GEOGRAPHIES
of the EAR

SIGN, STORAGE, TRANSMISSION

A SERIES EDITED BY JONATHAN STERNE AND LISA GITELMAN

GEOGRAPHIES
of the EAR

———

THE CULTURAL POLITICS OF
SOUND IN CONTEMPORARY
BARCELONA

Tania Gentic

DUKE UNIVERSITY PRESS   DURHAM AND LONDON   2025

Designed by Matthew Tauch
Typeset in Alegreya by Westchester Publishing Services

Library of Congress Cataloging-in-Publication Data
Names: Gentic, Tania, [date] author.
Title: Geographies of the ear : the cultural politics of sound in
contemporary Barcelona / Tania Gentic.
Other titles: Cultural politics of sound in contemporary Barcelona |
Sign, storage, transmission.
Description: Durham : Duke University Press, 2025. | Series:
Sign, storage, transmission | Includes bibliographical references
and index.
Identifiers: LCCN 2024057242 (print)
LCCN 2024057243 (ebook)
ISBN 9781478032076 (paperback)
ISBN 9781478028802 (hardcover)
ISBN 9781478061021 (ebook)
Subjects: LCSH: Sociolinguistics—Political aspects—Spain—
Barcelona. | Sociolinguistics—Spain—Barcelona. |
Multiculturalism—Spain—Barcelona. | Language and
languages—Political aspects. | Globalization.
Classification: LCC P40.45.S7 G46 2025 (print) | LCC P40.45.S7
(ebook) | DDC 306.4409467/2—dc23/eng/20250529
LC record available at https://lccn.loc.gov/2024057242
LC ebook record available at https://lccn.loc.gov/2024057243

For Dave, Miriana, and William

Contents

Preface

In 2006 I was in Barcelona for my first extended stay when I came upon a group of men and women dancing the *sardana* in the Plaça de Sant Jaume. I was thrilled; as a graduate student in Hispanic Studies, this was an example of Catalan culture I had been waiting to see. The nineteenth-century poet Joan Maragall had called the *sardana* "la dansa més bella / de totes les danses que es fan i es desfan" (the most beautiful dance / of all that come together and apart), and the travel writer Aurora Bertrana had written that the dance was both highly local and inclusive.[1] In this era before YouTube was the norm, it had not occurred to me to Google it and see what the *sardana* was all about: my knowledge to that point was entirely textual. I was surprised by how slow the music was. Given the passionate descriptions of the dance and imagining it to be somewhat like the Mediterranean folk-dance with which I *was* familiar, the Serbian *kolo*, which I had on occasion danced as a child, I had expected something upbeat. "Yeah," said the Colombian student I was with, who lived in Barcelona at the time. "It's pretty much for old people."

The memory of that moment came to me as I was walking through the Plaça Nova on another trip to Barcelona, in July of 2019. There, another circle had formed, this one made up of tourists with cell phones and selfie sticks from around the world listening to James Brown's "I Feel Good," expertly mixed with hip-hop rhythms, as a group of break dancers and gymnasts performed stunning acrobatic feats. The group, called Street Flow, or Fusión Callejera, was made up of men from Puerto Rico, Brazil, Venezuela, and El Salvador. At the end, they addressed the crowd mainly in English, throwing in the occasional Spanish phrase, as they asked for folks to spare some change in exchange for having enjoyed the show. Afterward, Elvis Crespo's "Suavemente" blared from their speakers. A young woman asked for a picture with the group. "You guys are so good!" she exclaimed in English.

The scene could have been anywhere in the so-called English-speaking world. After all, I realized, Barcelona, one of the most highly touristed cities on the planet, *is* the English-speaking world, at the same time as it is governed in Catalan and haunted by a history of Spanish dominance. This kind of cosmopolitan soundscape appealing to tourists was not new to the city; if I think back to my first backpacking trip in 1997 I can recall hearing the ubiquitous "El Condor Pasa" played on a pan flute, just as I did in Rome and Paris; this time around the pan flute was a couple of blocks away from Fusión Callejera, on Les Rambles, but it was playing ABBA's "The Winner Takes It All." As visitors continue to crowd the city, though, they are no longer just spectators: in the summer of 2024, the L4 metro car I was on, between Jaume I and Barceloneta, erupted in cheers as Lenny, a tourist from Liverpool who told me afterward he only rapped at home or on TikTok, begged the busker for the mic and engaged in a rap battle with him. Before walking away pretty much empty-handed, the busker, in Spanish, concluded by rhyming that, unlike "mi amigo de Liverpool" (my friend from Liverpool), he also rapped, at times, "en la línea azul" (on the blue line).

These scenes illustrate the way in which our aural imaginaries of a place are often at odds with the day-to-day soundscapes of them. As I contend in this book, these imaginaries are often reinforced by the way in which we continue to hear language as tied to territory or nation, but they are also present in how culture is packaged up and sold to audiences both local and global. When I went to Barcelona in 2006 I had studied a largely Hispanophone concept of Iberia prior to that point, and despite the fact that my doctoral studies had impressed upon me the ways in which Orientalism, colonialism, and globalization had created for Western eyes and ears what would later be called the Global South, the aural imaginary around Catalan as a repressed language still coming into its own held a different kind of appeal. Barcelona is a place shot through by aural imaginaries of national identity that are not unique in the West, though the tensions that arise from political conflicts about how the city should sound, and the varied, daily acoustic realities of the place, may have more resonance for Catalan constructs of national identity than such conflicts do elsewhere. By that I mean that, despite recent claims for Catalan independence, which are based at least in part on economic arguments that suggest the Spanish state is repressive, these are aurally played out through the politics concerning the Catalan language, and whether or not it is supported, promoted, or disseminated by cultural institutions. The linguistic soundscape in this context bleeds into musical culture, questions regarding immigrant voices in court systems, and how media is distributed in Spain and around the world.

Now that I have spent more time in the city, the differences between the aural imaginary I had of Barcelona on my first trips and its daily realities seem obvious, even a little banal. Few residents of Barcelona would ascribe homogeneity of any kind to the city, aural or otherwise. But that does not mean the colonizing aural imaginary that emplaces culture through language and sound does not still resonate: it hauntingly returns in encounters with accent, sounds of gentrification, music, and politics, daily specters that, like implicit bias, seem hard, if not impossible, to shake. The same Sunday evening that I was strolling through that cosmopolitan musical soundscape in 2019, protestors were confronting the *mossos d'esquadra* (Catalan police) about the eviction of a family in Sants who had lived in their home for sixteen years; the shouts of neighbors defending them were captured on cell phones and posted to Twitter under the hashtag #AbdelahNoSenVa (AbdelahIsNotLeaving). Later that week, I was meeting with Salvador Picarol, the founder of a free radio station that had never been allowed to obtain a radio license, despite being one of the first to broadcast in Catalan after Franco, but had helped circulate the early punk sound in the city. I was also set to follow an *okupa* (squatter) protest by the "poetically incorrect" group Bio-lentos, who use poetry as a tool of direct action against tourism and gentrification in places like Sants and Gràcia. What ear was I bringing to this place now that I had been thinking about the city for over a decade? Was it a Hispanophone ear, given my studies? An ear informed by my upbringing by two immigrant parents, one a heavily accented native speaker of Serbian who did not learn English until he was in his twenties, the other an English immigrant with a love of proper grammar raised in Toronto? Did it matter that I was not a native speaker of Catalan, though I could usually pass for a native speaker of Spanish (albeit an accented one, generally not from whatever place it was I was speaking)?

In his book *Barcelona's Vocation of Modernity*, Joan Ramon Resina writes that foreigners who come to Barcelona often interact with it as a nonplace, in the vein of Marc Augé, obviating in their appreciation of its architecture, or their expectations formed from stories about the city, the day-to-day realities and history of the place.[2] This fascination with new places that is the definition of travel, he suggests, produces a different form of knowledge of a place that, in effect, colonizes the place because the gaze is one of self and other.[3] Certainly, most analyses of travel writing since Mary Louise Pratt's *Imperial Eyes* have recognized that tendency, evidenced over centuries in texts as varied as Mungo Park's descriptions of West Africa or Joan Didion's chronicles of El Salvador. But, as Pratt also deftly showed in her seminal text, transculturation works in multiple directions, as observations of places made from afar, or by foreign eyes, often continue to inform how local eyes see their own space. I had grown up in the United States being told, repeatedly, by my parents that we were *not* American. But my ear certainly was. Wasn't it? Or could I overcome that ear through close listening?

Keeping these dynamics in mind, I tried listening through different filters: was Barcelona still the *ciudad nerviosa* (nervous city) Enrique Vila-Matas had described in his *crónicas* via a discussion of his own experiences in Chicago two decades earlier?[4] Was Barcelona the *Ciutat Podrida* (rotten city) it had been called by La Banda Trapera del Río in the 1980s? Could I even pretend to listen in, authoritatively, to a city that was not my own? When I heard Les Rambles as more congested with tourists than on my first visit in 1997, was I hearing the city as it was, or was I hearing my own ear differently? These questions sound rhetorical, but they reflect an auditory self-awareness that has come from years of reading Latin American postcolonial and decolonial theory and which I carry with me even though I am neither from Latin America nor training my ear on it in this book. How could I not hear my own out-of-placeness as I, funded by Georgetown University, silently attended the antigentrification protest of poets who considered themselves *poéticamente incorrectos* (poetically incorrect), staging out of the Ateneu Llibertari de Gràcia one Saturday afternoon and occupying local plazas with music and poetry? They sprayed graffiti on statues and shop fronts saying things like "Tourist go home" and were greeted by shouts, in English, of tourists from balconies yelling at them to stop complaining. Was I the tourist? An ally? A *flâneuse*?[5]

I asked these questions even though I came to this book about Barcelona after years of studying its culture, literature, and politics, and also after many trips made over the course of twenty years. I say this not to defend my analysis, but rather to admit that how and what we hear is always a contingent experience, but it is one in which we participate every single day, often without thinking about it. During one of my travels I realized that I have listened more closely to the soundscapes of certain neighborhoods of Barcelona than I have to those of the Washington, DC region, where I live. I feel—though I am sure any resident would tell you I am wrong—that I know how Gràcia sounds. But I could not tell you a thing about the soundscape of Adams Morgan.

This backdrop is important to keep in mind, because within the supposition that there exists a native way of listening that can be opposed to a foreign one, there is a shared imaginary of aural coloniality that supposes alterity is at the heart of all cultural encounters. Edward Said said as much in his suggestion that Orientalism was both self-defining for the West, and a definition of the East by the West.[6] As Jonathan Sterne has already explained in his decisive undoing of Walter Ong's notion of the sensorium, when it comes to sound, such modes of hearing alterity merely naturalize an understanding of sound that, by opposing it to vision, is itself ideological.[7] One of the primary determinants of the ideology of sound, I suggest here, are the conceptual geographies that produce the ear as it moves. These geographies—which change over time and through our travels and daily encounters with language, whether we hear them in person or through the media—are the subject of this book.

Introduction

Echoic Memories of Dispossession

There is a geography to the ear. Hearing is spatial, of course.[1] For his part, Caleb Kelly has argued that, unlike light, sound turns corners, allowing us to experience distant phenomena we cannot see.[2] But the ear—which is not just hearing but a convergence of physical, affective, and ideological practices and discourses—is also imagined, ideologically produced, and carried around with us, sometimes in the cell phone in our pocket or the newspaper in our hand, other times in our memories. As a geography, sound is a function not just of one's daily movements but of the discourses and media that produce the sounds of places as history, culture, and politics.[3] The geography of the ear informs how we know place and how we emplace ourselves—and others—in it.

Barcelona is an illuminating place in which to think through these aural complexities. Culturally and linguistically, Barcelona is both an amalgamation of local identities grounded in its *barris* (neighborhoods), each with its own history and characteristics, and a thriving, modern, globalized metropolis at the crossroads of multiple diagonal geographic relationships that cannot be easily defined as fully occupying North or South.[4] The echo of the transatlantic slave trade is present, albeit overlooked, in the lauded architecture of the city, built financially on the backs of Indigenous people and trafficked Africans put to work in Spain's colonies in the Americas and Africa. Constantly present, too, is the memory of forty years of dictatorship under Francisco Franco, in which the city's native tongue, Catalan, was officially forbidden in public settings although it was the everyday language of speech, even on the streets, and produced an entire musical movement of resistance to Franco's dictatorship, the *nova cançó*, prior to his death.[5] For over a century, Barcelona has been theorized by Catalan intellectuals as Mediterranean, in part to resist being included in a Spain that has been considered the south of Europe; for the last two centuries, the Pyrenees have been viewed on and off as a dividing line between Europe and Africa, despite Spain's great imperial wealth. In addition to participating in the colonization of the Americas, then, when it suited their imperialist needs, intellectuals and politicians in Spain, including in Barcelona, embraced their supposed "Africanness" to justify their right to colonial domination in Equatorial Guinea and the Sahara—not to mention an incursion into Morocco—in the early to mid-twentieth century. In fact, most of the colonial enterprise in Equatorial Guinea, which only won its decolonial struggle against Spain in the 1960s, had its business and ecclesiastical center in Barcelona. At the same time, there is a geographical south to Spain, Andalusia, which comes to Barcelona in the form of migration, in ways that, for almost a century, have informed a racialized economic hierarchy between the Catalan bourgeoisie and migrants that spills into a sonic and class difference marked by accent and the bilingual sounds of speakers' non-Catalan languages. Those same migrants, many of whom lived in shantytowns around the city before these were razed to make way for the 1992 Olympic venues, have been both courted by and excluded from a Catalan sense of nationalism, grounded in linguistic identity, which spans—and at times also divides—both working-class and upper-class Catalans.[6]

Since 1975, the defining cultural project of the city has been an aural one: the promotion of the Catalan language, a movement that both responded to Franco's repression of its use in public and was tied to Barcelona's desire to be a modern, European city with its own unique identity.[7] Books and comics aimed at children and adults were printed in Catalan, and television and radio stations consolidated rules for the sound of proper Catalan in guidelines for their broadcasters in an effort to erase Castilian Spanish "barbarisms" from people's speech.[8] As the 1980s wore on, neoliberal planners for the city not only supported an incipient Catalan rock, they ensured that the opening ceremonies for the 1992 Summer Olympics included a musical performance of Catalonia's national folk dance, the *sardana*, and a new song called "Barcelona" that would bring the city's famed Liceu Theater to a global pop ear when their own soprano opera singer Montserrat Caballé performed with Freddie Mercury. Yet at the same time, working-class punks and experimental musicians disillusioned with the new democratic system looked to England, Germany, and the United States for musical escape from the Spanish music industry. They reveled in the experimental, nonindustry noise they could make to offend sensibilities associated both with Spain and with the Catalan drive for a European modernity, which they heard as linked to the previous silencing of voice that defined the Franco regime. Occupying FM frequencies with unlicensed free radio stations and taking to nightclubs to create new sound art, they created micro spaces of sonic agency outside the public sphere. Starting in the 1990s, immigrants from the Americas, Africa, and the Middle East (especially Pakistan, Romania, and El Salvador) began coming to the city looking for economic stability, bringing with them multiple new languages, accents, and dialects; 23.6 percent of the city's population today is foreign-born.[9] Since the 1980s, squatters have converted houses and commercial buildings into sites for concerts and demonstrations in order to highlight the city's disregard for the precariousness of its marginalized populations. Combined with all of this, the sounds of tourists in the city's center—up to 15.6 million in 2023—increasingly overwhelm the daily sounds of the 1.7 million who live in the city, especially in the areas around the Gothic quarter, Barceloneta, and the immediate environs of the Sagrada Familia. Throughout the transition to democracy and beyond, then, Barcelona went from being a quiet, somewhat sleepy place to a noisy, polyglot city, with both local music and art culture rooted in specific neighborhoods and the foreign sounds of tourists and migrants resounding in public squares, all while its institutions remained committed to producing a Catalan soundscape through music, literature, and an ever-more-perfect sound of speech.

My argument in this book is that we can hear within these sounds (and particularly the often contradictory aural imaginaries that surround them) the echoic remnants of a now-globalized colonial ear. This ear was forged in the Americas but reproduced in transatlantic dialogue with Europe and Africa, and it resounds in how these sonic practices, often centered on language, are related to what it means to be global or modern, to have a community identity, or even to be a democracy. This may sound counterintuitive at first, but in Barcelona, as elsewhere, today's globalized ears hear today's cultures by processing them through understandings of selfhood, nation, colonialism, and democracy that often hark back to earlier moments of history, including conquest, even if they do so from within local contexts that do not seem to share a direct historical link to the scenes of the past that inform them. Because all types of media have normalized and often universalized notions of coloniality, inequity, and human rights, they are at times applied to seemingly incongruous scenarios, which call into question the line between colonialism as a historical production of space and coloniality as a structural, epistemological understanding of subjectivity linked to voice. The ideological structures of (accented) voice that materially produce space are crucial to understanding that relationship, as well as to questioning the theoretical construct of aural coloniality as a territorialized force. By this I mean that the varied relationships between the accented or bilingual sounds of voice, the meanings of language, and the sensed perceptions of music or protest have helped construct a modern(izing) geography of the ear that also emplaces sound in a particular way, often linking contemporary sounds of voice to a monolingual aural imaginary of national identity that, in turn, has its origins in a binary colonial epistemology.[10] As I will show here, sound circulates via what I theorize as an *echoic memory* of the perception and experience of language, music, and voice whose aurality is also grounded in a colonial way of listening. I will explore the concept at length later, but briefly, echoic memories are aural feedback loops that emerge in brief moments of sensation (humor, anger, discomfort, offense) that construct local spaces and identities every day, not just through listeners' daily interactions with sound, but through mediatic portrayals of it. The idea resonates with Jennifer Stoever's concept of listening ears, through which "sounds from the past come to us already listened to; they are mediated through and by raced, gendered, and historicized 'listening ears,' [which are] an embodied cultural process that echoes and shapes one's orientation to power and one's posture toward the world."[11] Still, I am interested in geographically complicating this notion by exploring how linguistic soundings of race, gender, and history, often through how we hear accent or voice, emplace sound—that is, attribute a historical, cultural, or political place to it—even as the tensions between the present material instantiations of sound and the at times untraceable histories of how perceptions circulate through communities often reveal the cultural politics of that emplacement. Because sound is echoic, stretching across memory and history as well as across borders, it can at times carry with it colonizing assumptions about places and peoples that were sounded decades or even centuries before. At the same time, the effects of Barcelona's modernizing project, as well as the global realities of migration, have produced pockets of dispossession around the city whose sounds also reflect an echoic memory of the colonial condition, one that is not discursive but lived. These echoic memories of a colonial aurality are present not only in acoustic settings but also in an intermedial cultural production of sound that, as I explore, includes fanzines and comic books, documentary films, popular music recordings, and local concerts, as well as antiglobalization protests and prodemocracy demonstrations, in which sound is spatialized and imagined in contradictory ways. In other words, at stake in this modernizing geography of the ear is what Ana María Ochoa Gautier defines as an aurality that "is not the other of the lettered city but rather a formation and a force that seeps through its crevices demanding the attention of its listeners, sometimes questioning and sometimes upholding, explicitly or implicitly, its very foundations."[12]

As anyone who has followed Spanish politics in recent years will know, the same institutions that have sought to produce a more vibrant Catalan aurality have also often claimed that Catalonia has been colonized by Spain, at least in part because of its language.[13] So strong is this sentiment that, in 2017, Catalonia held a referendum, deemed unconstitutional by the Spanish government, in which 90 percent of the 1 million people who participated (out of a population of 7.5 million) voted to secede from Spain; Catalan government officials, initially charged with sedition, were jailed or went into exile.[14] Still, as Raphael Minder has pointed out, Catalan is hardly fighting for its survival: In 2016, in a population of which 35 percent were born outside the region, 94 percent of residents understood Catalan, and 80 percent could speak it.[15] Wrapped up in this sound of language, then, are opportunities to interrogate how the voice as a democratic construct has become entwined with a colonial ear that has extended, through an aural imaginary of sound, language, and voice, back across the Atlantic today. Nationalist discourses about Catalan oppression by Madrid began to circulate in the late nineteenth century—both as Catalonia was coming into its own as a political, as well as literary, entity and as the Spanish empire was faltering. National celebrations in Catalonia invoke the historical date of September 11, 1714, which commemorates the day Barcelona fell to Bourbon Spain during the War of Spanish Succession. This ushered in the decrees of the Nueva Planta, which for the first time placed Catalonia under the control of a Captaincy General, the same kind of colonial governance structure used in the Americas for border regions; some independentists invoke this history as evidence of Catalonia's oppression by Madrid. When placed in a historical frame of *longue durée*, one which takes into account what Mary Louise Pratt has called a planetary consciousness,[16] however, complaints of Catalonia's colonization seem to ring false as compared to colonial struggles by Indigenous or Black communities in the Americas and Africa. After all, the money that drove Catalonia's modernization, allowing it to assert its autonomy from Spain, was derived from the slave trade and the colonial enterprise in the Caribbean, mainly Cuba. But locally, after Franco's death, assertions of colonial rule reflected a more recent lived experience of censorship and repression of identity that lasted for forty years in the confines of an isolated city space. That experience has produced a memory of dispossession that echoes, affectively and narratively, the sense of marginalization and alterity also experienced in Spain's colonial hegemonies across the sea. The government, financial, and media institutions that have pushed hardest for Catalanist modernization so that Spain—but primarily Barcelona itself—can finally be heard as part of Europe have simultaneously produced a capitalist dispossession of precarious migrant populations, which might actually be more consistent with the colonial condition understood in terms of class, race, and identity.

Although Barcelona is not usually described outside Catalonia as colonized in the historical sense, then, my attempt here to think through the disjunctures of sound and language in post-Franco Barcelona is my own way of broadening and complicating a decolonial epistemological critique by addressing the very spaces that have benefited from colonization and today feel its echoes in new ways, especially as migration and tourism radically change the living structures of language and accent, and as discourses of (colonial) oppression become a political staple in some Western societies. As Arturo Escobar has succinctly put it, the modernity/coloniality paradigm that undergirds the decolonial approach derives from "a new spatial and temporal conception of modernity" that rethinks the linearity of the historical paradigm that runs from Greece to Rome to Christianity to modern Europe. Instead, it considers "the foundational role of Spain and Portugal (the so-called first modernity initiated with the Conquest) and its continuation in Northern Europe with the industrial revolution and the Enlightenment (the second modernity, in Dussel's terms)."[17] This approach, as Dussel writes, recognizes that "modernity, colonialism, the world-system, and capitalism were all simultaneous and mutually-constitutive aspects of the same reality,"[18] and they began with the conquest of the Americas. Later scholars, furthermore, extend the concept to argue that the epistemological contours that accompany modernity as an economic structure across the Atlantic world also inform racial categorizations, the notion of rationality as in opposition to affect, and binary codings of gender.[19] As Walter Mignolo suggests, the modern world-system at stake in this understanding is not simply an issue of economic development and change, but "a spatial articulation of power."[20] This approach resonates with Fernand Braudel's theorization of the Mediterranean as a nonlinear space that exceeded linguistic, cultural, and national borders, at a time when linguistic identity and emplacement were not as fixed by the nation-state as they are today. In this book, then, I try to listen to Barcelona through aural geographies informed not just by local or European epistemologies but by a transatlantic form of dispossession heard in and through a longer Atlantic frame that recognizes how auralities formed in the colonial conquest echo in other places around the Atlantic world, even those that overlap territorially with the places that originated the modern world-system. Barcelona in particular is a place with a long history of both Mediterranean and Atlantic crossings, material and conceptual, that do not sit easily within any of the frames attributed to it. As José Luis Venegas has argued in his work on Andalusia (a source of migration to Barcelona for decades), for example, "before the Global South, there was the Mediterranean."[21] Although he is not writing about Catalonia per se, Venegas draws attention to the fact that geographical attempts to establish relationships transversally beyond state borders, in particular those like the Global South that seek to produce alternative geographies through a shared notion of periphery in the face of modernity/coloniality, often elide "the contradictions of capital and imperialism" that make it difficult to satisfactorily sustain these concepts once one looks at the granular frame of the local.[22] When one takes into account how aurality sounds within different frames—the Mediterranean, the city, the neighborhood, and the globe (as I do in each of the chapters in this book)—the echoic memories that produce the Atlantic allow us to hear how important geography is to framing the ear. In line with this idea, Catalonia has been embroiled in peninsular relationships, in particular with Madrid and Andalusia, that complicate what it means to be a center, the North, European, or modern. Moreover, the religious frames that often crop up in discussion of Spanish culture take some curious turns. In just one example, as Eric Calderwood has provocatively shown, al-Andalus in particular became a focus for Francoist imperial discourse that, paradoxically, produced contemporary notions of Morocco for Moroccans themselves, while also positing Catholic Spain as spiritually outside Europe and in touch with the wider Mediterranean/Arab world.[23]

Given this complex geography, I am not interested in perpetuating the tired binaries of colonized and colonizer, or their derivatives, which have too often been dehistoricized and misapplied to contemporary situations. Instead, I wish to use Barcelona to think through how the sound of voice, and particularly a monolingual aurality tied to notions of nation, produces a daily, spatialized, lived experience of the body in and through place in the late twentieth and early twenty-first centuries. The aural frame of the monolingual sound of language as identity informs the nineteenth-century European ideal of the nation, but as I show later, it begins to reverberate with the colonial ear produced across the Atlantic in the time of conquest by hearing language contact as conflict instead of communication. The colonial aurality that echoes into the twentieth century splits the sound of voice, transducing what Roland Barthes called the "grain of the voice" into a sensation that is contrasted to the meaning ascribed to voice as language. As I theorize below, this is a political structure of listening that, over time, normalizes a foreign ear as being unable to hear meaning, and thus treats it as out of place, in order to advocate for, or justify, authority over it. Yet, at the same time, the process takes place in an aural imaginary related to law that transduces the paralinguistic sound of voice into a spatial geography of belonging and exclusion, of borders and imagined contestatory movements, into territories-cum-geographies. This politics of listening produces an epistemological continuity between a colonial geography of the ear and a globalized Barcelona, but it also reaches further back in history than the colonial period to justify its listening practices. Although many times echoic memories harden into stereotypes about the sounds of place, ethnicity, gender, or race, the temporality of the echo as a creation of spatialized sound also disrupts these codings as it creates them. Perceptions of accent, which has its own complicated history in Barcelona and the larger Catalan region, are a key mover of these aural geographies, and one which I focus on throughout the book.[24]

The difference between territory and geography is crucial to this work: If territory refers to the land that is life, to reference scholar of settler colonialism Patrick Wolfe, geography is the imaginary about space that informs how we perceive and produce that land.[25] Moreover, this geographic spatiality is temporalized: As Joan Lafarga i Oriol puts it, geography brings memory to territory.[26] Like all national identities, Catalanism is an identity tied to a constructed geographical memory, be it of the so-called Països Catalans (the regions where Catalan is spoken), an ideal *Catalunya ciutat* (Catalonia city), or Barcelona as a cultural and political center that is at odds with Madrid.[27] And language—be it Catalan, Spanish, English, or otherwise—plays a role not just as a tool of meaning-making or a metaphoric description of experienced sounds but as a sense of geographical subjectivity captured by the ear (often through social conceptions of accent, but at times through tone or linguistic difference) that forces the listening subject to locate the contemporary presences or distant origins of sound in voice.[28] This makes echoic memory part of the perceptual coding of space through which we engage in what David Panagia refers to as "the aesthetic-political dimensions of democratic life."[29] Almost always, these codings become conflated with notions of *emplacement*—who belongs and who doesn't, which sounds are local or foreign, and which voices should be present in a space, and which should not.

At the same time, sound can shape knowledge differently through its material resonances. There is always a spatial production involved in the materiality of voice: Listeners are affectively moved by voices, even while, as Alex Chávez asserts, "voicing *takes place*—its material enactment constructs mattering maps that represent the ways social actors move through the world, or desire to do so."[30] The materiality of vocal movement through space thus intertwines sound with bodies, sensations, and the production of place from the local to the global, sometimes at the same time: Anarchist *okupes* (squatter activists) who squat houses or abandoned banks in the Barcelona neighborhood of Gràcia, converting them into sites of poetry readings or concerts, are materially enacting a differently sounded city than is the neoliberal gentrification or consolidation of Catalan media apparatuses for radio and television that define Barcelona's official soundscape after Franco. Yet often these enactments are not territorially separate phenomena. We might consider how the gentrification and investment in tourism has markedly changed the soundscape of Barcelona since the Transition. The Plaça Reial, for example, was once the home of José Pérez Ocaña, whose trans* occupations of the streets with vocal difference signified both migration and queerness.[31] His efforts are now reflected in the name of a restaurant in the plaça called "Ocaña," where his queerness seems mostly to be a marketing tool for selling overpriced drinks and tapas, as the square resounds with noisy laughter and polyglot languages from tourists around the world, many of whom would not know Ocaña's name. The way in which those mattering maps are produced and overlap in the cultural and political soundings that take shape—and are then reshaped—in Barcelona after Franco's death produce the competing geographies of the ear I address here.

The specificity of Barcelona's sound, produced not just by institutional decisions regarding broadcast language or education, but through the scale of the listening act, thus complicates some of the assumptions about geography and sound that have crept into sound studies over the years, particularly around the notion of the Global South. Steingo and Sykes, for instance, draw our ear to the binary geography implicated in the recent notion of the Global South, which they summarize thus: "Whether the relationship is dialectical, supplementary, or hybrid, sound and the South are the Others of the visual and the North. And like poles in any binary opposition, 'sound' and 'South' can easily be substituted for multiple 'Other' terms, including 'nature,' 'woman,' 'native,' 'Africa,' 'black,' 'queer,' and 'disabled.'"[32] As Susan Martin-Márquez has argued, however, Spain experiences coloniality in ways that complicate the old binarisms of East and West, colonizer and colonized, related to Edward Said's notion of Orientalism (which, arguably, is also reflected in the geography of the Global South): "Spain is a nation that is at once Orientalized and Orientalizing. . . . For Spaniards, this positioning on both 'sides' of Orientalism—as simultaneously 'self' and 'other'—may bring about a profound sense of 'disorientation.'"[33] For Calderwood, this idea allows us to recognize not only the geographical limitations of Said's project but just how contradictorily Spain's global colonial projects and national identities at home and abroad inform each other.[34] Within Catalonia, this dynamic is complicated further by the relationship between ideology and linguistic sound that emerges out of the end of the imperial period in which Spain and Catalonia both participated; the loss of empire in the Americas and the Philippines by Spain in 1898 provided Catalanists with an opportunity to bring their cause to the center of Spanish political discourse.[35] Soon entwined with the desire to "elevate" spoken Catalan to the status of a lettered language equivalent to Castilian Spanish, the Catalanist project looked both forward and backward, inward and outward, and North and South as it tried to simultaneously resuscitate the lost Caribbean colonial project in Equatorial Guinea and build Barcelona as a modern city.[36] This is the project that is recuperated after Franco's death, although in a way that forgets Catalonia's own participation in colonialism and focuses instead on producing media in Catalan and rebuilding Barcelona in order to recenter the city firmly on the European and global map.

Importantly, as the case of Barcelona illustrates, living in translation, or hearing languages or sounds we do not understand, is a common phenomenon in the Atlantic world (and beyond). As a bilingual—or even, as some scholars now argue, a socially trilingual—space, Barcelona shows us that the monolingual approach to identity and sound is in and of itself a residue of a nineteenth-century construct of nationalism imposed on lettered societies in ways that are inconsistent with the diagonal lived experiences of their communities.[37] We confront unknown, or partially known, languages, music, and sounds on a daily basis, whether we recognize it consciously or not; many of them are transculturated from elsewhere in ways that give them local meanings that illustrate the tensions between emplaced sound as an echoic memory and lived experiences of those sounds. As I show in the chapters ahead, the Catalanist ear, in striving to hear Catalan as a modern language, often defined that modernity through processes by which accented pronunciations or foreign languages were objects of derision or, at the very least, heard as being in need of correction. The origins of unknown languages and accents are not always clear to listeners; the geographies of the ear that echoic memories produce fill in the knowledge gaps we face when confronted with linguistic difference. Aural imaginaries of language as sound often suture these subjects and places to each other through ways of hearing accent, music, and noise as local or national, familiar or other, settled or in movement, or at times all of these things at once.

With all of this in mind, some of the motivating questions of this book are: How is sound perceived when it is heard through an ear that does not (fully) understand the language or culture in which the sound is inscribed? How does accent mean, especially when it intersects with questions of racial, national, or gendered identity? What relationship does sound have to the imaginary of place through which we "hear" nationality and identity? How does sound as sensation intersect with cultural and political notions of voice? What role do local music movements or radio transmissions play in reshaping how a place is heard, and vice versa? And if historically the colonial experience can be located in certain territories, how does a colonial aurality travel in the contemporary, globalized period, not just through music or accent but through mediatized notions of voice and dispossession as they relate to community and democracy? With what cultural and political consequences?

ECHOIC MEMORY AND THE SPATIALIZATION OF SOUND

Henri Lefebvre's 1974 *The Production of Space* is instructive for understanding the spatial shaping of sound and place, both as an epistemology and a material practice. That practice is embodied and sensed, which allows us to interrogate sound in terms of the decolonial possibility (and more often than not, impossibility) of knowing otherwise. For Lefebvre, ideology "achieves consistency by intervening in social space and its production, and by thus taking on body therein. Ideology per se might well be said to consist primarily in a discourse on social space."[38] But, despite this linkage of ideology to discourse, both participate in a broader production of spatial codes that are "not simply a means of reading or interpreting space; rather [they are] a means of living in that space, of understanding it, and of producing it."[39] These coded relationships between representations of space, what he calls representational spaces, and the practices that reproduce ideology through space involve both bodies and discourses. They also imply time. In a present moment of experience, a subject may perceive a space, including a representational space, such as a consecrated place like a church, which grounds social relations through the practices that emerge around it and also through the social discourses in which it is implicated. Yet within that immediate moment of perception the subject is also implicated in the discursive and spatial practices that produce the space ideologically and representationally in history. Consequently, representational space "is essentially qualitative, fluid, and dynamic" and ideology can no longer be separated from knowledge: "knowledge must replace ideology" as one of the tripartite means through which space is produced, the other two being lived practices and perception.[40] Although he does not put it in these terms, if we consider sound as a kind of relational knowledge, as Steven Feld does when he coins the term *acoustemology* (a sonic way of knowing), in a sense what Lefebvre allows us to conclude is that there is a feedback loop between language about, and even visual portrayals of, space and the aurality of a space itself.[41] Sound is never sound by itself, but a variety of echoic sensations and discourses that coexist in spatial experiences, situating the subject aurally in past, present, and future, sometimes simultaneously. The scale of the geographies of the ear we choose to engage, and how we emplace voice as sound and sensation matter here. They matter, moreover, not just for how they make language mean, but for how they construct the acoustic spaces—the "mattering maps"—in which sounds, music, media, and voice are aurally emplaced.

I will offer an example. One evening in June 2011, viewers of Catalonia's comedy sketch show, *Polònia*, were treated to one of many impressions of Spain's conservative Minister of Education, José Ignacio Wert.[42] Often portrayed on the show as a Spanish version of Austin Powers's Dr. Evil, in this episode Wert is dressed up as a conquistador, and bursts into a Catalan language class. While the teacher and the students speak fluently in Catalan, Wert speaks mainly in Castilian Spanish; the show is clearly directed toward a bilingual audience, a fact that already separates Catalonia's audience from much of Spain's non-Catalan-speaking population. His first word is a high-pitched, nasal "How," in English, meant to mimic a Hollywoodized North American Indigenous greeting, which he repeats with his hand held up in the air, before he asks—in Castilian (Spanish)—"Queréis hacer educación los indios, o qué?" (Do you Indians want to do education, or what?) When the class responds with an unenthusiastic "How" in return, he exclaims, "Ay, tranquilos, salvajes, ¡no me mordáis!" (Oh, calm down, you savages, don't bite me!) Speaking as if they cannot understand him or even hear him, he says loudly and patronizingly, in a faux-nasal tone: "Yo—Ministro de Educación. Vosotros, indígenas sin educación." (I: Minister of Education. You: Uneducated Indians). Wielding a scroll, he lays out the rules that from now on, in the classroom, Catalans will "only speak the language of empire," by which he means Castilian.

The show reflects the fierce political debates taking place at the time over the role Catalan should play as a language of instruction in schools in Catalonia. Although the debate about *català a l'escola* (Catalan in schools) had ebbed and flowed since the early 1970s, by the first decade of the twenty-first century, the argument about the language of education fed into a larger question of Catalonia's political and economic power within Spain, as well as Catalonia's national identity.[43] Shown on TV3, Catalonia's main television station, which is supported by the Catalan government, the Generalitat, the sketch frames the language question as an echo of colonial conquest, suggesting both that Spain hears the Catalan language as barbaric, and that Spain's attempt to control language use in the classroom represents an imperial position over a subjugated people. In this way, the program hears colonialism as aurality inscribed into national linguistic identity. It intentionally harks back to the conquest, when Spain not only violently overtook Indigenous territories in the Americas but also produced the first official Spanish grammar, Antonio de Nebrija's 1492 *Gramática de la lengua castellana*, which explicitly stated that language was a tool of empire. The scene thus reflects an imaginary of colonization in which the defining factor of conquest is not physical violence or economic exploitation, but the sound of communication: Catalonia is, the sketch suggests, linguistically colonized by Spain.

By using the English-sounding "How" to reflect Indigeneity as a Catalan identity that is undervalued by Spain, though, a gross aural stereotyping comes to play through the old binary of civilization and barbarism, one that moves the geography of Spain's imperial past into an amalgamated, Hollywoodized sound of "savagery." Visually portraying the conquest only through Wert's clothing and the scroll, the scene strips colonialism of its historical setting and the racialized bodies that would have been present. The show also ignores the long history of Catalan participation in imperial pursuits in the Americas and Africa, through which Barcelona made much of its modern wealth. Its critique of Spain's educational language policy thus works by extracting a certain aural imaginary of colonizer and colonized from the historical record and converting language into a sound of imposition and oppression that, because it is voiced by a parody of Dr. Evil, can be judged both true and absurd: Are Catalan students forced to learn Spanish really the victims of colonialism in a way that equates them with the violent slaughter of the Indigenous communities of the Americas five centuries ago?[44] To a wider, globalized public sphere, this scene could be a reason for rage, because it sounds an offensive misappropriation of Indigenous history. But for a strident Catalanist it confirms that, as a very localized language, Catalan can sound and feel "minor" because throughout Western history, "only a few languages [have been] deemed 'reasonable' for international communication," and Catalan is decisively not one of them.[45] Regardless, the scene suggests that colonialism and its critique take place both in the distant past and in the present, in a moving aural geography that spans continents, cultures, and the shifting sounds of multiple languages.

I present this rather uncomfortable scene as an example of what I am calling throughout this book echoic memory. In particular, I am interested in the transposition of a particular geography of the ear onto contemporary politics—specifically a democratic politics grounded in neoliberal economics and increasingly globalized media—in ways that extract the sounds of language from their specific spoken contexts in order to attribute social meaning (identity) to it. At times, as is the case in the example above, such geographies are created by opposing the sounds of local languages and voices to ideologically produced memories of other emplaced sounds.

Echoic memory, then, is a multipronged feedback loop that moves among the local, national, and global ideologies and cultural practices that circulate in a mediatized, transatlantic milieu (at times simultaneously), in which the sound of elsewhere can be produced in any number of ways. Filtered through daily experiences as sensations, as well as mediatic representations of the spaces one occupies, these memories (re)produce how we hear sound in the present and how we expect it to sound in the future.[46] In that sense, echoic memories are transductions, which, as Stefan Helmreich has argued, are not about adopting a point of view, but about sounds "tuning in to surroundings and to circumstances that allow resonance, reverberation, echo—senses in brief, of presence and distance, at scales ranging from individual to collective."[47] In my reading, the echo especially foregrounds both rupture and simultaneity in the acoustemological knowledge of space that obtains in contemporary democratic societies; often that rupture is evident in how discourse informs and interrupts our perceptions of sound. Steven Feld argues that the conjunction of acoustics and epistemology through the relationality of bodies to networks and spaces occurs over time: "Knowing through relations insists that one does not simply 'acquire' knowledge but, rather, that one knows through an ongoing cumulative and interactive process of participation and reflection."[48] Yet these accumulations of knowledge often settle into ideology, becoming the regimes of truth that sustain our thinking. Echoic memory, however, need not be cumulative, working toward a final goal of completion: Rather, it may be *restless* or continually moving, like musical experimentation or improvisation that has not yet solidified into a recording or final version. It is iterative, yes, but as Amit Pinchevski has argued, echo is distinguished from reverberation and resonance: Rather than a simultaneous fullness and dying away of sound determined by the size of the space in which it moves (reverberation), or a vibration that begins in one object and causes another contiguous one to vibrate in turn (resonance), echo is a dislocation of sound. It "resounds in contradistinction to the origin. It returns belatedly enough to be noticed independently, hence heard as both replica and response."[49] This doubleness as replica and response destabilizes a binary model (even a mutually transductive one), because even as echo is a repetition, it is one that implies difference: It is always "potentially divergent."[50] Echoic memories are marked by temporal delay, returning to the past and projecting into a future. They are also separated from perceived origins by both space and time, and by the differentiation through which each sounding becomes its own. Because they are always being reheard through a present moment of sound, moreover, echoic memories have an elasticity that allows them to be recuperated discursively at the service of seemingly opposing political and ideological projects, but without settling into either. Far from mere instances of misrepresentation, echoic memories are the means by which the (accented) sound of voice, or an unfamiliar musical genre, returns to dislocated origins (which need not, in fact, be the actual starting point for the sound's production). They are also how the contemporary ear responds to those perceived origins, iteratively reproducing geographies of distance and proximity in contemporary spaces geared toward the future as well.

In this echo, as Don Ihde has argued, there is also a relationship between our perception of the shape of sound, and its visual and tactile contours: "With the experience of echo, auditory space is opened up. With echo, the sense of distance as well as surface is present."[51] Thus, we can hear echoic memory as what Nina Sun Eidsheim calls a thick event, in that it is simultaneously tactile, spatial, material, and vibrational.[52] Yet because these repeats are echoic, moving back and forth from imagined origins into contemporary reiterations that re-create the geographies they define in the process, these scenarios are anything but static, engaging instead in a constant remaking of the soundscapes, and lived distances from them, in which a subject hears itself and others. For example, in the early '80s, when the alegal, free radio station Radio PICA began broadcasting the Ramones and the Sex Pistols, bands whose music was otherwise unattainable in Barcelona and rejected by the Spanish music industry, from the neighborhood of Gràcia, the station created an aural origin point for punk rooted in both the new tones and chords of electric guitar that had never been heard in the city before, and also in the sound of the English language as a free, antisystem aurality that was experienced in Barcelona as a kind of underground knowledge. The punk music culture that developed in the city at the same time was not an imitation of English music but a way of hearing, through both local and foreign bands, a longer echoic memory of Barcelona sound as resistance against Spain.

In a broader geographical example from a few short metro stops away, we may consider a moment from August 2017, when a Moroccan-born ISIS (Islamic State of Iraq and Syria) member drove through Les Rambles, killing thirteen people and injuring 130 more. He fled on foot directly past the Palau de la Virreina, which is situated next door to the city's famed Boqueria Market, at the touristic center of Les Rambles, and only a block away from the Catalan National Library. The building, which is now a free museum of sorts, the Centre de la Imatge, was acquired by the Ajuntament (Barcelona City Council) in the 1940s. It was built, however, in 1772, by one of Spain's most ruthless viceroys—a Catalan named Manuel de Amat y Junyent, with wealth acquired while he was in Peru representing the Spanish crown. If we take Achille Mbembe's coloniality of necropolitics into account, we may imagine the screams and police presence caused by the 2017 attack as part of an echoic memory of *longue durée*, in which the colonial relationship between Europe and Africa, and the United States and the Middle East, echoes in intertwined yet at times politically opposing ways, through and along the Barcelona city spaces that were built through violent encounters in the Americas that were, in turn, justified by a (Spanish) linguistic and legalized geography of the ear.

My work is also different from Feld's argument that this sort of relational epistemology of acoustemology is a "cornerstone of decolonized Indigenous methodologies."[53] As I interrogate in my chapters, an echoic acoustemological approach is not always decolonizing in and of itself, because sonic knowledge is also formed through discourse and its ideological valences. In this sense, too, I differentiate my work from any decolonial presumption of being able to represent "worlds and knowledges otherwise" as forms of thought that, in the contemporary period, are completely outside Western epistemologies.[54] After all, the *Polònia* sketch that presents bilingualism as colonizing relies on a sensation of oppression that harks back to an aural imaginary of coloniality as universalizable, notwithstanding the historical realities of its production. Language, both as a sound that means and as the way through which we mediate our "listening to listening" through meaning, produces the ear that hears geography in conjunction with the sonic aspect of voice—the tones, affects, and sensations—which Mladen Dolar calls the "material element recalcitrant to meaning."[55] Yet the seeming alignment between voice and body, as Michel Chion has shown (and which I discuss in Chapter 2), is often false. Although I draw on decolonial theory to frame my understanding of sound, then, I do not presume my work to have the same subject-subject relationship as that of Dylan Robinson, for example, through which his own Indigenous subjectivity allows him to practice a "resurgent" listening that hears that which has been erased. Aware of my outsider's position, I do, however, attempt a critical listening positionality, which "engages how perception is acquired over time through ideological state apparatuses at the heart of subjectivation," all the while reading those positionalities (my own included) alongside the materialities of sound that emerge in practice.[56] I therefore dialogue with a decolonial/transmodern approach that originated in Latin America in order to recognize its echoic geography within Europe, and how the abstraction of the colonial *as* an imaginary often delinks from the material experiences of daily sound. Historically, as Etienne Balibar has pointed out, as a subject-making and un-making phenomenon that is "cognizable," language, the most immediately obvious sound of national subjectivity, has been framed over the last century as an affective sense of origins in which subjects can feel themselves as individuals and part of a community.[57] However, the sensed notion of language is also present in how affect, tone, and corporeal sensation derive from the sound of language as a modern construct of democratic voice, and it is here where the contemporary aurality of dispossession, echoing the colonial, is produced.

SLEIGHT OF EAR AND COLONIAL AURALITIES OF VOICE

If, for decolonial theorists, modernity and coloniality are two sides of the same coin, joining the Atlantic space as a single unit of epistemological production, it behooves us to interrogate how current conceptions of voice as sound have been perpetuated throughout that space and over time. Veit Erlmann has productively read into the Western philosophical canon to unearth notions of sound that have been subsumed by vision; in just one example, he contrasts René Descartes's modes of thinking understanding as a reflection grounded in the visual perception of the mirror, to Denis Diderot's notion of resonance as both the intimacy of an idea and the acoustic quivering strings that form the core of the enlightened self.[58] To be sure, this contrast draws our ear to the role of sound and its conjunction with vision in the West. But in the interest of rethinking this production of sound as transatlantic, and using a decolonial frame to do so, I want to go back to a founding document, the *Requerimiento*, which is echoed in the aurality of the *Polònia* sketch I discussed a few pages ago. Doing so demonstrates how the mediatized and sensed sounds of language as voice that influence daily a contemporary Catalan(ist) geography of the ear—and its tensions with other aural geographies—play out across an echoic transmodernity that situates the sound of language and accent within a construct of the ear that is both a listening to sensation and an aural production of law.

The *Requerimiento* is a one-thousand-word text, drafted in 1513, that proclaimed Spain's right to seize Indigenous lands and goods, legally justifying any violence that would occur were the Indigenous people not to comply immediately with what the conquistadors demanded. When those acting on behalf of the Spanish crown encountered new communities, a member of the conquistadors' party would read the text in Spanish to the Indigenous populations prior to overtaking them. A seemingly participatory text, it "ask[s] and require[s]" its audience to accept the Catholic Church and the Spanish monarchy as their rightful rulers. But within the appeal to participation is coercion in the form of a threat:

Y si así no lo hicieseis o en ello maliciosamente pusieseis dilación, os certifico que con la ayuda de Dios, nosotros entraremos poderosamente contra vosotros, y os haremos guerra por todas las partes y maneras que pudiéramos, y os sujetaremos al yugo y obediencia de la Iglesia y de sus Majestades, y tomaremos vuestras personas y de vuestras mujeres e hijos y los haremos esclavos, y como tales los venderemos y dispondremos de ellos como sus Majestades mandaren, y os tomaremos vuestros bienes, y os haremos todos los males y daños que pudiéramos . . . ; y protestamos que las muertes y daños que de ello se siguiesen sea a vuestra culpa y no de sus Majestades, ni nuestra, ni de estos caballeros que con nosotros vienen.

(But, if you do not [submit], and maliciously make delay in it, I certify to you that, with the help of God, we shall powerfully enter into your country, and shall make war against you in all ways and manners that we can, and shall subject you to the yoke and obedience of the Church and of their Highnesses; we shall take you and your wives and your children, and shall make slaves of them, and as such shall sell and dispose of them as their Highnesses may command; and we shall take away your goods, and shall do to you all the mischief and damage that we can; . . . and we declare that the deaths and losses which shall accrue from this are your fault, and not that of their Highnesses, or ours, nor of these gentlemen who come with us.)[59]

This legal document is, historically, perhaps one of the most important instances of what J. L. Austin would consider a performative speech act. As Paja Faudree succinctly puts it, "the very act of uttering the text was intended to fundamentally alter the social relations between Spanish and natives."[60] Still, the actual effectiveness of the document was questionable. Various historical texts make clear that, when it was read at all, it was read aloud to Indigenous peoples who did not understand the Spanish language.

What most interests me about this founding document of imperial conquest is the specific *imaginary* of a legalized construct of voice as a sound that it employs. The fact that the Indigenous listeners will not understand the meaning of the text underlies the entire premise of the *Requerimiento* as a founding document of Ibero-American Atlantic relationships. On the one hand, the document is a classic example of the kind of imperial authority that multiple scholars have linked to the lettered-orality divide.[61] It helps build what Ana María Ochoa Gautier calls a Western "power-knowledge nexus" that validates certain perceptions of sound by inscribing them, in writing, in "an acoustic regime of truth . . . in which some modes of perception, description, and inscription of sound are more valid than others in the context of unequal power relations."[62] However, as José Rabasa has pointed out, "contrary to the commonplace that presumes that the opposition between orality and writing is transhistorical, I would not only argue that it assumes different values in different historical moments and cultures but also insist that this [oral-lettered] binary was hardly central to sixteenth-century relations."[63] After all, most of those on the ships were likely also illiterate. In other words, as Ochoa Gautier has further argued, the letters-orality dyad is, in many ways, an a posteriori echoic memory of a situation imagined from well within an established colonial frame, allowing lettered elites to construct the notion of literacy as constitutive of the modern.[64] As she writes, the field of orality thus "functions as a mechanism through which the subaltern is simultaneously named as having a voice, yet such a voice is subordinated by the very same principles through which it is epistemically identified as other."[65] The opposition between orality and literacy is a reflection of a modern(izing) Western aurality that denies the lived experience of those participating in the production of the events later inscribed in written memory.

But I am interested in the sound of language itself. That being the case, I would like to suggest here that the performative circumstances imagined by the document, despite its likely material ineffectiveness, position the sound of language as a sensation tied into its legal structure in a way that is still deeply enmeshed in how we hear voice and language today.[66] Legally, the document participates in a European sense of textuality in which the written word transmits authority. As a performative act in the Americas, however, this expression of legal voice is addressed to an ear that will not hear its meaning: Neither the text nor the conquistador reading it aloud hears the voice of the law as meaningful. Rather, the law is a double enunciation in which the paralinguistic sound of voice *as* a sensation of sound prevails over its meaning and does so in the interest of war, not communication. The transduction of law into a sensation that cannot linguistically mean serves as a founding principle of its rule. As Barry Truax has argued, in any context in which a language is not understood, the ear aurally processes the paralanguage of the other; that is, the tone, pitch, volume, or timbre of its interlocutors' voices, and ascribes some kind of interpretive meaning to the sound.[67] The sound of voice as both an acoustic and a discursive event is present in what is imagined by the *Requerimiento*'s creators to be the performative aspect of the document itself. The phrase "y si así no lo hicieseis o en ello maliciosamente pusieseis dilación, os certifico . . ." (But, if you do not [submit], and maliciously make delay in it, I certify to you) linguistically presumes understanding, but in the aural imaginary present in its articulation, it reveals a desire by the speaker (and the apparatus he represents) for the listener to *not* understand what is said, since that could leave room for discussion or resistance. Moreover, the sound of the language perceived to be dominant in this model acquires an affective weight on both listener and speaker that does not necessarily map to linguistic meaning. The document and its readers *desire* the act of communication to be primarily about noise, not comprehension, even as its reading aurally imposes a monolingual geography of the ear (a Castilian Spanish one) onto a territory to which it does not belong. The delay in comprehension by Indigenous listeners can therefore be interpreted as a delay in compliance and justification for pillaging whole communities, changing the very places in which conquering voices are heard and creating a new, transatlantic geography of the ear grounded in the epistemological tensions between the sound and the meaning of legal language.[68] The acoustic occupation of the territory by the voice of the Spanish conquest reallocates that space as a property owned by the sound of Spanish, whether it is understood or not; this dynamic will reverberate, with English and Catalan as well, in the acoustics that obtain in the gentrification of Barcelona as a newly globalized space centuries later.

In this case, the notion that the law's voice is *not* heard in any meaningful way by the public to whom it is read, that the ear of the "foreign" person being addressed *cannot* make it mean, is what makes the law all-powerful. For here, "foreignness" is heard not in relation to territorial ownership but as a reflection of an ear that either does or does not understand Spanish. As the sound of language is imagined in the text, the natives' inability to hear a different language as meaning and behave accordingly justifies the legality of the colonial project. We might say that, in this model, the work that has gone into trying to decide whether or not the subaltern can speak does not really matter, since it is whether or not the subject hears properly, according to the rules of those with voice (grounded in the particular sound of their language), that is of concern. Thus, while it may be commonplace to assert, following Gayatri Spivak, that the subaltern subject has no voice, we may reply that, according to the colonial aurality that already supposes the other will be unable to understand language as law, here she is perceived also as having no ear.

I am not arguing that this is a founding document for Barcelona or Catalonia's self-image. The first obvious point is that it is written in Castilian, not Catalan. Moreover, whether or not Catalans participated in the conquest has been a contentious point among some historians.[69] Rather, I am arguing that this is an early instance of an aurality of voice, territory, linguistic identity, and law that ties monolingualism to an assumption of voice as split between sound and meaning. Moreover, it signals the complications of a geography of the ear that hears language as emplaced by a territorial origin: Clearly, languages that originated in the Americas are silenced by the sound of Spanish, which converts territory into geography. I am suggesting that the idea that the Amerindians were legally responsible for the loss of territories and violence that was to be wrought upon them because they had heard, and not understood, the *Requierimiento* creates a deceptive aurality of noncommunication, rooted in a sensation of voice, in which the "unintelligible" *sound* of voice supersedes any meaning of language it could communicate, even as the Spanish language seemingly justifies itself as the only sound worth listening to. And it is this deceptive aurality of noncommunication, which takes place in the bidirectional transduction between legal (or otherwise socially authorized) discourse and the sound of voice as an affective sensation, or tone, for ideological or other purposes, that I am calling *sleight of ear*. That is to say, voice is not simply a corporeal relationality that can somehow overturn the metaphysics of logocentrism by drawing our ear to our interlocutors' humanity, as feminist philosopher of voice Adriana Cavarero has argued. Nor is it simply the aural counterpart to what Diana Taylor has called a scenario—the repeated acoustic staging of a colonial encounter that "numbs us with familiarity," the aural equivalent to images of Indigeneity and conquest becoming a "paradigmatic system of visibility [that] also assures invisibility" because we do not even see the historical violence it represents anymore.[70] Rather, I am suggesting that voice, as sound, itself becomes inscribed in a Western ear that associates sensation with nonmeaning and uses that sensation as a force that validates the semantic sound of (usually a single) language as reason and modernity. This sound of voice (what Barthes famously called the "grain of the voice") is not limited to sound as rhythm or resonance but extends to timbres and affects that create spaces—including the negative affects Sianne Ngai has theorized as emerging through tone: "Tone *is* the dialectic of objective and subjective feeling that our aesthetic encounters inevitably produce."[71] She is referring to the tone of literary texts, but the idea of an affective dialectic in the subject-object (voice/ear) relationship that has produced a transatlantic colonial aurality also obtains here.

Within this context, the *Requerimiento* is a productive tool for understanding how coloniality became embedded in an aurality of voice tied to law. In the scenario of the *Requerimiento*, the tension between the legalistic/discursive voice that can be understood and the affective sensation embedded in both the sound of the conquistadors' speech and the listener's ear (conceived as faulty) produces a whole new imperial Atlantic acoustic geography. It is a geography that hears the ear of the colonized as deaf to the acoustic reason wielded by the colonizer but still attuned to tone, timbre, and the paralinguistic sounds that make up its regime. At the same time, the Spanish can use their own deceptive hearing of enunciation as sound as a new form of spatial appropriation under the guise of what they would begin to call, in 1573, "pacification," rather than conquest.[72] This colonial aurality, as an epistemology of voice, has continued to reverberate not just around the West but in Orientalist approaches to other parts of the world, like Africa and the Middle East, as well.

The geography of the ear that echoes into sensations of dispossession today, in fact, depends on this aurality of deception, the *sleight of ear* I have referred to. In the *Requerimiento*'s foundational dialogue, language as system is subsumed by language as sensation—as embodied feeling—which in turn constructs the deaf ear that can be heard as inferior and in need of training. This exemplifies sleight of ear, an operation that produces the ear as split between voice as sound and voice as meaning *for political purposes*. This is a politics that travels echoically across the sea; it creates a geography of the ear tied to reason and sensation where both are necessary for constructing a shared aurality but that also requires the tension between them to be repeatedly produced. Theories that seek to keep the discursive out of sound, proposing daily aurality as some sort of universalizable experience, repeat this gesture when they extract language from sound.

Moreover, echoic memories heard into this aurality of voice have real effects on the spaces in which they circulate, cycling into political actions with material consequences. In the *Requerimiento*, the ear of the other is already heard as deaf to meaning by those who control the political sleight of ear. When those in control address someone who does not understand their language as capable only of a sensation of the ear, incapable of hearing meaning, or reason, the listener's inability to understand seems to justify any action taken to make the rest of their body the site of subjective construction, since neither letters nor voice can suffice to produce a subject who can hear properly. And with this comes all of the uses and abuses of punishment by the state that Foucault theorized, and which, as Achille Mbembe has shown, colonialism has wrought on the minds and bodies of those it conquered and enslaved, producing a geographical, if not always territorial, distinction between subjects with rights, theorized from Europe, and the "living dead" who worked the plantations, whose bodies, tied to voices heard as meaningless and ears heard as incapable of hearing meaning, were expendable.[73]

As such, the sleight of ear that often defines contemporary political aurality has consequences for studies of democratic voice in Spain and elsewhere. The voices of immigrants, political opponents, women and LGBTQI+ people, and other voices of alterity are all subject at varying times to the sleight of ear which excludes them from the production and sharing of meaning. This is especially true when, in the most extreme cases, these voices or other thick sounds are dismissed as gibberish.[74] Rita Segato has interrogated the same epistemic principles in order to argue that, in the Americas, the production of minority subjects as other through the colonial episteme reshaped a reciprocal, dual Indigenous social structure into a hierarchical, binary one through which the masculine public sphere became the domain of the universal One, which has the strongest impact on gender: "Thus understood, the history of the public sphere is nothing less than the history of gender. The public sphere, that state agora, thus becomes the locus of enunciation of all politically valued speech."[75] Segato has in mind multiple forms of alterity—"feminine, nonwhite, colonial, marginal, underdeveloped, deficient"—when she defines the patriarchal public sphere as an epistemic structure. But within this frame, we may then consider the multiple gendered repercussions the sleight of ear produces when it splits the sound of voice as sensation from its linguistic meaning: the raped woman whose "no" sounds like "yes" to her rapist or a judge; the trans person whose sense of gender does not sound "correct" when voiced and who thus may be dismissed as not hearing even themselves correctly. The voices of those who are heard as incapable of speaking "properly" due to their accents or the visual interference of skin tone become even more vulnerable because these traits make it easier for the material sound of the tongue's difference to be identified and narrated as out of place.[76] Sleight of ear is not simply present in the vocalization of letters as representation and authority; it normalizes the notion that it is the inability of the listener's ear to process meaning correctly, or of their voice to perform the law or the norm as it is "meant" to be heard, that is the unspoken justification for a dominant voice's authority over a space, whether domestic, local, national, or global. Queerness, racial difference, and class are all susceptible to the sleight of ear when not just politicians but everyday people hear accented or "foreign" languages as outside the realm of "intelligible" voice, often while also hearing the idea of democratic voice as tied to a particular monolingual structure of the law.

In today's media age especially, these geographies can migrate echoically from one territory to another, creating imagined aural relationships that cross borders, both spatial and temporal, often creating new political or musical connections in the process. For some scholars, such as Luis Cárcamo-Huechante, broad invocations of "coloniality" are problematic because, unlike embodied Indigenous approaches, which may be situated in specific histories and lands, they perpetuate "a state-tied view [that] reflects the Althusserian tendency to 'theorize' the colonial question as a matter of 'relations of power' within the horizon of a disembodied and ahistorical superstructural sphere."[77] I agree with Cárcamo-Huechante, especially within a Latin American context in which, as he has pointed out, the colonial period has never ended for Indigenous communities.[78] But in a situation like Barcelona's, in which the argument for Catalanism is directly linked to the notion of a native tongue and is at times made by media companies, politicians, and businesspeople who wield incredible political and cultural power over the mattering maps of the city, it seems important to interrogate the theoretical and the material together to understand how this extraction of the materiality of sound from language as voice has taken place over time. Doing so allows us to recognize the daily disjuncture between, first, the epistemological and economic projects of modernity qua coloniality that might inform perceptions of belonging and exclusion in contemporary democratic spaces, and, second, the material sensations of sound as noise or alterity that are evident within the aurality of accent, voice, music, and noise that obtain in polyphonic situations.

As a political discursive tool, I might even venture to say that the geography of the colonial ear encapsulated by the *Requerimiento* has so saturated contemporary media in the Atlantic world that echoic memories of coloniality can be harnessed to produce political performances of colonial oppression even by those who have historically and economically been empowered (as the *Polònia* representation of upper-middle-class Catalan students as oppressed suggests). This occurs when a Catalan ear can hear itself as colonized by Spanish, or when a Spanish ear refuses to hear a Catalan complaint of repression. It also takes place when the Catalanist project of defining voice by the proper sounding of the Catalan language in turn dispossesses immigrants who are unable to speak with the sound of a native or who cannot speak the language at all. These constructs of voice presume a monolingual context that has never been adequate to the mattering maps that circulate daily in almost any modern space, but especially in Barcelona. So listening in through a paralinguistic context, as I do in the chapters ahead, allows us to hear how sensations of varied linguistic or musical sounds complicate nationalist, ethnic, or even territorial origin stories of sound and identity. In geographically triangulating my theoretical approach to sound in Barcelona by linking Spain to Africa and the Americas, though, I do not in any way wish to suggest that we can hear Catalan in the same way that we hear Indigenous languages in the Americas, like Quechua, Gitxsan, or Mapudungun, or that their political struggles are the same in historical, socioeconomic, or geopolitical terms. Quite the contrary: What I am concerned with is how the sound of language as voice comes to play a role in producing not just a geography of the ear, but a political sleight of ear. In this case, it is one in which some Catalans can affirm Catalan's minority status with respect to Castilian Spanish in order to gain political power, while in the process perpetuating a modernity/coloniality framework with respect to populations in the city who are economically and culturally dispossessed. And by dispossessed I mean not just immigrants who do not speak the language but even the Catalan working classes whose voice is the materiality of their expression as Catalans, but whose accents may be heard as different with respect to linguistic norms; thus even their ears may be perceived as faulty. In Barcelona, where the felt experience of oppression vis-à-vis Spain at times coincides, paradoxically, with economic and territorial power, the political sleight of ear that adopts colonialism as a rhetoric of political exclusion also at times dispossesses the very marginalized populations who are already heard within economic and legal frameworks as incapable of voice, whether they speak Catalan or not.

THE THICK SOUNDS OF BARCELONA

In the chapters ahead, I interrogate just some of the geographies of the ear that have obtained in and around Barcelona after Franco's death, and how their echoic relationships emplace and move sounds to produce mattering maps of the city that are present simultaneously, at times overlapping, at times producing quite different acoustic spaces from one another. I do not concentrate on the tried and true analyses offered by institutionally backed Catalan culture, which frequently extol the city's modernity, celebrate the literary and political forefathers of Catalanism, or denounce the linguistic repression suffered under Franco.

Instead, while respecting those perspectives, I listen in to the smaller soundscapes of the city's *barris* and communities that coexist daily with the grander narratives of Catalan identity since the Transition, and about which much has yet to be written. The accented immigrant communities. The sonic occupations of the city performed by anarchist protestors. Queer presentations of traditional *coplas* performed publicly in drag. The working-class musicians and sound artists who were punk before punk was mainstream, and whose free radio experiments have participated in the Barcelona underground for over four decades. The hidden-in-plain-sight colonial imaginaries of race that continue to be part of the Catalan mainstream media even as African migrants are now an everyday part of the Barcelona soundscape. The construct of the "global war on terror" that reverberates in the geopolitical ear and often taints how immigrant voices are heard throughout the West today.

I want to make clear from the outset that although I am centering my work on Barcelona, I recognize the city is just part of a wider swath of the voices and accents throughout the autonomous community of Catalonia, the broader construct of the Països Catalans—including the Balearic Islands and Valencia—and the polyglot space of Spain more generally, which includes not just Castilian and Catalan as native languages, but Basque and Galician, as well as Asturian, Valencian, and Aranese.[79] I do not want to repeat the gesture of presuming Barcelona is somehow a superior or more Catalan space than those I do not directly address. However, because it is such a polyphonic place, it is attractive to me as a way of thinking through the geographies of the ear that produce urban spaces.

As with Barcelona itself, the progression of the chapters reflects a gradually changing city that went from being a very local place at the time Franco died to one that today is highly globalized. At times the colonial ear is my focus (Chapter 1), but at other times it is subsumed in the flow of other ways of hearing dispossession, such as through gender and migration (Chapter 2), disputes over access to the airwaves (Chapter 3), or protest (Chapter 4).

Chapter 1, "Travel, Race, and the Colonial Sleight of Ear," uses the vast media production around a classic of Catalan children's literature, Josep

Maria Folch i Torres's 1910 character Massagran, to interrogate how a colonial sleight of ear originally figured through aural depictions of African voices echoes into comics, television cartoons, records, and stage plays that return in the production of a Catalan aurality in the 1980s and 1990s. While these newer media extend an imperial geography of the ear grounded in the transatlantic context of European colonization in the Americas and Africa into a globalized setting, they also produce a racialized notion of what I call a *toothless voice* to sustain a Western logic of monolingual intelligibility.

In Chapter 2, "Of Immigrants and Accents," I turn my ear to the intersection of gender and (im)migration in Barcelona, both during the early days of the Transition and in the decades since. Speaking to the role that linguistic, rather than state, borders play in producing geographies of belonging and exclusion in Barcelona and Spain, this chapter shows how José Pérez Ocaña's public, trans* subversions of Catalan spaces with the sound of Andalusianness and other representations of accented immigrant language in several contemporary novels entwine with sounds of gender and queerness. Together, they *redress* extant theories of voice and vococentrism, as, I argue, the migrant accent produces *vocal chords* that constitute a critique of voice as representation.

Chapter 3, "Radiophonic Restlessness," turns to music and radio, examining the underground punk and experimental sound movements that emerged in the early 1980s, around the same time as the Catalanist project for Linguistic Normalization was coming into its own. Exploring the radial microrevolutions that echo into and out of the neighborhood of Gràcia via the free radio station Radio PICA, I show how punk bands and sound art grounded in a libertarian or anarchist ideology tap into a local echoic space with a long history of aural resistance, through which the scale of national sound as a dominant aurality is called into question.

Last, in Chapter 4, "Protest and the Acoustic Limits of Democracy," I focus on a constellation of so-called *algarabías* (noisy rackets) that took place in Barcelona before and around 2017, including pro-independence demonstrations, a terrorist attack, and antigentrification protests. Exploring how democracy is sensed as well as practiced through sound, I demonstrate how a long temporal geography of the ear fuses with the tangle of sensations and affects, including joy and rage, that circulate through the sound of protest to repudiate an aurality that figures legitimate democratic voice as silent. Through the etymological reach of *algarabía*, we hear how the contemporary democratic soundscape is shot through with fractal echoic memories that bring the past into the sound of the present.

Finally, in the Coda, I narrate my own experiences listening to Barcelona as someone who is not from there, who listens with an imperfect ear, and who is herself a first-generation immigrant, raised in a bilingual family with two very differently accented parents, and who is now raising her children in the United States in Spanish, a language that is not natively her own, but is now theirs, despite their non-Hispanic heritage. I consider how my mattering maps of Barcelona have been forged over years of travel, mediatic consumption, and archival study in order to examine how, in that very act of listening to identities other than my own—as we all do daily—I am thrust into, and continuously navigate, multiple geographies of the ear at once.

Travel, Race, and the Colonial Sleight of Ear

In the opening pages of her now-classic *Imperial Eyes: Travel Writing and Transculturation*, Mary Louise Pratt writes, "Empires create in the imperial center of power an obsessive need to present and re-present its peripheries and its others continually to itself. . . . Travel writing, among other institutions, is heavily organized in the service of that need."[1] Focusing on narratives about Africa and Latin America, she attempts to understand a gaze that originates in Europe, travels abroad, and returns home in the form of a "planetary consciousness" that "create[s] a sense of curiosity, excitement, adventure, and even moral fervor about European expansionism."[2] In this groundbreaking rethinking of Ángel Rama's notion of transculturation, the European gaze not only reinforces itself through the colonial encounter, it repeatedly produces the reader's knowledge of the places he inhabits, whether at home or abroad.[3] This construction of colonial identity for a European audience also produces the readers who are the objects of the travel narratives' discourse, as they come to know themselves through what they read.

Once a lettered knowledge available only to a few, travel narratives such as those Pratt describes became more common in the late nineteenth century, when the massification of the newspaper led to an explosion of transatlantic travel *crónicas* (chronicles) by writers like Rubén Darío and Enrique Gómez Carrillo, who published descriptions of the lands they traveled to in newspapers on both sides of the ocean. Within a few short decades, though, travel narratives were no longer strictly textual: When the writer Ramón Gómez de la Serna became the first orator on Spanish radio to take his *micrófono ambulante* (walking microphone) to the streets, broadcasting the live sounds of the Puerta del Sol for listeners, he re-created the experience of the travel chronicle for a local audience, defamiliarizing the daily life of the city through the intervention of this mechanical ear. As the twentieth century progressed, other mediatic portrayals made interactions with travel a common daily experience for increasingly globalized publics. The geographical complexity of these daily sights and sounds, seen and heard through mediatized notions of travel, allows us to recognize that, even as ideologically a center/periphery binary seems to undergird the economic divide between the Global North and the rest of the world, the daily interactions that take place because of travel and migration, whether real or imagined, are rarely so simple.

Indeed, Pratt's concept of an imperial vision relies on written texts, but with radio, television, and the internet it becomes prudent to ask what role sound plays in producing the planetary consciousness that obtains in the geography of the ear. As John Mowitt has suggested, there is no sonic equivalent to the gaze. Instead, he proposes the *audit* as a kind of hearing with an aesthetic perspective. This is neither perception nor a sensible event, in the way Jacques Rancière understands it, but "a fold where perception turns back over on itself, traversing the faculty of hearing with the angle, the posture of listening."[4]

This angle, this "posture of listening," I argue here, emerges not just in the exploratory or ethnographic kind of writing that informs lettered knowledge of distant places, but in the daily experiences of sound that engage with the materiality of local spaces but also veer into the virtual, often through media that fictionalize cultures at home and abroad. In the twentieth century, mediatized constructs of nation, in particular, are of paramount importance. Paul Giles has argued that nations create their virtual imaginaries through exposure to the literary and cultural representations other nations make of themselves, allowing us to consider narratives of national identity in terms of "familiarity and alterity, domesticity and estrangement" that look abroad as well as within: "The point here is that national histories, of whatever kind, cannot be written simply from the inside. The scope and significance of their narrative involve not just the incorporation of multiple or discordant voices in a certain preestablished framework of unity, but also an acknowledgement of external points of reference that serve to relativize the whole conceptual field, pulling the circumference of national identity itself into strange, 'elliptical' shapes."[5] The virtuality of these modes of perceiving the nation, he suggests, converts material objects into mirror images, "in a process of aestheticization that highlights the manifestly fictional dimensions of their construction."[6] The mirror imagery he invokes, with its distortions and refigurings, carries a similarity to the reverberations of what I refer to as *echoic memories*. As Steven Feld writes of the relationship between sound, time, and space that obtains in that term, "Acoustic time is always spatialized; sounds are sensed as connecting points up and down, in and out, echo and reverb, point source and diffuse. And acoustic space is likewise temporalized; sounds are heard moving, locating, placing points in time."[7] The "strange, 'elliptical' shapes" of the virtual nation are much like the fold that Mowitt describes as part of the audit, a material change that is not just a change in perspective, but a modification in the very material that echoes. In Barcelona, these shapes are heard in the sound of the Catalan language as an echoic space in the world that can both travel and be traveled to, identified both with place and with a certain way of being. These discordant transfigurations of the national imaginary are the mechanisms by which the listening ear produces the sound and voice of others: Perceptions that enter the community or the mediascape at a certain point become reified as aural imaginaries, in a constantly adjusting feedback loop between the sound that impacts the ear and the ideologically informed spatial understanding of how to interpret the local, the national, and the global. When linguistic sound strikes the ear, the temporal and geographic aural imaginaries that imbue sounds with affect, judgment, and meaning hear them as familiar or strange, orderly or chaotic, civilized or barbaric within a broader aural imaginary of planetary space.

My contention in this chapter is that the supposedly accented sound of language is fundamental to these echoic memories of nation and global space, and that even in written texts accent forms part of how the traveling ear is built. Language is often the primary and most constant daily experience of sound travelers have with local communities. It is bound up with concepts of the mother tongue as the founding element of national identity, which, as Etienne Balibar has discussed, elides the difference between what Ferdinand de Saussure called *langue* and *parole* (the system of language as a structure versus the daily spoken iterations of it) to produce a shared sense of identity, even though in so doing the sound of language itself confirms unspoken differences among members of the community: "Belonging to the linguistic community—chiefly because of the fact that it is mediated by the institution of the school—immediately re-creates divisions, differential norms which also overlap with class differences to a very great degree . . . [such that differences in linguistic competence] function as caste differences, assigning different 'social destinies' to individuals."[8] Indeed, most linguists will tell you that accents are, largely, the social understanding of dialectical variations.[9] But accent is not just pronunciation or the presence of certain lexical or morphological features in a language. As Ana María Ochoa Gautier has further argued through engagement with biopolitical theory, the sound of language forms part of a lettered concept of aurality that is not only associated with progress, it goes so far as to distinguish between the human and the animal in order to create a polity: "Eloquence involved . . . a grammaticalization of the voice in order to create the distinction between a proper and an improper human, a way of 'directing the human animal in its becoming man.'"[10] Even before the radio, she argues, a lettered voice is distinguished from an unlettered one by a process of rhetorical training in which the onus is put on the speaker to be heard as eloquent, and the incorporation of nonstandard sounding language into literature is a challenge to normative institutions.[11] Similar dynamics emerge in the portrayals of language and accent that originate in Barcelona around the turn of the twentieth century, when reforming the language was tied up with aspirations for modernization and progress, as well as a drive for political autonomy from Spain. The Catalan literary establishment, including newspaper and magazine culture, focused on developing Barcelona as a political and economic powerhouse within Spain, even as Barcelona's lettered and political elite also turned their attention south and west to former and current colonies in Latin America and Africa as they confronted the loss of Spain's empire, and north to France, Germany, and England for models of modernity. In portrayals of Africa especially, racialized constructs of language became a prominent way of asserting Catalonia's modernity over "barbarous" territories in need of further civilization, whether abroad or at home. Local and foreign language was therefore ever present in the planetary consciousness, informed by empire, that shaped the lettered fin-de-siècle geography of the ear in Catalonia.

This tendency to listen simultaneously outward and inward that defined the late nineteenth century for Catalonia emerged again seven decades later. After Franco's death, Barcelona became both Spain's great hope for inclusion in Europe and its representation on the global stage, from which it had been separate for so long. All the while, it remained a territory whose economic prowess, alongside its memories of political repression, ended in arguments that Catalonia merited more autonomy within, and eventually independence from, Spain. Those longing to overturn Catalonia's repression and censorship by the dictatorship thus felt that the correct sounds of Catalan voice had to be taught and the Catalan soundscape had to be actively produced. Their goal was to combat not just the dominance of Castilian Spanish but the increasing influence of French, Italian, and other European television programming and film after Spain opened up to the rest of the world.[12] The result was the production of a linguistic norm. Thus, in the early 1980s, the presenters at the newly formed Catalunya Ràdio and TV3, as well as those who wrote for the newspapers *Avui* and *Diari de Barcelona*, began to follow linguistic guidelines known as the *Orientacions lingüístiques*, based on linguist Lluís López del Castillo's 1976 *Llengua standard i nivells de llenguatge*. The guidelines aimed to standardize a public language that had not had a full media presence since the Second Republic, such that it would model proper pronunciation, vocabulary, and grammar for listeners. As the former director of *El matí de Catalunya Ràdio*, Antoni Bassas, put it, "El nostre català s'havia anat embrutant de barbarismes gairebé sense adonar-nos-en, i un parell de generacions més tard d'acabada la guerra civil dir bona tarda era una extravagància. O sigui, que tots, emissors i receptors, vàrem tenir molta feina a acostumar els uns la parla i els altres l'oïda" (Our Catalan had been sullied by barbarisms without our even realizing it, and a couple of generations after the end of the civil war saying *bona tarda* [good afternoon] was an extravagance. That is, all of us, broadcasters and audience, had a lot of work to do: some to attune their speech and others to attune their ears).[13] With other languages publicly considered "barbarisms," deviance from the new norm could easily be heard as a sign of an insufficiently or improperly Catalanized speaking subject. The move to normalize the Catalan language through television and radio, in schools, and through public campaigns that supported Catalan culture also sought to reassure Catalan speakers that they could freely speak the language in public. It might, in turn, create new speakers to augment the Catalan public sphere, which would itself have political effects. If, prior to the dictatorship, Catalan intellectuals' focus was on normalizing the language as a lettered form of expression that would elevate a largely oral language to the status of other European languages, afterward emphasis was on reviving and reforming that newly lettered sound as the dominant everyday sound of the region.[14]

In what follows, I interrogate a particular set of travel tales that began at the turn of the century but have been a persistent presence in the Catalan mediascape since then. These tales revolve around an adolescent character called Massagran, developed by Josep Maria Folch i Torres in 1910. An early popular example of how voice is sounded for the public in Barcelona, these novels figure the tangled relationships among language, nation, and colonization both at home and abroad, in ways that draw our attention to the role fictional constructs of sound play in producing aurality as a vocal construct that is also visual. Importantly, this visual aspect is tied intimately to race, which comes to represent both the underside of modernization and the need for linguistic reform. Moreover, this racialized portrayal of voice links the colonial ear forged in the early twentieth century to the debates about nation that would take place during the Transition decades later. As Balibar writes, linguistic projects around nation produce a racial and ethnic construct around the sound of voice that amounts to "class racism": "'Foreign' or 'regional' accent, 'popular' style of speech, language 'errors' or, conversely, ostentatious 'correctness' immediately [designate] a speaker's belonging to a particular population. . . . The production of ethnicity is also the racialization of language and the verbalization of race."[15] Writing from a Marxist lens highly influenced by Louis Althusser, Balibar's emphasis is mainly on class as an aural instance of hierarchy. Yet Fran Tonkiss complicates this understanding of the national ear by illustrating how skin tone and accent become intertwined in how publics perceive and discuss immigration: "Some people . . . sound stranger than others; certain voices jar to certain other ears. The immigrant, it has been said, is *audible*, and indeed those forms of race thinking that cannot bring themselves to speak of skin often are happy to talk of language."[16] In other words, for Tonkiss, the sounds of accent or mispronunciation in language allow for political discussion of belonging in ways that cannot be articulated via direct discussions of race. I like this approach because it allows us to consider the fact that perceptions of accent do not always relate to pronunciations of one's own language: In fact, as we see below, not all listeners can immediately identify the features that make up a language's specific pronunciation, especially if the language is an unknown one. Indeed, Homi Bhabha argued in his now-classic text "DissemiNation: Time, Narrative and the Margins of the Modern Nation" that colonized subjects who return to the metropolis as migrants after decolonization challenge the pedagogical model of national identity through the performative differences they enact daily. For Bhabha, these interstices between the pedagogical and the performative are the site of "writing the nation."[17] As I will show throughout the Massagran series, however, this normative and normalizing character draws the ear of a reader or viewer to the faulty speech of non-Catalan speakers through a seemingly humorous portrayal of accent that maintains the pedagogical aurality of standardized language as a cultural norm, but it does so by extending it first to an African and then to a global setting. In these tales, it is in the play between the mouth and the ear, as embodied producers of language, where the sounding—as well as the writing—of the Catalan nation really takes place.

Tonkiss's suggestion, that one can silence skin by talking about language instead of race, is largely the focus of the Massagran tales I explore here. Massagran was invented in 1910 as a children's book character who travels to and linguistically conquers Africa. Over time, he has been reproduced in stage plays, comic books, records, television cartoon shows, and even on websites, in scenarios that put him in touch with "Eskimos," "redskins," Pacific Islanders, and other *salvatges* (savages) that allow him to model a Catalan ear as a necessary part of civilization, progress, and development. These constructs form part of the Catalan imaginary of what Nina Sun Eidsheim has called "sonic blackness."[18] Here, though, this Blackness is figured not through actual listening to the timbres and tones of the voices of Black people, as is the case in Eidsheim's analysis, but rather through an elliptical, fictional portrayal of Black and Brown aurality that justifies the colonial project. As Iñaki Tofiño has pointed out, "En el imaginario colectivo español, África es musulmana, no negra. De ahí que el discurso sobre la antigua Guinea española haya sido en general escaso, fragmentario y minoritario" (In the collective Spanish imaginary, Africa is Muslim, not Black. For that reason, discussion about the former Spanish Equatorial Guinea has been, in general, scarce, fragmented, and marginal).[19] More to the point, Joshua Goode signals the tendency across the contemporary political spectrum to declare Spain different from its European counterparts, unaffected by racism or even racializing thought, despite the fact that, as he and others like Julia Chang have shown, race is ever present in the discourses that produce the construct of nation in Spain (including, I would add, Catalonia).[20] Indeed, the stereotyped assumption that Blackness does not form part of Iberian history reverberates in the portrayal of religionless Black aurality in the Massagran stories, setting the stage for a broadening out to an imaginary of non-Catalan voice in general as uncivilized, a trope that extends beyond the tales and into ideas about immigration and tourism to Barcelona. After forty years of insularity caused by Franco's dictatorship, the Catalan fin-de-siècle geography of the ear informed by an aspirational colonial project returned in the 1980s, taking a global turn to address several troubling aspects of Barcelona's history that had lain dormant in the national narrative for decades. In the updated storylines that populate the comic books and television cartoons of the 1980s and 1990s, then, Massagran participates in his own version of the Olympic games in Athens, addresses the history of Catalonia's lucrative participation in the slave trade, and discovers the perils of globalization, as Catalonia's textile factories are shut down by the exportation of labor to cheap Southeast Asian markets. The fictional sonic Blackness developed in the early texts shifts the aural imaginary to a nationally framed setting of migration and tourism through a cunning sleight of ear, an aural structure that, as I explored in the Introduction, plays on the relationship between language as meaning and language as sound in order to hear one's own voice authoritatively. In these later texts, accent is not simply a metaphor of an imperfect language that needs to be lifted to the level of established national languages through education, as it is in the early twentieth century; in a new mediatized era, it becomes a lived acoustemology directed toward racialized and traveling sounds in a newly globalized Spain.

Despite its popularity, few critics have analyzed this children's series in depth; those who have tend to point out that the stories are humorous and that, therefore, they avoid the seriousness of other adventure tales, such as *Robinson Crusoe*, or those of Jules Verne, Mark Twain, and Robert Louis Stevenson.[21] Alternatively, the character is compared to the Belgian Tin Tin, not in a way that recognizes that *Tin Tin in the Congo* is a colonialist and racist depiction of Africa, but merely a popular media phenomenon. To me, Massagran is a compelling object of study because for a century the character has helped forge in young readers, viewers, and listeners an aural imaginary of Catalanness that informs so many of the tensions around identity, language, and nation that have dominated Catalan-Spanish politics over the past three decades. I wish to clarify from the outset that this is not an interrogation into the actual colonialism perpetrated by Catalans in Cuba, Equatorial Guinea, or Morocco.[22] I am interested, instead, in how the series interpellates listeners again and again into the ideology of the colonial ear by linking echoic memories of race to contemporary questions of travel and immigration. In the process, it extends a geography of the ear grounded in the fictional and philosophical engagement with the early transatlantic context of European colonization in the Americas, into fictionalized African, Indian, and Pacific Islander contexts that a more precise ear would recognize as different, but which here instead privilege a Western logic of monolingual intelligibility over the territorial specificity on which the fictions of language and sound seem to be based.

As I explore below, the sound of accent is at the center of the texts—and, later, the audio recordings' and television productions' depiction of Catalan culture and the "barbaric" characters who present Massagran with the opportunity for producing not just a stronger Catalan voice, but an ever-more-perfect ear. By displacing loaded concerns about empire and modernity to humorous, and, I would argue, offensive, situations involving an adolescent abroad (or, after the Transition, interacting with foreign travelers and immigrants at home), the weight of the cultural interactions can be brushed off as child's play. But behind the humor is an *audit*, an aesthetic figuring of sound that reflects the colonial aurality that helps sustain the national imaginary of Catalonia within Spain, paving the way for discussions of globalization and immigration that are still playing out in the city today.

BUILDING THE COLONIAL EAR

For over a century, children in Catalonia have been reading or watching the tales of Massagran, an intrepid young Catalan explorer who sails from the port of Barcelona and has rousing adventures at sea and in far-off lands. Author and playwright Josep Maria Folch i Torres first conceived the character in 1910, when he published *Les aventures extraordinàries d'en Massagran*, illustrated by caricaturist Joan Junceda. The Massagran tale is the first book for children written in Catalan and might even be considered the first novel in the language.[23] It was so popular that Folch i Torres wrote a sequel in 1913, *Noves aventures d'en Massagran*, and a stage play based on the books in 1921.[24] Publication of the Massagran texts was prohibited under Franco but, due to the fact that censorship laws about the Catalan language did not apply to sound recordings (and perhaps because it aligned with Franco's support of colonization in Equatorial Guinea), in 1958 the author's son, Ramon Folch i Camarasa, produced a short recording on vinyl of the story, voiced by the well-known Ràdio Barcelona voice actor Isidre Sola. After the death of Franco in 1975, the Folch i Torres franchise once again acquired a robust presence in Catalan society. Although the list is long, it is worth briefly outlining some of the multiple ways in which the character has been recuperated since then. On the one hundredth anniversary of Folch i Torres's birth in 1980, the serial *Nadala* (an annual publication by the Fundació Jaume I distributed to Catalan institutions with the intent of establishing a shared Catalan intellectual culture) published an issue in his honor, titled *Josep M. Folch i Torres: per a una cultura catalana majoritària*.

Much as previous issues had celebrated the five hundredth anniversary of the first book printed in Catalan, or the hundred-year anniversary of Enric Prat de la Riba's birth, this issue used Folch i Torres's memory to suggest the importance of recognizing Catalan as the national language in print, on the radio, and on television. Shortly thereafter, in 1983 TVE Catalunya (Televisió Espanyola Catalunya [Spanish Television Catalonia]) produced a live-action miniseries based on the play, *Les aventures d'en Massagran*, which featured actors in blackface playing the African characters Massagran encounters.[25] The author's son also published a series of Massagran comic books, illustrated by J. M. Madorell, that ran from 1981 to 2002, and in the 1990s a television cartoon came out based on them. 2010 was named the *Any Massagran* (Year of Massagran), and the latest edition of the original two novels, complete with Junceda's drawings, was announced in the long-running magazine for children, *Cavall Fort*, and published in 2018.[26] Today, all the cartoons are available on a website called Super3, a page for children under fourteen associated with *Televisió de Catalunya* and the last vestiges of the Club Super3 show that defined childhood for many starting in the 1990s and ending in 2021. The website, which has video games and mobile apps, assures parents of the good quality of the content.[27] As his son once wrote, Folch i Torres "taught three generations of children to read."[28] Now his works serve to introduce children to televised content online as well. There is no doubt that, despite its suppression by Franco, the Massagran character constitutes an enduring Catalan franchise.

Through all the texts and programs, Massagran is presented as "un noi de casa bona" (a boy from a well-off family), who is an adventurous traveler and often runs into trouble.[29] In the first novel, looking for something fun to do, he sneaks onto a ship. After he is forced overboard for stealing a chicken from the ship's galley and washed ashore on an island, the rest of the tale focuses primarily on his encounters with threatening animals—mosquitoes, a whale, a giant condor, a hippopotamus. In the sequel, *Noves aventures d'en Massagran*, we learn that "despite his modesty," Massagran harbored desires of domination.[30] Thus, the second novel revolves around the conquest of several Black tribes in an unidentified African land: the peaceable *kukamuskes*, who make their savior a general for defeating the warmongering *buskabronkes*, and the *karpantes*, cannibals whom Massagran helps the *kukamuskes* outsmart with his keen ear. As the tale comes full circle, the captain and cook who had thrown Massagran off the ship in the first novel are washed ashore and captured by the community that Massagran has successfully colonized.

When the books were first published, Catalan intellectuals of the period commonly argued that, following the loss of Spain's last American colonies in 1898 (Cuba and Puerto Rico), Catalonia needed to "civilize" itself in order to compete with the global empires that were still going strong. Catalonia thus needed to become modern, forming what one of Folch i Torres's contemporaries and founder of the aesthetic branch of Catalanist ideology known as *noucentisme*, Eugeni d'Ors, would call a *Catalunya ciutat* (Catalonia city).[31] In this idealization, the modernization of the city would coincide with a celebration of the pastoral, folkloric origins of the Catalan peasant. In the process, Catalonia would also recuperate an imperial position in the world that Spain was no longer capable of. An example of this ideal, the unassuming, quick-witted Massagran manages to outsmart his elders while also successfully producing a new Catalan community, one that displaces to an imagined territory the *noucentista* suggestion that the Catalan populace can be civilized by language, culture, and law. Massagran in fact renames the community of African people he has colonized as "Katalatribu": "En Massagran ho va aprofitar per a parlar-li del seu projecte de formar un gran poble de negres amb el nom de Katalatribu, en el qual hi hauria escoles, orfeó, casino i camp de futbol." (Massagran took the opportunity to describe his project of forming a large community of Blacks with the name Katalatribu, in which there would be schools, a choir, a casino, and a soccer field.)[32] The book also associates the uninstructed "savages" with the children reading the book, as those yet to be civilized into Catalan society, by specifically calling his new subjects *katalanins*, a term he explains refers to their status as not yet fully formed Catalans because they lack understanding: "Perquè acaba en *nins*? No veus que teniu tan poc enteniment? I al meu poble, als que no tenen enteniments, se'ls diu *nins*, de manera que *katalanins* vol dir, ben traduït, 'nins catalans'" (Why does it end in *nins*? Don't you see you have little understanding? And in my town, we call those who have little understanding *nins*, such that *katalanins*, translated well, means "little Catalans") (311). By making lack of understanding what unites the reader (a child) with the *katalanins*, the book makes clear that both are Catalans in development. And as we will see shortly, language and a particular kind of ear are central to this civilizing project, which aimed to extend a Catalan geography of the ear not just to the new *Catalunya ciutat* but to the wider Mediterranean world.

Before getting to the text, though, I wish to outline some of the colonial and national history that sets the stage for the ideas of travel that are at stake in the original texts, and which recirculate at the end of the twentieth century. The creator of Massagran, Josep Maria Folch i Torres, was one of five brothers who were extremely influential as Catalonia came into its own politically around the turn of the twentieth century. In addition to the Massagran stories and the now-traditional Christmas play *Els pastorets*, Folch i Torres also produced 1,906 editions of *En Patufet*, a children's magazine that reached sixty thousand subscribers—a remarkable distribution for the period, greater than that of any newspaper or magazine for adults.[33] Enric Jardí has argued that Folch i Torres's prolific publishing was the patriotic gesture of a devout Catalanist, his writing dedicated to "la formació moral de nens i adolescents, d'acord amb la moral cristiana. . . . féu adquirir a uns públics joves l'hàbit de la lectura en català, en una època malauradament massa prolongada en què l'ensenyament de la nostra llengua ha constituït casos excepcionals" (the moral formation of children and adolescents, in accordance with a Christian morality. . . . he made young publics acquire the habit of reading in Catalan, in a period—unfortunately far too long—in which [acts of] teaching our language constituted exceptional cases).[34] The popularity of his children's series far exceeded that of the literature by Jacint Verdaguer, Narcís Oller, or Joan Maragall, who have come to represent, often as much for their extraliterary status as for their writings, the origins of Catalan literature.[35]

This new emphasis on children's literature echoed other attempts of the period to convert Catalan into a "llengua de cultura" (language of culture), which, as Joaquin Fuster put it, would help Catalan speakers "eludir el complex de patois a què semblaven condemnats" (avoid the *patois* complex to which they seemed to be condemned).[36] The lettered classes throughout Spain wrote and spoke primarily in Castilian, so to elevate the Catalan language from the orality of the masses into a daily, lettered interaction required training in a new kind of speaking and listening. Of particular importance was the work of Pompeu Fabra, who through the Institut d'Estudis Catalans (Institute of Catalan Studies) published the first official orthographic norms in 1913, and in 1932 the *Diccionari general de la llengua catalana*, on which the official Catalan dictionary today is still based, in an effort to standardize Catalan and thus elevate it to the level of other national languages, despite its being stateless.[37] But the poets, editors, and publishers associated with the institutions run by the Lliga Regionalista also contrasted their brand of lettered Catalanism to the popular Catalanism associated with anarchists, proletariat groups, and working-class citizens who might either support the Spanish federal government or internationalist understandings of the class struggle.[38] Children's literature was fundamental to the *noucentistes*' attempt to solidify and spread this new lettered Catalanism, and thus, as the author of the first Catalan novel for adults, Josep Maria de Sagarra, once wrote, Folch i Torres's efforts went beyond the pedagogical: "en Folch i Torres ha ensenyat als literats del nostre país que, escrivint en català, un no s'hi mor del tot de gana, i per això tots els que tenim una ploma als dits hem d'agrair-li-ho" (Folch i Torres taught the writers of our country that, writing in Catalan, you won't die of hunger, and therefore all of us who hold a pen between our fingers should thank him for that).[39] Given his respected place in Catalan literature and culture, it is not surprising that as recently as in the summer of 2019, when I was at the Biblioteca de Catalunya (Library of Catalonia) in Barcelona, the second Massagran comic book, *Aventures encara más extraorinàries de Massagran*, was on the display of recommended classics for children. What was surprising—even shocking to me—is that, in its visual and aural depictions of race, the story is among the most offensive of the series. But I will return to that later.

Around the time Folch i Torres invented Massagran, the Barcelona elite were navigating a touchy relationship with the central government in Madrid, their loss of colonies in Latin America, and a new turn to Africa as a possible site of future economic development. Travel, as both a popular activity and a possibility for future expansion of Catalan prosperity, was tied both to the colonization of Africa and the production of space locally within Catalonia. In the nineteenth century, watching Latin America gain increasing independence, Antonio Cánovas del Castillo, the future prime minister of Spain, had redrawn the imaginary end of Spain at the Atlas Mountains, conveniently including Morocco in its territorial imaginary. He cited Roman historians as justification for this imperialist geography.[40] Similarly, the influential *Regeneracionista* (Regenerationist) intellectual Joaquín Costa, whose politics were otherwise largely a negative response to Cánovas's restoration of the monarchy following the failed First Republic, also justified Spanish colonization in Africa; he argued that for two millennia Spain and Africa had maintained contact and thus shared an "ibero-bereber" culture that meant Spain was better able to understand Africa than other European colonizers could.[41] After 1898, most discussion about colonialism even within Catalonia still concerned the future of a Spanish state that had lost revenue and identity when it lost its colonies in Latin America and the Philippines, the sentiment echoing Isaac Muñoz's 1913 lament, "Will this inert and rigid Spain ever be convinced that the future is an African future?"[42] Therefore, as Brad Epps and Gustau Nerín i Abad have shown, even as Spanish intellectuals and politicians worked to reconstruct ideological connections to postindependence Latin America by promoting the idea of a Hispanic brotherhood, they also justified several colonialist wars in Morocco by drawing on the above-mentioned romanticized understandings of Spain as sharing a Hispanic history and geography with Africa. The Treaty of Paris that put an end to the Spanish-American War had also placed part of Equatorial Guinea, Río Muni, under the rule of the island Fernando Po (today Bioko), which had been a Spanish colony since 1848.

Spain's colonization of Equatorial Guinea outlasted the Moroccan project and came to form part of what Abad calls a *hispanotropicalismo* that extended the presumed Iberian sensibility also seen in Portuguese colonialism further into the African south and west. Unbeknownst even to many in Spain, Equatorial Guinea remained under Spanish colonial rule until 1968. Spanish colonists, most of whom hailed from Catalonia, put the rival Fang and Bubi tribes to work in deforesting, cacao, and coffee production, even as the Black upper class in the territory, the *fernandinos*, drew on their earlier contact with British colonials to claim elevated social status. The primary colonizing force in Equatorial Guinea were the Claretian missionaries, founded by a priest from the Catalan town of Vic whose power within Spain was such that he was at one time the personal confessor to Queen Isabel II. The main agricultural company, the Compañía Nacional de Colonización Africana (ALENA), was largely run by Catalan businessmen, and the agricultural union had its headquarters in Barcelona, creating relationships in the city with Guinean plantations that supported their business interests.[43]

In the wider Africanist effort that reflected Spain's continued attempts to play a role on the continent, then, Catalonia occupied a unique position. The Catalanist campaign at home, meanwhile, emphasized the importance of autonomy, which often linked national politics to an idea of language as identity: Even as it monopolized control of Equatorial Guinea, Catalonia rejected the conscription of its men into armies whose purpose was to expand Spain's presence in Morocco, not because they were against the war per se, but because they felt such a move represented central Spain's imposition of power over them. Such arguments thus rejected the construct of *Hispanismo* that justified Spain's interventions in Africa as some kind of consolidation of the *madre patria*.

Moreover, the conservative political party, the Lliga Regionalista, and the Catalanist publishing industry supporting it—such as the newspaper *La Veu de Catalunya* and the satirical magazine *Cu-Cut!*—focused on consolidating local political power, advocating for Catalan autonomy from Spain. In 1909, what is known as the Setmana Tràgica massacre of Catalan workers protesting Spanish presence in the city took place, accompanied by a violent shuttering of the *Cu-Cut!* magazine, whose satire had drawn the ire of the Spanish government. It is worth noting that the editor of the magazine, Josep Baguñà, is the same person who acquired the *En Patufet* children's series written by Folch i Torres and made it profitable. Moreover, the illustrator for the Massagran series, Joan Junceda, also drew caricatures for the magazine.

As far as travel was concerned, not only was there a constant movement between Catalonia and Equatorial Guinea for economic and missionary reasons, a newfound national pastime in Catalonia, *excursionisme*, was reaching peak interest. The counterpoint to the *flâneur* walking the city, the excursionist explored the countryside, affirming for himself his connection to Catalonia's identity as a newly modern place that also reflected a rural, naturalized—and nationalized—concept of the past: "Caminar implicaba un proceso de territorialización de la cultura, donde los mapas, las tradiciones y las epxresiones orales de la lengua fueron los ejes del primordialismo cultural que hacia posible imaginar la nación" (Walking implied a process of territorialization of culture, in which the maps, traditions, and oral expressions of language were the axes of the cultural primordialism that made it possible to imagine the nation).[44] As Ricardo Cicerchia has explored, the turn-of-the-century obsession with hiking—which began in France and soon spread to many parts of Europe—valued the scientific aspect of exploration, as hikers carried out what we may consider a touristic appropriation of Alexander von Humboldt's methods for their own personal enjoyment. In Catalonia, the excursion was presented as having the pedagogical utility of producing good Catalan citizens. As Manuel Milá i Fontanals, of the Associació Catalanista d'Excursions Científicas (Catalan Association of Scientific Excursions), wrote in 1879, discovering the rural sites of Catalan culture that had been neglected would help educate the Catalan public about its history, elevating the national culture.[45] The Centre Excursionista de Catalunya (CEC; Excursionist Center of Catalonia) and its monthly publication, the *Bulletí*, moreover, linked the excursion to literary celebrations of Catalan lands, like Verdaguer's *Canigó*, as well as *costumista* attempts to record local speech, and general celebrations of ruralism.[46] In some cases, this ruralism carried with it anarchist underpinnings, as local *ateneus* devoted to promoting internationalist thought among their members, also promoted excursions meant to escape the increased urbanism that accompanied the city's development.[47] A thousand years after its founding, the Santa Maria de Montserrat abbey, whose Virgin the pope named the patron saint of Catalonia in 1881, became a frequent destination not just for religious believers, but a new form of pilgrim, the tourist.[48]

The national(ist) project of *excursionisme* was, however, also tied to a planetary consciousness through shared institutions, clubs, and publications that critiqued French, Portuguese, and other European colonialist enterprises in Africa, not because they were seen as morally repugnant, but because they were considered insufficient in their civilizing efforts. Archaeological excursions and an incipient tourism also overlapped with industrial interests in discovering opportunities to further or better exploit foreign resources or develop areas abroad.[49] In 1909, the year before the first Massagran tale was published, a distinguished member of the CEC founded the Turisme Marítim (Maritime Tourism) section, under the umbrella of the Asociación Nacional del Fomento del Turismo (National Association for the Promotion of Tourism). Concurrently, members of the CEC who had begun with hiking and now added maritime tourism to their definition of patriotic activities also saw it as their civic duty to convert Barcelona into a tourist destination. One of the Folch i Torres brothers, Ramon, was the general secretary of the Societat d'Atracció de Forasters (Society for the Attraction of Outsiders), which was dedicated to the promotion of Barcelona to tourists.

If Ramon was focused on attracting tourists to the city, Josep was not only publishing children's books (as well as "practical" tourist guides to Barcelona and Catalunya for his brother's organization), he was writing travel guides to Spain's African territories. In 1911, two years before he published the tale of Massagran successfully colonizing the "katalanins," he wrote a nonfiction book called *África Española*. A unique mix of photographs, drawings, history, and anecdote, this rare text begins with an "excursión por las poesesiones españolas de África Occidental" (excursion through the Spanish possessions of West Africa) that frames the travelogue with a dialogue between a child eager to know about Africa and his uncle Antonio, "comisionado por el Gobierno español para recoger datos respecto a las obras que deberían realizarse en aquellos territorios para ponerlos en condiciones de salubridad" (commissioned by the Spanish government to collect data regarding the work that needs to be done in those territories to bolster their health conditions).[50] Yet the editor's note, as well as several commentaries throughout the text, relate that at stake is the investment potential of the territories in need of development. Part of that development means dealing with local cultures, which Folch i Torres elaborates "in picturesque detail," as the editor writes. Thus, the generalized "moro de marea" (coastal Moor) of the Sahara has an expression in his eyes that "revela una extremada hipocresía" (reveals an extreme dishonesty); he "es cobarde y traidor, humillándose rastreado ante las más absurdas y brutales imposiciones de los moros del interior" (cowardly and treasonous, humiliating himself before the most absurd and brutal impositions of the Moors of the interior).[51] In contrast, the *pámue* (Fang, a Bantu ethnic group in Equatorial Guinea) is a cannibal, but "por razón de su fuerza no es hipócrita como la mayor parte de los africanos" (by dint of his strength he is not fake [*hipócrita*] like the majority of Africans) (60). After speaking with both a French anthropologist and a Catholic missionary, Folch i Torres concludes that "los pámues son los más fáciles a la civilización, y prueba de ello es que las escuelas de las misiones son frecuentadas por una proporción de pámues muy superior a la de otras tribus" (the Fang are the easiest to civilize, and the proof of that is that a higher proportion of Fangs go to mission schools than do those of other tribes) (60–61). The section on Equatorial Guinea in particular foreshadows the sensationalized imagery and themes that will appear in the later Massagran stories: The body of a Bubi child is "más mono que humano" (more monkey than human) (18), the tribesmen have "limitada inteligencia" (limited intelligence) (66), the Bubi king sits at a table made from a human skull (12), and the Spanish and Catalan missionaries are striving to produce in the natives "el amor al trabajo" (the love of work) (29).

Within this cultural framework of travel as both a leisurely development of the nation locally and an economic investment in progress abroad, Massagran's original journeys participate in the construction of a national imaginary brought not nostalgically to a land that is represented as naturally Catalan, but to new, unexplored places that can be produced as such. As we will see, a particular cultivation of the ear is fundamental to that process.

SLEIGHT OF EAR AND THE TOOTHLESS VOICE

In the Introduction, I argue that what I call *sleight of ear* is central to the way the colonial ear conveniently uses the rupture between language as sound and language as meaning to hear the ear of the other as faulty. This model of hearing is present in both of the original novels written by Folch i Torres, which present Barcelona as a port that opens the city to the wider world and represent other cultures as in need of development. In both original novels,

Folch i Torres demonstrates a comparative epistemology typical of the colonial eye and ear, replicating early modern chronicles of the new world by seeking to understand new places through Barcelona as the central point of reference. An alligator's mouth is like the gaping maw of Antoni Gaudí's Sala Mercè theater.[52] The trees are compared to the plantain trees on Les Rambles, while a fruit Massagran finds is similar to Valencia oranges (80). The first island he lands on is as quiet as "un solar de l'Eixample" (a vacant lot in the Eixample) (79), and the *kukamuskes'* music, which they perform in Massagran's honor after his defeat of the *buskabronkes*, is compared to the Murga Gaditana, a form of street music from the city of Cádiz performed during Carnaval. In a way, the texts also echo the sixteenth-century Valladolid debate on the subjectivity—and even humanity—of Indigenous people.[53] In a more disturbing comparison, then, sounding more like Juan Ginés de Sepúlveda than Bartolomé de las Casas, the text repeatedly uses animal imagery to describe the Africans Massagran encounters.[54] When Massagran first sees the *kukamuskes*, he is still out in the water, and they swim out to him. At first he is happy to see people—"encara que fossin negres" (even if they were Black)—but he then decides he will need to discover if they are "bona gent o si, al contrari, són d'aquells que els agrada la carn blanca" (good people or if, on the contrary, they are the types that like white flesh) (175). Between the binary of "good people" and cannibals, though, the narrator quickly presents a third option that will allow Massagran to feel safe around them: They are "salvatges" (savages) who are like the animals he had encountered before. Thus, they swim out to him "com si fossin peixos, obrint contínuament la boca" (as though they were fish, continuously opening their mouths), with "dents més blanques que una rajola de València" (teeth whiter than a Valencian tile) (177). Massagran thus begins to throw raw hippo meat at "els negres, els quals, més contents que un gos amb un os, van començar a saltar per l'aigua" (the Blacks, who, happier than a dog with a bone, began to jump around in the water) (177). As if to solidify the relationship between them and the animal kingdom, when they arrive back on shore, the Africans begin to dance around Massagran, and his dog Pum "va posar-se a lladrar de la manera més desaforada" (began to howl in the most outrageous way).[55] Although the sound scares the *kukamuskes*, who run away, this allows Massagran to conclude—in yet another animal metaphor—that "Si que són gallines, aquests salvatges! Per un lladruc, tot això?" (They really are chickens, these savages! All that because of a bark?) (178).

Indeed, sound, and especially the sound of language, perceived by an ear attuned to perfecting speech, is central to the text's portrayal of a civilizing force. In *África Española*, Folch i Torres presents clues to the colonial aurality that is at play in his children's literature by arguing that the Bubis speak a language that is "sonoro y agradable al oído" (sonorous and pleasing to the ear) because almost all the words end in vowels, which creates a softness of expression.[56] This softness is visually reversed in the children's tale, as the dominant sound of African voice is demarcated by the letter <k>.[57] Although the sound /k/ is present in Catalan pronunciation, it is usually written as <qu> or <c>; the use of <k> as it is written here seems to signal a harsher, more guttural sound, which would be unfamiliar to the Catalan mouth and ear.[58] Thus, in the first instance of dialogue, the reader is struck by the preponderance of the unfamiliar letter throughout the phrases uttered by the *kukamuskes'* leader, Penkamuska: "—Kukanova: Benarri bat tusi gasen kasan ostra. Sinofas kosa malanotin dràsfat ics; aprosifas kosadol en tallavores tela karrega ras" (Kukanova: Wel cometo ourhouse. If you donoth ingbad, noth ingbad willha ppen to you. But if you be have badly, you will pay for it) (182).[59]

As written, the sentence is at first glance unintelligible, because in addition to making an orthographic change, the narrator has truncated syllables and rearranged phonemes in ways that create new words entirely. With no other information to draw from, Massagran has no other recourse than to attempt to pull from the unfamiliar sounds some kind of meaning, which he cannot do correctly: "Com que en Massagran no sabia el llenguatge d'aquell país, no va entendre res del que li deia en Penkamuska, si bé li va semblar que parlava d'ostres, de teles i de tallar vores" (Since Massagran did not know the language of that country, he didn't understand anything Penkamuska said to him, although it seemed he was talking about oysters, cloths, and cut edges) (182). The oysters, cloths, and cut edges are an example of the messy phenomenon of parsing, through which listeners identify morphemes, the units of language that convert sound into meaning, or when they try to make sense of a language they do not speak. Parsing occurs when a child—or, as Marit Westergaard suggests, an adult in a multilingual situation—processes an unknown, or otherwise "external," language, through their own *I-languages*, where *I* stands for both internal and individual.[60] Both an attempt to organize external sounds into a recognizable system, and a reflection of one's own exteriority to spoken sound, parsing attempts to make meaning from noise.[61] Massagran's inability to understand the language of the *kukamuskes* suggests that his ear is not fully developed. In fact, this scene of aural misunderstanding restages the scenario of the *Requerimiento* I described in the Introduction. As I argued there, in the

Requerimiento a directive is ordered in a language that is not understood, which allows the Spanish conqueror to hear the ear of the Indigenous listener as faulty, and thus without a voice; only through a well-trained ear can one obtain a voice that is intelligible. Here, however, the joke is that Massagran has the faulty ear incapable of understanding and thus is at risk of violence, specifically by cannibals, who, consistent with most stereotyped portrayals of colonial travel, make an appearance later in the text.[62] Yet because the text places the reader on the side of the future conqueror, it also produces the conditions by which he must somehow prevail. And prevail he does, because in the end the novel makes clear that whatever sounds the African tribes might make, Massagran has a standardized and proper Catalan on his side.

In fact, the text creates the unknown language by using truncations of words and rearrangements of phonemes that, when read aloud, demonstrate that they are simply a rewriting of Catalan itself. Lexically, what differentiates Catalan from the language of the Black tribes Massagran encounters is merely a different word segmentation, as well as a frequent use of the grapheme <k> to distinguish the African voices from Catalan ones. The inserted or eliminated orthographic breaks thus change the perception of words from "casa nostra" to "kasan ostra," "te la" to "tela" and "cosa dolenta llavors" to "kosadol en tallavores." A reader who reads aloud will realize that the only difference is the placement of the breaks between sounds, but Massagran is presented as being faced with an unknown language he cannot process. The text thus suggests the importance of a standardized written form, even as it imagines an incorrect voicing as indicative of a whole group of speakers represented by the *kukamuskes* whose faulty ears have reinforced each other in producing a barbaric community. The graphic representation of the *kukamuskes'* voices, with the <k>, shocks the eye, and although the grammar and vocabulary of Catalan and their language is the same, the pace and segmentation of their delivery, as it is imagined here, shocks the (civilized) ear. That being the case, a few lines later, the narrator translates the text for the reader, a practice that continues for a while until the reader can translate for herself: "En realitat, el que Penkamuska acabava de dir no era més que una salutació que, traduïda al català, diu: 'Senyor novament vingut: Ben arribat siguis a casa nostra. Si no fas cosa mala, no et passarà res de mal; però si et portes malament, te la carregaràs'" (In reality, what Penkamuska had just said was nothing more than a salutation which, translated into Catalan, was, "Newly arrived sir: Welcome to our house. If you do nothing bad, nothing bad will happen to you. But if you misbehave, that is on you [you will carry that]") (182). By infusing the written word with both correct and incorrect pronunciations, based not on the phonemes themselves, but on a phonosonic aurality that depends on the reading ear's ability to recognize the breaks between them, the text draws the ear to the importance of proper pronunciation, as well as the relationship between the vocalic and what Mowitt refers to as the "nonvocalized vocalizations (sounds that occur in or on speech but that lack phonemic value)."[63] For Mowitt, these are whistles, whispers, and gasps, but here it is the break between phones, the slight pauses we learn to hear between words, that creates a new sound of language. Indeed, the chapter, titled "Parlant la gent no s'entén" (Talking, people don't understand each other), plays on the Catalan proverb "Parlant la gent s'entén" (Talking, people understand each other) in order to illustrate not just the difficulty and necessity of translation between different cultures, but more importantly, the fact that even within the same language, perceived mispronunciation—what we might broadly call an accent—hinders communication by drawing the ear to what is missed rather than what is shared. At the same time, the text draws attention to the cultural expectations one brings to a newly encountered culture, even if its language is the same as our own:

> —Japres pos sessió deka sas eva—li va dir en Penkamuska.
> "Ara em deu presentar la dona," va pensar en Massagran, en sentir l'últim mot "eva."
> . . . En realitat el que li havia dit en Penkamuska era que ja havia pres possessió de casa seva, com ja hauran comprès els meus lectors. (183)

> (Alreadytake npo session of yer'ouse—Penkamuska said to him.
> "Now he's going to introduce me to his spouse," thought Massagran, as he heard the last word "'ouse."
> . . . In reality, what Penkamuska had said was that he had already taken possession of his house, as my readers will have already understood.) (183)

Supposing a shared communal structure with a home and a Biblical background, the joke in which Massagran hears "Eva" and assumes that means "wife," as opposed to "kasa seva" (your house), highlights the acoustemological homogeneity underlying Massagran and his readers' cultural knowledge. While this structure might reinforce a particular moral hierarchy for the children reading the text, it is the unspoken yet underlying debate around the standardization of Catalan that adult readers could have recognized in the text: Written at the same time as Pompeu Fabra was standardizing Catalan spelling for the first time, 1913, the text is a not-so-subtle commentary on the need for unity among the varieties of Catalan speech and spelling that at the time coexisted in what today we might call the Països Catalans, or the regions where Catalan is spoken. By substituting <k> for <c> or <qu,>, or creating caesurae that interrupt the flow of speech and produce new rhythms in the language, the text represents barbaric language not as other but as sounding jarring and unintelligible simply because it is accented. I recognize that, linguistically, accents are largely social perceptions of the many varieties of a language that may exist; but it is the sound of difference, of not pertaining to an imagined or desired norm, that I am signaling with the term. In fact, <k> aside, the *kukamuskes*' voices seem to trend more toward the Balearic or Valencian pronunciation of Catalan, rather than the Eastern centralized one that is spoken in Barcelona.[64] That being the case, the accent at stake in the "barbarians'" voice is not only meant to represent Africa as in need of colonization; it is meant to represent the outlying edges of the Països Catalans as in need of linguistic standardization, an idea that was also present in Pompeu Fabra's and his contemporaries' *barcelonista* approach to "civilization," which favored the urban dialect over the pronunciations found in other regions.

From a linguistic standpoint, the production and reception of the sound of accent reveal the breaks between the phone as sound, the phoneme as a sonic production of language as system, and the written text as phonetic representation of phones and phonemes. In other words, this scene uses graphic representation to disrupt what Nicholas Harkness has called the "phonosonic nexus . . . a medium through which we orient to one another, not directly, but through phonic engagements with sonically differentiated frameworks of value that shape our social interactions."[65] For Harkness, the literal voice, or what he calls a "voice voice," allows a speaker to "situate herself in worlds of significance, which makes the configuration of social lives . . . congruent and in dialogue with presupposable social configurations of some sort, at different scales."[66] But only so much is in control of the speaker, for whom, in most circumstances, not being heard as having an accent equates to passing, to forming part of an imagined linguistic community that is ideologically tied to a shared phonosonic value associated with the so-called mother tongue. What I am calling accented speech, meanwhile, diverts the listener's imagined construct of linguistic community by revealing the precariousness of the shared phonosonic relationship. In speech heard as accented, the material production of the phone—the sound realized by a given mouth and tongue—already haunts the listener's imagination of the phoneme before it is spoken. Here, the "accent" is written, but the ideological effect of demonstrating to the reader the instability of Catalan as a written, aural, and oral language that can be normalized remains.[67]

By repeatedly interpreting for the reader, moreover, the narrator draws attention to the unstable relationship between the sound and the letter of language. At the same time, the gesture suggests that with repeated exposure to how accented language sounds, eventually a lettered ear will recognize this other language as Catalan in disguise, a Catalan that is at best funny and at worst threatening, but which can be reformed by a smart listener: Once Massagran learns to listen, he will also be able to communicate with those he encounters. There is a double process of learning that must take place for dialogue to occur, in other words: The subject of colonization must be educated into speaking properly, while the Catalan who already speaks well must be educated into hearing how to correct the language when it is mispronounced and where the faults of their own ear lie.

For Massagran, this project is made more difficult because he can only listen, while readers have the graphic representation of sound printed in front of them. Readers affirm the dominance of the colonial ear when they learn to laugh at Massagran's inability to hear the language because of the *kukamuskes'* different pronunciation. But by laughing, they also affirm that they are complicit in perpetuating this kind of aurality that hears "good" language as perfect and accented language as inferior. Indeed, whether we read the *kukamuskes'* enunciation as a representation of a vulgar Catalan populace who needs to learn to speak properly, or as a colonial geography of uncivilized Indigenous people in need of European civilization, in either case the colonial aurality revolves around the construct of accent and what I will now describe as the *toothless voice* as an ideological product relying on the construction of the faulty ear of the other.

In a much darker echo of the scenario of conquest that the novel imagines, it is not just the ear of the child or the ear of the "savage" that must be trained. In the case of the cannibalistic *karpantes*, the whole body of the person being colonized by the Catalanizing mission must be reformed. Massagran easily befriends the *kukamuskes*, and, his ear having matured to a point where he can understand their accented language, he is able to communicate with them enough to become their leader. Yet Massagran's biggest ingenuity, according to the book, is a "projecte genial" (great project): to "senzillament, conquerir i civilitzar els . . . horroritzeu-vos! . . . els karpantes. . . . Volia fer-se'ls seus, conquistar-los, educar-los i ensenyar-los la sardana, el joc de trencar l'olla i la llengua catalana" (simply conquer and civilize the . . . oh horror! . . . the *karpantes*. . . . He wanted to make them his, conquer them, educate them and teach them the *sardana*, the game of breaking the pot [*trencar l'olla*], and the Catalan language) (303).[68] In a brutal act of trickery, Massagran and his loyal subjects, the *kukamuskes*, lay a trap in which they construct a large group of statues made of mud and river stones to look like men. The *karpantes* are so hungry for human flesh that they rush at them and begin to eat them, breaking their teeth in the process. Consistent with the moralizing function of the text, the deception through which Massagran attracts the *karpantes* puts the onus of the *karpantes*' suffering on them for indulging their ravenous desires. Their pain is so great that they give up fighting. Massagran later explains to them that cannibalism is a vice, a sign of poor education that "no est[a] de moda" (is not in style) in Barcelona and "[c]om que jo vull educar-vos, he pensat que el millor que es podria fer per a treure-us aquest vici, era deixar-vos sense dents" (since I want to educate you, I thought the best thing that could be done to get you out of that vice was to leave you without teeth) (311).

This violence against the mouths of the cannibals is presented as a pacifying force, but aurally, the text suggests, it does not affect their speech: They continue to pronounce words in the same incorrect way their noncannibalistic counterparts do, with a preponderance of <k> and incorrect word truncation or segmentation defining their speech patterns. As Steven Connor writes, "A particular kind of mouth is needed to speak a particular language, or dialect. . . . But languages produce different mouths—or what, adapting the term 'voice-body' . . . , might be called mouth-bodies."[69] Here, the mouth-body is completely changed: Losing one's teeth would make any dental production of sound impossible. Yet the narrator's representation of the sounds of these reshaped mouths remains the same. "Civilization" is not just a silencing of voice, then, but a disciplining of the body whose violence is further elided when these damaged bodies are heard speaking with the same voice as before: Although their teeth are removed, both the "noble savages" who appreciate Massagran's presence and the "barbaric" ones he has to tame through violence are heard as speaking in the same manner, despite what must be a difference in sound produced by the difference in body. The inability of the narrator, the character, or the reader to hear the violence behind voices that would, in fact, sound different than before indicates a willful mishearing (a sleight of ear) of the role the mouth plays in producing voice. This scene is useful for thinking through an epistemology of the ear that obtains in colonial encounters and continues to be prevalent today. I am referring to the tendency by scholars, most notably Walter Ong, to think orality in terms of what Jonathan Sterne calls the *oral-literate dyad*, a series of binary oppositions that suppose sound is always other to vision, reason, capitalism, and other hallmarks of modernity.[70] Ong writes, for example, "Many, if not all, oral cultures strike literates as extraordinarily agonistic in their verbal performance and indeed in their lifestyle. Writing fosters abstractions that disengage knowledge from the arena where human beings struggle with one another. It separates the knower from the known. By keeping knowledge embedded in the human lifeworld, orality situates knowledge within a context of struggle."[71] Yet I would like to use voice to consider a different kind of struggle, one grounded in the materiality of the sound-making body. As Brandon LaBelle puts it, voice can be thought of as "a tension—a tensed link, a flexed respiration, and equally, a struggle to *constitute* the body."[72] With this idea in mind, thinking embodied knowledge as orality that is somehow separate from literacy—or more specifically, the notion that literacy removes one's ability to embody sound as (sensed) language—represents a nostalgic fantasy of Indigeneity that wishes to preserve an image of the non-Western subject as more pure, or natural.

However, when we take into account the sleight of ear that contextualizes Folch i Torres's work for its specific audience, it becomes clear that the Massagran construct of the cannibal voice, both before and after the violent removal of the speakers' teeth, is itself a construct of the *same* Western voice that presumes an ontological purity in vocal sound, because both are reinforced by lettered depictions and representations of the same semiotic system. That is to say, the accented sound of difference is reinscribed into a Western notion of voice as representable. This erasure of the violence against the barbarous voice in order to represent civilization as proper speech, simultaneous with the inability of a Western ear to distinguish that sound from a previous "uncivilized voice," perpetuates a virtuality of voice as lettered sound that upholds the aural imaginary of a newly envisioned Catalan society, silencing the actual sound of the mouth in the process. At the same time, it legitimates the fiction that physical discipline is not behind the colonial production of sonic voice as orality that will bring the uncivilized aurally into line.

In other words, neither Massagran nor the narrator distinguishes between language with teeth—language that is heard as a threat because it is voiced by a "cannibal" who seems to present a physical challenge to one's safety—and language that has lost that threat because the teeth have been eliminated, but which is still heard as "foreign" and in need of discipline because it is voiced incorrectly. This is the fine line that divides the sound of voice as accent within a known language and accent as indicative of externality as a threat. Hearing the accent of a Black speaker as simply different, without identifying the material scope of that difference—its forced manifestation on the lips, tongue, or teeth by violence—defines the colonial ear that overwhelms and justifies a conqueror's violence against the body of a "barbaric" other. Hearing racialized accent thus performs a virtualization of the object of the audit, producing an echoic memory of voice as a bodiless form of representation that does not need to take the materiality of lived experience, like colonial racial violence, into account. This toothless voice is related to the fiction of what Douglas Kahn calls the *deboned voice*, which in the next chapter I explore with respect to how accent shapes space. For Kahn, deboned voice reflects how speakers hear in their interlocutors' voices a difference in the texture of the sound, since a speaker hears their own body resonate in the bones of the ear as they speak, while the listener hears that voice "deboned."[73] Here, though, we have not a deboned voice, but a toothless voice, a racialized subjectivity that is produced by violence inscribed into a virtual sound of voice not tethered to the materiality of the body, but rather to an idea of identity that can be "heard" through written language. Civilization, as a project of modernity, occurs by removing voice's teeth, producing an aural concept of voice that is a virtual image or echo of the lack of skin tone implicit in Western subjectivity, which also ignores the material conditions of conquest that produced this construct rooted in the subject-object epistemology in the first place.[74] For LaBelle, in its sounding, voice is like a cord that connects a body to the promise of its subjectivity; voice "does not move away from my body, but rather it carries it forward . . . it drags me along, as a body bound to its politics and poetics, its accents and dialectics, its grammars, as well as its handicaps."[75] I am positing here, then, that the toothless voice, as a universalized construct that results from violence and homogenizes voice as a language that can be standardized, cuts that cord. Voice, as a representation of subjectivity, thus becomes virtual to the ear when it is stripped of the material conditions of its production. Even so, as an echoic memory that becomes recycled in

Western literature and media, this civilizing or modernizing function of voice as toothless, one that ignores the possibility of accent or mispronunciation, is perpetuated in the aural imaginary as the key to community.

The other side of this written, toothless voice of civilization, the one that rules at home and informs the colonial endeavor, is law. For successful colonizers not only harness the sleight of ear to civilize society, they work around law to achieve desired results by hearing it in their favor. In the 1910 novel, for example, Massagran outwits his elders by hearing the ship's lawbook in a way that liberates him from the confines of international maritime law. Early in the story, Massagran is unable to resist the smell of a freshly cooked chicken in the galley, and eats it. Accused of stealing, the captain has him tied to a plank and plans to throw him overboard. In a vocal performance of the law, as Massagran is being hoisted down, the captain reads aloud article 249 of the ship's lawbook, which states that any passenger or shipman who is caught stealing will be punished with death. The captain is just giving the order to throw him overboard when Massagran asks him to reread the passage aloud. Excited upon rehearing the phrase—and having forced the captain and all around him to listen again—Massagran explains that technically they did not catch him in the act of stealing; he was found out after the fact. The captain protests that they know he was the thief, to which Massagran replies, "Tot el que vulgui, però la llei és la llei, i com que allí diu que sols es donarà pena de mort al qui s'*atrapi* robant, i a mi em sembla que està ben clar que ningú no m'ha atrapat, si volen ser justos i bons complidors de la llei, ja m'estan deslligant sense perdre temps" (Whatever you may like, but the law is the law, and since there [in the rule book] it says that only someone who *is caught* stealing is punished with death, and since it seems pretty clear to me that nobody has caught me, if you wish to be just and comply with the law, you won't waste any time untying me).[76] In this exchange with the law, Massagran demonstrates an exquisite ear, an ability to hear legal language so keenly that he does not merely sign off on the law, as the captain does: He rehears it in a way that makes it his own.

In that hearing, a Catalan ear is not about a form of expression to be translated or mistranslated by a keen listener, but rather a tool that one can always learn to wield more perfectly. Far from proposing the law itself needs to be changed, the scene suggests that the written law must be heard as already perfect in order for a community to function, while it is the task of listeners to hear into it what they need. Moreover, when, in the sequel, *Noves aventures d'en Massagran*, the tables are turned and Massagran has a chance to punish the same captain and cook who had exiled him from the ship, Massagran explains that he can't really be upset with the captain for throwing him overboard since "l'home complia la llei, i ja se sap que la llei és sempre la llei" (the man complied with the law, and it is known that the law is always the law) (324).

Language is tricky, the novels suggest, but it is the ear that can and must define it. Both Massagran and the reader must learn to recognize the greater lie behind the African speaker's language—that it always already was Catalan, just a Catalan that needs to be corrected, by force if needed. Moreover, a keen ear recognizes that the natives' voices are noise that is meaningless, except when they are heard and interpreted by a knowledgeable ear that reasserts the authority of a Western aural imaginary of language as the foundation of empire and civilization. A keen ear also hears into the law a means of appropriating it for one's own interests, to individualize the authority of the community rules to suit a different civilizing project than the one being carried out by those in power. Moreover, Catalonia's colonial presence in Africa, as it is fictionalized in the early Massagran tales, sounds the elision of violence that is the foundation of the colonial ear. That elision supports a universal construct of voice as a linguistic representation without a body that shrouds the echoic memory of a perceived toothed threat to the Western ear, replacing it with a universalizable toothless voice that neglects to hear the deception that created the loss of teeth. As we will see below, this hearing of voice as bodiless produces a geography of the ear in which travel is always a colonial endeavor, brought forth after the Transition through Barcelona's late twentieth-century engagement with globalization, immigration, and tourism.

SOUNDING RACE (ECHOES)

I have been arguing that the sleight of ear that takes place in the colonial endeavor renders voice toothless, at the same time as it perpetuates a sound of difference in which mapping voice to race constitutes a threat to civilization. Indeed, notwithstanding its metaphoric relationship to reforming Barcelona and Catalonia, the Massagran tales base their idea of barbaric voice on a racialized body. If *África Española* used photographs and realist drawings of the places and people the traveler could meet in Spanish Africa, in the children's series the illustrator Joan Junceda seems to draw on the wider European figure of the Golliwog, dark black in skin tone and with grotesque facial features, such as white circles for eyes and wide white mouths, to produce the *kukamuskes, buskabronkes,* and *karpantes.*[77] Several of the drawings in the children's books reflect the supposed savagery of the characters with whom Massagran comes into contact, as they are depicted eating the leg of a white man on a plate, or, in one particularly disturbing scene, are drawn with bodily proportions and poses that recall those of chimpanzees. Moreover, late in the second book, Massagran discovers that the king of another tribe is actually a Catalan poet who had been shipwrecked and had painted himself black because he feared the tribes he encountered would eat him if they saw he was white (294). This character in blackface, Baldufa, joins Massagran in his civilizing project.

Perhaps one of the best-known images from the tale is one of Junceda's drawings, a depiction of a large Black man with the words "Taca Caca" (Poop Stain) inscribed on his chest, which serves as a blackboard for educating the rest of the group. (This is the image I found so offensive in 2019 when I saw it on display in Catalonia's national public library.) The novel insists time and again that Massagran should be lauded for "ensenyant de llegir i d'escriure a aquells pobres salvatges, fent-los aprendre la dolça llengua catalana, ensenyant-los a ballar la sardana i a fer la sortija" (teaching those poor savages to read and write, making them learn the sweet Catalan language, teaching them to dance the *sardana* and make the ring).[78] Yet this picture highlights the disconnect between an orality that must be affected by the violent reshaping of the mouth and its translation into written text where the body is absented, converted into a blank slate on which the words seem to sound a "soft" vowel—<a>—all while naming the Black body a dirty stain. The picture thus encapsulates the way the "toothless voice" exerts a violence of its own on the Black body through the phonosonic nexus of language.

As Nina Sun Eidsheim has written, what she calls *sonic blackness* is associated with vocal timbre that allows racism to emerge in how voice is heard, even as it has become politically incorrect to be racist on the basis of skin tone. This "sonic blackness" is "not a single phenomenon, but might be a combination of interchangeable self-reproducing modes: a perceptual phantom projected by the listener, a vocal timbre that happens to match current expectations about blackness, or the shaping of vocal timbre to match current ideas about the sound of blackness."[79] Historically, at least in the United States, the idea of Black voice plays on what sociologist John Cruz has referred to as *ethnosympathy*, a nineteenth-century construct of Black voices heard by white ears, who can assume Black slaves are human enough to suffer and portray that suffering through spirituals and other forms of Black music. I would add that a crucial part of this sonic Blackness is the sleight of ear. In Catalonia, this sonic Blackness is produced through tales like the Massagran novels, which not only produce the stereotype of the Black ear as faulty because it cannot reproduce a lettered sound of voice, but which also hears the Black voice as problematically accented, even threatening, or otherwise offensive, and whose body must therefore be reformed, through violence if necessary. The supposed humor of the drawings is based on the violent aspect of Europeanization that both produced the sound of mispronunciation *as* a threat and then rendered it inaudible through the very act of colonizing and conquering that makes the voice, as voice, toothless.

As I described above, this echoic memory of race in the form of voice preexists any acoustic sounding of the Massagran tales because it emerges through writing. Yet it acquires specific acoustic support once the Massagran tales are performed on stage, recorded on vinyl, and portrayed on television. For example, the 1958 audio recording of the Massagran story adapted by Ramon Folch i Camarasa, titled *Les aventures d'en Massagran. Cuento infantil*, was voiced by a white man, one of Ràdio Barcelona's most famous actors, Isidre Sola.[80] At the time, Spain was in the midst of the larger decolonial struggle in Africa. Many colonial states, including Algeria, the Congo, Angola, and Kenya, not to mention Equatorial Guinea, were lobbying for independence, and as French and Belgian colonies gradually obtained their freedom, Spain and Portugal struggled to hold on to their African lands. That same year, the so-called Guerra Olvidada in Ifni, a now-former Spanish province in Morocco, was taking place, as the Moroccan Army of Liberation sought to free territories from Spain's rule. Although Spain managed to retain its possession of the Sahara, ideologically the defense of Spanish imperialism overseas fell to nineteenth-century, European constructs of nation that perceived African Indigeneity as in need of spiritual and economic salvation. On June 3, 1961, Franco himself declared Spanish colonization efforts to be superior to those of the past because

no puede confundirse la noble empresa de colonización, la elevada tarea de alumbrar pueblos nuevos, entregándoles generosamente—en una verdadera transmutación espiritual—toda la propia herencia de cultura, con este concepto peyorativo y actual, encarnado en dolorosas realidades de hoy que se ha llamado colonialismo o coloniaje.

(one cannot confuse the noble enterprise of colonization, the elevated task of conceiving new peoples, generously delivering to them—in a true spiritual transmutation—all of one's own cultural heritage, with this current and pejorative concept, incarnated in painful realities that today have been called colonialism or the colonial period.)[81]

Given Spain's censorship mechanisms, it is not surprising that the Catalan recording from 1958 leaves out all references to Massagran wishing to instill in the African tribes any sense of Catalan identity. Instead, the story focuses on Massagran's capture by the *kukamuskes* after he lands on their island, and how he saves himself after the leader, Penkamuska, sentences him to death. Massagran escapes by employing a singsong tone and repeating "calla, bonica" (be quiet, beautiful) to tame a lion—his later constant companion, Pam. After training Pam to roar on command, he then uses the roar to scare off the cannibalistic *karpantes*, who are about to wage war on the *kukamuskes*. As a result, Massagran is crowned king of the tribe. Focusing on Massagran's ability to manipulate the *karpantes* through his use of a sonic scare tactic rather than the cruel means through which he tricks them into losing their teeth in the book, the record overlooks the violence of the colonial endeavor to focus on the tale as a mere childhood adventure. But in so doing, it perpetuates a generalized echoic memory of race as accented, where race is associated aurally with skin color, and not the violence against the rest of the body that has historically accompanied its production. In other words, the record audibly realizes the idea of the toothless voice I described above.

In the process, the record appropriates and inverts the notion of ethnosympathy that produced sonic Blackness in nineteenth-century North American music by attributing an affective portrayal of voice to the characters that contrasts the tone and pitch of Massagran to the voice of the leader of the *kukamuskes*, Penkamuska.[82] As in the books, the record establishes a phonic differentiation between a correct and an accented form of language, aurally replicating the written representation of the *kukamuskes'* imperfect Catalan in the novel. Penkamuska's words are peppered with mispronunciations. And like the written text, Penkamuska's staccato and guttural voice inserts pauses between the phonemic building blocks of words, and joins the endings with the next word in the sentence. Sola also voices Penkamuska as having a deep-pitched, almost robotic delivery devoid of emotion. This contrasts with the smooth and fluid, high-pitched and emotionally varied vocal prosody of Massagran. The exaggerated distinction between the pitch and prosody in the two voices creates a kind of counterpoint between white and Black, civilized and uncivilized, feeling and unfeeling. The aural cues in the recording thus correlate the sound of emotion to a perceived humanity and intelligibility innate to the European character, even as Massagran's voice suggests a physical weakness that makes his ingenuity all the more necessary. In a sense, then, unlike North American publics who valued the emotive sounds of Blackness for their portrayals of suffering, here that sound is co-opted by a Spanish civilizing force that sounds weak even as it prevails. Massagran's tone rises and falls constantly as the voice actor adds *ayyyys* to reflect his moments of cowardice, or uses a higher timbre than that of Penkamuska to communicate his youth. The tribe, on the other hand, chants and sings along with a steady drum beat that suggests their society as a whole lacks emotion, incorporating the lack of fluidity of their leader's voice into an evenly rhythmic chant. Vocally stripping them of humanity, the record replicates aurally the hideous drawings of African characters as animals incapable of human emotion, while Massagran is an exaggerated portrayal of excessive feeling who can only be steadied by the law. In both the text and the audio recording intended to reflect it, it is not merely the orality of the language, nor its difference in grammar or vocabulary from Catalan that makes the natives' voice less intelligible than a European one. Rather, the affective tone of voice's sonic construction, communicated through accent, rhythm, and lack of emotion, also is meant to be heard as demonstrating the unfeeling barbarism of the *kukamuskes*.[83]

By eliminating direct references to Catalonia from the story, moreover, the recording slips the colonial ear into a tale of Africa that supports Franco's claims that Spanish colonialism was different from that of the rest of Europe, that Spain had a legitimate claim to colonization, perhaps because of its shared spirituality—a notion Franco had used during the war to legitimate his appropriation of North African non-Catholics into his cause— or simply because of the geographic division between Europe and Africa through which he could hear Spain as African. In fact, a sleight of ear could hear Spain into Africa: In the record, the *kukamuskes* remain savage but open to salvation, while the *karpantes* are silent, ravenous cannibals whose ears are better attuned to animal sounds than to language.

The record also resonates with a popular culture of the time that redirected the ear toward other conflicts in Africa, distancing Spain from Equatorial Guinea's decolonization process. The Catalan singer Dodó Escolà, for example, used tasteless humor similar to that of the Massagran tales to hear the colonies involved in struggles for liberation as a soundscape of *algara-bía*, or noisy gibberish. In his 1960 song "¿Qué pasa en el Congo?" (What's Happening in the Congo?), the names of the Congolese politician Patrice Lumumba and then–Army Chief of Staff Motubu Sese Seko, as well as the short-lived breakaway Republic of Katanga, blend together into a quick string of sounds lyrically compared to a bongo beat. In this hearing, the language and music of any African culture are interchangeable to a white ear, for which all are noise:

Y no es para menos,
pues estos morenos
les dan a los blancos
'mastica café.'
Que si son los balubas
porque están como cubas,
que si allí mister Hache
se ha metido en un bache.
Lumumba, Katanga,
Katanga Lumumba,
la selva retumba,
repica el bongó.
Katanga, Lumumba,
Lumumba, Katanga

(And not for nothing,
but these Blacks
make whites
"chew coffee."
If it's the Balubas,
it's because they are smashed,
and because Mister Hache
has fallen in a hole.
Lumumba, Katanga,
Katanga Lumumba,
the jungle booms,
the bongo pounds.
Katanga, Lumumba,
Lumumba, Katanga)

Since the sound of the chorus mimics the noise of "rabble," the line "¿Qué pasa en el Congo? / ¡Qué blanco que pillan / lo hacen mondongo!" (What's happening in the Congo? / Any white guy they find / they make sausage of him!) further blends linguistic alterity into an echoic memory of savage cannibalism that is both threatening to overtake the white man and easily ridiculed at the same time. Just as the Massagran series does, Escolà's record accompanies this sound with a visual atrocity, using humor as a shield: The cover features the singer in blackface, wearing a top hat and holding a spear.

THE EDGE-SHAPE OF THE TRAVELING EAR

Massagran's most long-standing recuperation, though, came in the 1980s and 1990s, just as the Catalan Normalization project was reaching its peak. During forty years of dictatorship, with its accompanying censorship machine, the legal use of Catalan as an official language that had been briefly allowed during the Republic was halted. All public street signs, publications, and other material circulating in the Catalan public sphere had to be written in Spanish. As a result, the city developed a linguistic mixing that has often been referred to as diglossia—in which speakers use a different language depending on the circumstances in which they find their interlocutors—but which recent studies have redefined as bilingualism, since, as Xavier Vila has argued, there is not always a strict division of the use of Catalan or Castilian according to whether one is speaking in a domestic or institutional setting, as used to be the argument among proponents of diglossia.[84] In point of fact, to hear the Barcelona soundscape as fully Castilianized prior to Franco's death or even as diglossic is itself a product of political narratives that forget the many ways in which the language *was* heard, publicly and privately, during the dictatorship.[85]

During the long decade following the transition to democracy, Folch i Camarasa and illustrator J. M. Madorell published fifteen comic books based on the Massagran character, and the television series, which ran for a season in 1993, had fourteen episodes. Consistent with the original stories, the audiovisual relationship in the comics and programs intertwine the appearance of race with the sound of accent. If the early texts reflect the turn-of-the-century culture of travel developed and promoted by the Centre d'Excursions Catalans as they engaged the possibility of spreading civilization and national identity through language, the later comic books and television programming engage with a cosmopolitan Barcelona preparing itself for inclusion in a globalized, Europeanized project of economic development, even as Catalan publishing houses, radio stations, and television producers recognized the need not only to recuperate Catalan voices silenced by Franco, but also to produce content in Catalan as part of the broader Normalization process after Franco's death. Of special concern now was the power of the media to train children's ears in language. As Joan Fuster put it in the 1980 *Nadala* journal dedicated to Folch i Torres, "Davant un televisor, un catalanoparlant deixa de ser catalanoparlant per a ser televident en castellà, o en francès" (In front of a television, a Catalan speaker stops being a Catalan speaker to be a viewer in Castilian or French).[86]

Central to the effort was the character Norma, who by the very sound of her name, embodied the new Catalan sound of grammatically correct speech. Figured as both a cartoon character and played by an actress in live action shorts, this adolescent girl modeled not just good Catalan, but good manners regarding how to react to the sound of imperfect speech. The title phrase of the soundscape-shaping effort, "El català és cosa de tots" (Catalan belongs to everyone), suggested that everyone had a right to speak Catalan and, moreover, that everyone had a responsibility to do so. By staging scenes of everyday affairs such as looking for an apartment, flying on an airplane, cooking a meal, or speaking about feelings, the show sought to revive Catalan as a living, public language. But within those performances of exemplary Catalanness were assumptions about social and cultural class dynamics. Thus, in one episode, Norma, a blond girl of the middle to upper class, goes to the Picasso Museum with a friend and asks the security guard why he does not speak Catalan. When he replies that he feels that Catalans always make fun of those who speak with an imperfect accent, she urges him to try again, and reassures him with a kiss on the cheek that his efforts are appreciated. Although presented as a cute interaction between an earnest young girl and a kind older man, the interaction also imbues Catalan linguistic competence with a patient and moralistic ear, repeating the civilizing aspect of the colonial approach to hearing voice. By making Norma a young girl who is not afraid to question those who are older than she, the videos suggest that speaking Catalan is a project of futurity rather than one merely rooted in the past.

The Massagran television series based on the comic books is also meant to combat the threat of Catalan children aurally privileging other languages over their own; television is heard in this scheme as creating young subjects who will be active participants in an ever-more-robust Catalan soundscape. Yet by returning to Massagran—whom we could envision as the cheeky other half of the too-perfect Norma—these programs also return to the echoic memory of race produced through colonization. This was also a moment in which Barcelona was just beginning to contend with immigration from other parts of the world. As Rubén Domínguez Quintana points out, early on in the Normalization efforts the production of Catalanness as a linguistic identity derived its force from defining itself in opposition to the *xarnego*, a highly derogatory term for Castilian-speaking Andalusian migrants. But with the 1992 Olympics, the city became open to visitors from around the world, and thus to other languages and accented voices.[87]

Although they are still set in the nineteenth century, the Massagran comics and programs restage the colonial ear to reflect this new travel reality, defined by migration, tourism, and economic development, as well as the normalization of both voices and ears that was seen as necessary for producing good Catalan speakers. In the more recent tales, it is not merely that Massagran travels to far-off lands, but rather that the world comes to him as he unsuccessfully attempts to relax at his *masia* (a rural estate). The texts suppose that readers and viewers already possess a knowing ear that assumes the supremacy of Western language and European culture, they just need to be made further aware of the possibilities for Catalan as it confronts non-Western cultures. But the gaze and the audit portrayed here echo almost exactly the colonial perspective from a hundred years earlier. While Madorell updates Junceda's illustrations of the novel by aligning them with the comic book genre, the images continue to reflect the stereotyped drawings of the Golliwog, and the television cartoon portrays all the Black characters in that tradition.[88] In fact, the visual use of the Golliwog and blackface persists into the twenty-first century, and it is not limited to the comics and cartoons. Live theater productions from as recently as 2010—the celebrated *Any d'en Massagran* (Year of Massagran), marking one hundred years since the novel's first publication—featured full-body black costumes with big red mouths and giant white eyes.[89] The drawings in the comic versions include other stereotyped visual cues when Massagran travels abroad, which work in conjunction with the affirmation of Catalan language to normalize Massagran and even his pet animals, while dehumanizing the people in the other cultures he encounters.

Indeed, the aural continuity between the original stories and the television and comic book series from almost a century later doubles down on the representation of accent as barbaric, carrying an echoic memory that seems to transcend the global recognition, however begrudging in some cases, of potential shifts in perspectives on race brought about by decolonization and civil rights movements outside of Spain. The metaphor of barbarism as animalistic, in point of fact, persists in the relationship between sonic and visual otherness in these tales, especially in the comic strips. Onomatopoeic representations of sound and the repeated portrayal of other languages as accented Catalan put the soundscape at the front and center of the adventure experience. In the first comic, for example, *Les aventures d'en Massagran*, which mainly deals with animals, there are plenty of instances of the visualization of sound. When a giant condor drops Massagran in the water, the panel shows a resounding "XAP!"[90] When he encounters a large fruit that strikes him as both like an orange and a watermelon, which he names "taròndria," the monkeys yell "Nyiiiic!" and "Huic," while the fruits themselves make sounds like "ZIMM" and "PLAP" (25), and the rocks flying and hitting them down from the trees go "FLAC" and "TRAC" (24). Graphically, some pages replace any illustrations of visual images with depictions of sound as voluminous and overwhelming, as when a panel shows only the large, cartoony words "MUU MUU MUU" to signify a sound Massagran hears before he sees its origin; he and the reader later realize that it is the sound of a hippo yelling (36). This panel, like others, performs a kind of visualized linguistic acousmetre in which the audit occupies space so fully that it obscures any image whatsoever, even as the sound, drawn into words, is both visually and linguistically represented. Paving the way for the engagement with accent and other languages that takes place in subsequent comics, these associations of loud noises with unfamiliar, exotic natural phenomena contrast with the novel's steady writing of the narrator and the voices captured in his hand. The series also adds a pet parrot to the dog and the lion Massagran befriends in the original series. Together, they are called Pim, Pam, and Pum, associating sound as noise, or as a drum beat, with animality. This construct will repeat in the way in which race is portrayed as being not simply about skin tone, ethnicity, or background, but as a sound of language that marks the dividing line between the human and the animal.

Perhaps the most egregious depiction of this notion comes in 1983, the same year as the Catalan Parliament passed the Llei de Normalització Lingüística, with *Massagran i el quadrat màgic*. The comic was also turned into a cartoon ten years later. As with most of the television programs, it adheres fairly strictly to the comic book; my commentary here is on the television program. The story begins with Massagran in his village in Catalonia, where he has been living for a year with Kokaseka and Bokazas, two African servants he has brought back with him. Bored, as he is at the beginning of each new adventure, he decides to take them to see a museum exhibit on "Raçes negres" (Black races). In a shocking example of the ease with which the series dehumanizes Black people, neither Massagran's dog Pum nor his African servants can enter because a sign outside the museum states that animals are prohibited. This is a recurring joke throughout the comic book and television series, in which Massagran's animals complain of being discriminated against on the basis of racism, a gesture that completely dehumanizes the nonwhite subjects and simultaneously seems to imply that it is impossible to accuse the series of being racist, since it invalidates the term entirely.

Given the rule stated on the museum door, Massagran smuggles his servants in, pretending they are museum exhibits. They thus become both the migrants who come illegally through the metaphorical back door of Spain, and the very exhibit on race Massagran wanted to see. Once they are smuggled into the orderly realm of Catalonia, represented as a museum, though, the African subjects must be silent figures, objects of amusement on display. Their Black bodies' preferred status as silent objects, tolerated and perhaps even the cause of some fascination and *burla* (derision), repeats the nineteenth-century exposition of Indigenous people in World's Fairs or circus side shows.[91] In this contemporary, yet simultaneously colonial, way of seeing and hearing race, African culture can only be present if it is perceived as mute, frozen in a primitive time that is prior to voice and, thus, prior to the human. In one of the scenes that causes Massagran to laugh heartily, for example, a museum visitor chastises her husband for trying to speak with the African men since "no veus que son estàtues?" (can't you see they are statues?). Instead, she takes her photograph with them. The lesson from Folch i Torres's original story, that "Parlant la gent no s'entén" (Talking, people don't understand each other), which made a case for "civilization," is replaced here with an expectation of complete silence by "the Black races" that are the subject of the exhibit. If Massagran's great accomplishment a hundred years prior was successfully instructing the *katalanins* in language, and in creating a toothless voice and modeling what Catalonia could become, in an incipiently globalized Catalan society of the 1990s, the program suggests the futility of hearing difference as human or engaging in dialogue while at home, and a desire instead to silence Black voices entirely, while keeping their bodies on display in both the museum and the media.

If in 1983 this storyline was already offensive, it becomes even more so when we consider that the television cartoon came out in 1993, a year after the Barcelona Olympics, which was at least in part defined by the *negre de Banyoles* controversy. A more public example of the endemic and willful overlooking of racist attitudes in some sectors of Catalonia, it was widely reported in 1991 that the Darder Museum, in the Catalan town of Banyoles, still displayed the body of a stuffed San man in its taxidermy exhibit. The man, whose body was purchased in the late nineteenth century from a French anthropologist who had taken it out of Africa, was labeled as a bushman, El Betschuanas, and had been blackened with shoe polish to make him appear darker. Although museum officials had begrudgingly moved the body off display for the Olympics (the rowing events were being held nearby), they returned "el negre de Banyoles" to the museum afterward. When international pressure finally became too great to ignore a few years later, the body was repatriated to Botswana, even though many Catalans argued that this was an injustice because he was "theirs"—he belonged to them.[92]

This portrayal of nonwhite people as bodies to be silenced or ridiculed is not limited to African migrants, however. Just as sonic Blackness is produced as a virtual echoic memory defined by its opposition to whiteness, it is transferable to other nonwhite cultures, real and imagined, as well. In *El bruixot blanc* (The White Wizard), Massagran rescues a Catalan doctor from a fictitious group of brown-skinned Pacific Islanders, the Balambas, who "parlen l'alababalà, una llengua que no és monosil·làbica sinó labio-palatal" (speak alababalà, a language that is not monosyllabic but rather labial-palatal) and whose language thus revolves around made-up words like "Balubalú" and "Obilabá bali."[93] Just one of many jokes about language that pepper the books, here the people holding the doctor are defined by a language that references a phrase derived from the Arabic *Alà Bâb Al-Lâh*. Although it originally signified "as God wills it," the Institut d'Estudis Catalans (Institute of Catalan Studies) defines the phrase, as used in Catalan, as signifying "sense posar l'atenció, el seny, necessaris en allò que hom fa" (without paying the necessary attention or common sense to what one does). Playing on the trope of *seny* (as common sense, an attribute Catalans often associate specifically with their identity), here the natives' voices are presented both in terms of their phonology (their language is somehow entirely labio-palatal and not monosyllabic, an assessment that mixes the analysis of the production of phones with the production of word units) and as indicative of their inability to think as they need to. Perhaps not surprisingly, then, the islanders are presented as cannibals even before the trip begins, as Massagran and his companions make numerous jokes about their being about to be eaten. This trope of white fear is repeated when they discover that the doctor has been made king because, in a takeoff on the narrative of Hernán Cortés's arrival in the Americas, the natives' religion foresaw the arrival of an emissary who would prepare them for the "fills de la gran serp" (children of the great serpent) (41). In addition, in an inverted echo of Baldufa's need to pass as Black so he could take over the *kamaliks* in the second novel, here the doctor has covered himself with flour to exaggerate his whiteness since "Al cap de pocs dies estava negre de pell com un turista i la gent començava a desconfiar" (after a few days, I was as black as a tourist, and people started to not trust me) (42). The idea that a darker-skinned body is that of either a savage or a tourist, neither of which is trustworthy, suggests a universal agreement in the superiority of whiteness. Indeed, the tale ends with another successful lesson imparted by the Catalan group who visits the island: The doctor's wife teaches the Balambas to knit, since "aquests salvatges són molt traçuts i fent mitja podrien confeccionar mil coses que els serviran per a fer intercanvi amb les tribus veïnes i les de les altres illes" (these savages are very skillful, and knitting they could make a thousand things they could exchange for the products of the tribes of the neighboring islands) (43). The narrator later explains that, within a week, the Balambas were being productive, engaged in a modern, transoceanic economy. The positive relationship between colonialism and neoliberalism is evident here, as the civilizing project articulated by Folch i Torres's *África Española* seems to finally come to fruition.

Still, Massagran does not have to travel abroad to perpetuate the colonial ear; often the opportunity comes to him. In the 1991 comic *Els Jocs Olímpics d'en Massagran*, prepared in anticipation of the Barcelona Olympics, Massagran suffers from the same imperial eyes and ears when he is summoned by a Greek messenger and then travels to Greece with him.[94] Catalan intellectuals have long associated Catalonia with a Classical Mediterranean past, and the Olympic games in Catalonia had been conceived by the writer Xavier Rubert de Ventós specifically to reflect the connection between Greece and Catalonia.[95] In the comic strip, however, by constantly engaging in a comparative approach to what he sees, Massagran paints a stereotyped picture of Greek culture that becomes an opportunity for the texts to reinforce a view of Catalan culture as superior.

The comic begins by portraying Greek visitors to Catalonia as confused and foolish. Folch i Camarasa uses his father's technique of adding <k>

to common words to illustrate the linguistic noisiness of accent as a humorous indicator of a non-Catalan's confusion and inability to understand one's surroundings:

FILOMENOS: Kemdius? Onsók? Kisou?
MASSAGRAN: Noi, no t'entenc de res!
FILOMENOS: Kima dutakí?
MASSAGRAN: Res a fer jove! Per mi com si parlessis grec!
FILOMENOS: Jogrek! Sígrek! Komhas endevinat?

(FILOMENOS: Wha're yousa' ying? Where'm I? Who're you?
MASSAGRAN: Kid, I don't understand you at all!
FILOMENOS: Who brung me here?
MASSAGRAN: Nothing, impossible to understand you, kid! To me, it's like you're speaking Greek!
FILOMENOS: Igreek! Yesgreek! How'd you know?)[96]

The double hearing of Greek—as a way of dismissing accented language as noise ("it's all Greek to me"), but also treating the visitor's identity as unintelligible—encapsulates the duality of the colonial ear in the contemporary period.

Nevertheless, the same rules do not apply once Massagran arrives in Greece, with his visit once again constituting an opportunity for conquest, even though he does not show even the same level of linguistic competence as Filomenos had. For example, after dinner at a wealthy Greek business owner's house, everyone throws their plates on the ground, prompting Massagran to quip "d'aquests grecs sí que no es podrà dir que no han trencat mai cap plat ni cap olla" (you could not say about these Greeks that they have never broken a plate or a pot), a reference to the Catalan custom of "trencant l'olla" (breaking the pot) (26). Likewise, when after dinner the Greek hosts dance the *sirtakis*, Massagran says, "A casa en diem la sardana de l'avellana" (At home we call that the *sardana de l'avellana*) (31). The reference to a children's song infantilizes the Greek tradition, even as it asserts a Mediterranean relationship that allows Catalonia to claim a shared history with Classical culture. Still, non-Peninsular language once again becomes the subject of tasteless humor that affirms a Catalanized ear as necessary for travel. In several other panels, the comic plays on the role of the translating narrator from the original stories, depicting a character saying something in Greek, and then including a footnote at the bottom of the page, "Això vol dir 'moltes gràcies' com és ben sabut" (That means "thank you very much," as is well known) (9). A few pages later, Massagran responds to another Greek word with a word bubble and a large exclamation point inside, with the footnote reading, "Es veu que en Massagran no sabia que això vol dir 'bon dia'" (One can see that Massagran did not know that means "good morning") (10). These rhetorical strategies preserve the narrator as translator, but the joke here—that it's all Greek to me—is that Greek is not "ben sabut" (well known). Unlike the accented Catalan of the *kukamuskes* that can be civilized and thus understood, there is no reason a Catalan would bother to understand Greek, even while in the country, since the non-Catalan language is minimal in its usefulness.

Such a strategy inverts the influence of Greek history over European culture to normalize Catalan as a dominant rather than minor language, but it also turns cultural difference into an opportunity for affirming the superiority of the Catalan ear in a global setting. Moreover, the comic uses the vocal silence of Massagran's pets to push this idea further without directly ascribing it to Massagran: As they are taking a tour of Athens, one panel shows Pum, the dog, urinating on the columns of the Acropolis and joking, "Els gossos grecs no necessitaven fanals" (Greek dogs did not need lampposts) (20). And when, in the end, Massagran participates in a private version of the Olympic games being held in Greece, his lion, Pam, gives Massagran the push he needs to win the long jump. Initially the public cries foul, but, in an echo of his reading of the law on the ship, Massagran appeals, stating there is no official rule prohibiting that a lion push you. The judges concede.

In the end, Massagran wins the Olympics by cheating the system, but he does it with a moral justification, in this case so that the "evil" shipping magnate at the center of the tale will be put in his place: Just as in Catalan missionaries' approach to Equatorial Guinea or Franco's understanding of colonization as a positive, spiritual force, Massagran subverts the rules and performs the sleight of ear in the name of a perceived higher good. Now, though, that higher good is Spain's integration into Europe, advocated for by Catalan governments and businesses that saw the development of Barcelona for foreign investment as the way out of Spain's past and into its modern future. In line with the Catalanist values that have defined the texts for a century, what the Catalan ear brings to the Olympic games is a modernization that even descendants of the Greek tradition could not achieve. The last page, wishing to establish a link between the games of the story and those about to be held in Barcelona, ends with a banner reading "Honor a En Massagran" (Honor to Massagran). But a strong wind rearranges the letters of the banner to read "En honor a en Samarang" (In Honor of Sama-rang), a reference to Juan Antonio Samaranch, former *falangista* and president of the International Olympic Committee (IOC) from 1980 to 2001. The panel shows Massagran saying, "I el cor em diu que un dia tindrem uns jocs olímpics a Barcelona, i que aquest nom hi tindrà alguna cosa a veure" (And my heart tells me that one day we will have Olympic games in Barcelona, and that this name will have something to do with it) (45). Given that it was later discovered that IOC officials—including Samaranch—had been bribed in order to fix the winner of the Salt Lake City Olympics in 2002, the finagling in the Massagran story strikes the contemporary reader as prophetic in another, unintended way. The actual outcome, moreover, would confirm the suspicions of those I discuss in Chapter 3, who saw the alignment of Barcelona's post-Transition government with the Olympic project as an example of Spain's capitulation to neoliberalism, which dispossessed local communities in the interest of economic gain.

Indeed, the larger question of globalization that the opening of Spain to migrants, and the spread of market forces, entails, emerges in one of the later comics, *En Massagran i la diadema robada*, from 1994, which comments on the black-market exploitation of textile workers in Southeast Asia, at a time that Catalonia's clothing industry was facing the export of textile production overseas. In the story, a turbaned man from India named "Panxatanta"—whom Massagran calls "Tantapanxa" (Big Belly)—arrives to speak with "Massagranah" in "Khatalunya." The gran rajah has sent him because a crystal ball has shown that the most powerful magician in the world, Tothosabah (Know It All), lives there and can help find a stolen diamond. Here, the non-Catalan voice is represented as aspirated and without pause—and once again the cause of humor. For yet again, Tothosabah is actually Joanet Baldufa, the *rei dels kamaliks* (King of the Kamaliks), in disguise. This is the same character who had first appeared in blackface in the 1913 novel. In this case, the Indians' pronunciation of his name, meant to be Tot ho sap (know-it-all), acquires its accent with the additional <ah> at the end that marks their pronunciation throughout the text. The punny joke is once again linguistic and classist: Baldufa reminds Massagran that he is the son of a shoemaker. In a flashback, we see the shoemaker and his son, with the sound of the word for shoe—*sabata*—resonating with the play on *saber* (to know) in Tothosabah.

Not surprisingly, much as the African characters are negatively stereotyped, here the wealthy gran rajah who has called for Tothosabah thinks that Pim, Pam, and Pum (Massagran's pet parrot, dog, and lion) are a gift for him to eat. The dog, riffing on the alliterative sound of their names, argues they must be a poetic "metàfora oriental" (Oriental metaphor). But the grotesque joke reveals an economic critique as well: Massagran discovers that the rich in India burn their outfits every day and keep the poor working to make more so they won't have a revolution. A Catalanist representation of overseas sweatshop labor, Massagran becomes a hero when he realizes that the rajah's second in command, Brahmdhasa, was actually hiding the clothing and selling it to foreign ships, thus both abusing the people of his country and feeding Western need with their exploitation.

In fact, some of the later comics seem to wish to atone for the earlier ones' emphasis on linguistic conquest through violence by seeming to critique some of the inequities of the past. In *Massagran i els negrers*, for example, Massagran manages to stop slave traders from carrying out their mission, despite continuing to portray enslaved people as incapable of saving themselves.[97] But even there the epistemology of conflictual colonialism remains in the stereotyped depictions of non-Catalan cultures on which the adventure narrative depends for its presentation of Massagran not just as an ideal Catalan speaker, but as an ideal listener.

As I suggested in the Introduction, echoic memory reflects a temporal presence of the old colonial ear in a contemporary space that is both immediate and long-term. Through the duration of sound, one acquires knowledge of one's ambient space; and as Don Ihde has further shown, the result is a kind of aurality that also gives that space a shape:

> The all-at-onceness to the outline shape [of a sound occurring] . . . is a matter of *temporal instantaneousness* or of *simultaneity*. But when I return to those experiences which give me shape-aspects I find that the one given is not a matter of instantaneousness but of a *sequential* or *durational* presentation. If the ball is dropped and does not bounce, I may not get more than a "contact point" as a vague and extremely "narrow" signification. But if the ball is rolled for several instants, if the rolling endures through a time span which is quite short, I get a sense of its shape as an *edge-shape*. This shape is presented not in terms of temporal instantaneousness but in terms of temporal duration. In both cases there is a need for some "time," as even visually the object presented in too small an atom of time remains equally spatially indiscernible.[98]

In a sense, we can consider the Massagran tale, with its elliptical relationship to national identity and movement throughout more than one hundred years of mediatized Catalan culture, an edge-shape. This edge-shape is an echo that moves through, but also continuously reshapes, a colonizing and globalized Catalan soundscape over time, figuring the sound of language as a sound of racialized belonging or exclusion that shapes the aurality of space into a recognizable place. In its constant reverberation, the echoic memory of the colonial ear embodied by Massagran returns through the different forms the character takes over the course of a century, all of which sound relevant to the period at hand, even if they are located in a nineteenth-century model of Catalan society based on lost empire and desire for future national glory produced through a colonial ear. The result for subjects experiencing the sound of echoic memory is a sort of Doppler effect as the sound moves between lived aural experiences, their entextualizations, and their sonic reproductions, in which prior ideologies of noise, race, and nation echo differentially in the "new" sounds the character encounters as he is recirculated, echoically, for audiences whose experiences of travel change with the period. The legacy of this sleight of ear as it plays out with respect to immigration, heard in what I consider an emplaced understanding of voice and accent, is the topic of my next chapter.

2 Of Immigrants and Accents

In the previous chapter, I explored the role travel plays in producing a colonial sleight of ear. Concentrating on the tales of the popular children's literature character Massagran, I focused on how visual and linguistic portrayals of Black voices in the early twentieth century helped normalize a Catalan aurality that returned to the mediatized public sphere after Franco's death. In this chapter, I wish to push the discussion of accent further to theorize what I consider an aurality of emplaced voice, which, I argue, plays out in the sounds of immigrant voices that reshaped Barcelona's soundscape beginning in the mid-twentieth century, especially after the Transition. Putting the cultural context of immigration in Barcelona in dialogue with critical theories of voice, I interrogate how the sounds of immigrants' embodied presence put increasing pressure on the aural imaginaries of voice I discussed in the previous chapter. I do so by drawing our ear to the sounds of the body in the city, and what I theorize below as *vocal chords* that occupy multiple territories at the same time. This attention to the relationship between body, voice, and emplacement allows us to consider how aural imaginaries of accent, gender, and race intertwine daily, which in turn offers us potential ways to redress a Western locus of audition that informs the colonial ear.

To explore this imagined emplacement of accent, I would like to begin by considering a very different kind of media production from the one I considered in the last chapter, one not embedded in the Catalan establishment, but tied to the counterculture. In 1977, just as Spain was transitioning from dictatorship to democracy, Catalan filmmaker Ventura Pons directed his first film, *Ocaña, retrat intermitent*. The film intersplices an informal documentary-style interview of the flamboyant Andalusian painter and performance artist José Pérez Ocaña with footage of him in drag as he stages performances and happenings throughout Barcelona.[1] Wearing a dress and nothing underneath, she struts down Les Rambles, flashing onlookers, while Juanita Reina's "Yo soy esa" plays over the scene. All alone in a cemetery, save for the camera, she performs an impassioned *flamenco* number dedicated to Federico García Lorca. He strips while singing during a rock festival at Canet de Mar. But at home in his bedroom, he speaks softly with an Andalusian accent, explaining with alternating sadness and glee the experience of growing up poor and queer in the south of Spain in the years following the Spanish Civil War. In addition to describing his early life, he emphasizes that his public stripping and dressing in drag are not stripteases or evidence that he is a *travesti* (transvestite), but rather provocations born of the fact that "a mí me gusta romper cosas" (I like to break things) and that "soy un teatrero, pero puro" (I am a drama queen [*teatrero*], but a pure one). These scenes are framed by title and closing credits that set the stage in Barcelona. The film opens by overlaying the sound of the *tenora*—the main instrument of the musical ensemble that accompanies the Catalan national dance, the *sardana*—over visuals of the Plaça Reial.[2] Similarly, the end of the film shows Ocaña walking through the desolate, presumably early-morning streets, with the same *tenora*, played by Aureli Vila, evoking a single solitary note.[3]

Barcelona itself has not been much commented on in descriptions of the film. Instead, critics and the public alike have been drawn to the scenes of Ocaña in drag, the image of which has come to represent the film itself.[4] At the time, so-called *travesti* performances (accent on the last syllable) were theorized within the counterculture as a way of transgressing all that Franco had wrought. As Toni Puig put it in the 1977 edition of *Ajoblanco*:

> El travestí aboga porque cada uno represente en cada momento el personaje que más le apetezca. Que más refleje su estado. Sí. Machote ahora, coqueto luego. . . .
>
> Hoy, el travestí, continúa siendo la crítica más subversiva . . . Y una subversión en exceso . . . Significa denunciar y barrer las máscaras de la comedia, los fingimientos, lo archifalso de nuestras relaciones sociales, políticas, jerárquicas. . . . ¿Cómo? Revistiéndonos, usando, agrandando, militando en la utopía.
>
> (The transvestite advocates for everyone playing whatever character they want to in the moment. The one that most reflects their state. Yes. Super macho now, a flirt later. . . .

Today, the transvestite continues to be the most subversive critic . . .
And a subversion in excess. . . . It means denouncing and sweeping
away the masks of comedy, the pretenses, the arch-falseness of our
social, political, hierarchical relationships. . . . How? Redressing ourselves, using, enlarging, militating in utopia.)[5]

As a practice of critical subversion that is also a kind of bricolage, a building and dismantling of self, the scenes of Ocaña in drag capture this idea
of *redressing* as both a performance and an expression of queer self that, as
I show below, is also a critique of geographical voice as representation. At
the same time, *Ocaña* questions the sexual norms enforced under Franco
and, as Josep-Anton Fernàndez has written, portrays characteristics of a
camp performance: "a somewhat paradoxical interplay between (in)visibil-
ity, representation, and recognition, ridden with ambiguities and ambiva-
lences that echo the double binds of the epistemology of the closet."[6] Sharing
a cinematic tendency to recycle Franco's kitsch into camp as a means of
resistance, the film is indicative of a cultural moment in Spain in which
normativity itself was being questioned.[7] The Catalan writer Terenci Moix,
who had shaken the Barcelona literary scene with his explicit depictions
of queer sexuality, called the documentary the first authentic representation of post-Franco Spain.[8] Today, the film could be considered a classic of
Spanish Transition cinema, in large part because it seems to encapsulate
so well the counterculture of the 1970s and its often-sexualized resistance
to Franco's repressive policies, as well as its recognition of queerness as a
public rejection of sexual normativity.

Yet I am drawn to the tagline with which the film was originally marketed to its audience, which claimed that this was "una película catalana
hablada en andaluz" (a Catalan film spoken in Andalusian). In many re-
spects, although the film is by a Catalan director—in fact, a director known
now for making the largest number of films in Catalan—it is not surprising that the film is in Spanish; much like speaking Catalan in public was
forbidden, *hearing* it through film was similarly difficult until after Franco's death. The official censorship machine was not dismantled until 1977,
a year before Pons made the documentary. And even since then, Catalan
language films have failed to achieve any sort of status within Spain or
even Barcelona, being primarily relegated to international film festivals, as
funding for Catalan works remains largely confined to television and theater.[9] Moreover, being Catalan does not necessarily mean speaking in Catalan, despite some of the rhetoric around identity; in fact, debates around

language and identity regarding "immigrants" to Barcelona had been going on since at least the 1930s, and *Ocaña* fits into that tradition. To outsiders, the very notion that people from the southern part of Spain were—and still are—considered to be immigrants to Catalonia, which is constitutionally still part of the same country, may seem odd.[10] But by focusing specifically on a supposedly Andalusian sound of Spanish, the phrase "hablada en andaluz" speaks to the role linguistic, rather than state, borders play in producing geographies of belonging and exclusion in Catalonia; these aural constructs of voice, which, as I explore below, interweave the sound of what I am calling accent with belonging, pave the way for how immigration from outside Spain and Europe more generally are considered in later decades, as particular sounds of voice become markers of difference.[11] In the 1960s, Francesc Candel coined the term *els altres catalans* (the other Catalans) to describe Spanish-speaking Andalusians in the city, a term that has continued to be used since, albeit with some wariness. Politically, the fact that *Ocaña* focused on a poor Andalusian in Barcelona was in some ways ahead of its time, since later the Partit Socialista Unificat de Catalunya (PSUC; Unified Socialist Party of Catalonia) would complain that *cine catalán* (Catalan film) or *cine hecho en Cataluña* (film made in Catalonia) tended to leave out leftist approaches that would validate the presence of immigrants in Catalonia.

Still, if queer sexuality and counterculture seemed to represent the "authentic" Spain after censorship and repression were lifted, so too did the use of accented language in the film, speaking as it did to the lack of linguistic uniformity in Spain, despite Franco's attempts to produce a monolingual aurality of Spanishness, at least in the sense that public performances needed to be in Spanish. Recognizing a polyglot aurality is itself a redressing of Spanish sound for the post-Franco era. Yet some critical approaches that appreciate the film's capacity to challenge heteronormativity have also suggested that, even if Ocaña's queering of voice through drag is a performance and a political critique, his accented pronunciation of Spanish is a stable representation of identity, and it marks him as out of place: "La presencia andaluza de Ocaña . . . termina imponiéndose, reflejándose esa sabiduría y esa idiosincrasia de pueblo viejo. . . . Ocaña se siente a gusto en Cataluña . . . pero él sigue allí, ceceando, sin perder para nada su identidad" (The Andalusian presence of Ocaña . . . ends up imposing itself, reflecting the wisdom and idiosyncrasy of the old *pueblo*. . . . Ocaña feels at home in Catalonia . . . but he remains there, lisping, without at all losing his identity).[12] In this interpretation, as much as Ocaña questions normative sexuality or narrates the difficulties of poverty, his lisp—heard not as queer, but as Andalusian—trumps other ways of hearing difference, as though one's linguistic identity occupied a higher place on a hierarchy of personal traits: accent is both a totalizing representation of identity and, at the same time, somehow separate from the rest of the person. Hearing accent this way grounds Ocaña in an aural origin associated with place, even while his performances illustrate the fluidity of gender.[13] In a way, one could argue that this aural imaginary echoes Nina Sun Eidsheim's research on listeners' perception of singers' or speakers' racial identity in the United States, where, she writes, both legally and socially "voice is established as unambiguously representative of a stable body," such that the sound of a Black voice's timbre becomes a way to project visual or other markers of race onto the singing or speaking body.[14] Yet there is a difference in Ocaña's sound of voice: Because his Andalusian-inflected singing voice is wrapped up in camp performances, it also draws the geography of the ear into dialogue with the temporalities of drag that, as Elizabeth Freeman has shown, is an archive that turns between the past and the present in ways that interrupt the idea of progress and the future as a linear model of time.[15] Much as *Ajoblanco* and other counterculture magazines were extolling drag as social critique, Freeman defines what I would consider to be the echoic temporality of drag as "a *productive* obstacle to progress, a usefully distorting pull backward, and a necessary pressure on the present tense."[16] Recuperating immigrant voices, as well as recognizing them as a challenge to a Catalanist linguistic ideal, pulls the ear back to the time of the Second Republic, whose Constitution allowed for multiple national languages, even as recognizing acoustic drag as temporally flexible warns against thinking language as a stable sound of the body politic in the future.

In a Spanish context in which race was itself a recent construct of imagined national identity, built by the Francoist media industry and unquestioning in its presumed whiteness, Juan-Fabián Delgado and others hear the sound of Ocaña's accented voice as producing a stable acoustic construct of vocal identity on which to ground his gender play. But this presumption of the acoustic "realism" of Ocaña's accented voice misses the point. As this film and several other later texts I explore show, listeners in Catalonia—and elsewhere, I would venture—are attuned to hearing certain vocal traits as identity, whether the speaker identifies with them or not, something which the performativity of drag calls into question. I position this assumption that identity can be ascribed to voice as a supposition that voice is emplaced, reflecting a certain notion of echoic memory tied to origin that I problematize below. In the case of this film, *el andaluz* (Andalusian variety of Spanish) is tied up with a long cinematic history of associating Spanishness with Andalusianness, defined through song and sound. But as the writer Quim Monzó argued in 2011, in the context of contemporary Catalonia, hearing Andalusian-accented Spanish as identity also consigns the speaker of the language to a lifetime of migrant status, never able to escape not just his own but his family's origins.[17]

In contrast to this supposition of an emplaced voice, I consider the temporal and spatial movement of what I call Ocaña's *vocal chords*—their occupation of several city spaces and cultural moments with voice—as the artist moves through the city, questioning the echoic memory of accent imposed on Spanish geographies of the ear. These chords challenge the emplaced voice that is the cornerstone of modern Western aurality because they draw our ear to an acoustic personhood that is always in movement and crossing boundaries, even as it confronts an ear that hears voice as rooted in a place or in a body with an origin. Like Jack Halberstam's construct of *trans**, where the asterisk implies "refusing to situate transition in relation to a destination," here accent is both a geographical performance of the ear and an embodied experience of daily movement that marks the "disaggregation of identity politics predicated upon the separating out of many kinds of experience that actually blend together."[18] Later in the chapter, I explore how three novels, Juan Marsé's *El amante bilingüe* (1990), Maria Barbal's *Carrer Bolívia* (1999), and Brigitte Vasallo's *PornoBurka* (2013), further critique the accent-identity geography of the ear by illustrating how globalized migration changes the vocal and linguistic soundscape of an increasingly polyglot city, challenging the emplaced voice—and the geographical boundaries heard into it—in the process.

REDRESSING VOICE

I wish to consider first the almost nonexistent role accent has played in critical conceptualizations of voice in literary and cultural studies in the twentieth century. We might argue that, informed by democratic constructs of representation, most modern Western cultures aspire to what I believe could be called a vococentric subjectivity. For Michel Chion, vococentrism refers to a natural tendency of the ear to pick out the sound of voice over other sounds, a phenomenon which the film industry mimics by highlighting voice while reducing other sounds to background noise.[19] As an ideological structure, however, in the construct I am putting forth here, vococentrism refers to the way in which listening subjects presume that the articulation of identity necessitates a representative concordance between subjectivity and voice, emplacing voice in (a socially defined construct of) the self. From the Enlightenment on, this vococentrism has largely avoided engaging the role of the body in enunciating or hearing the subject, as I argued in the last chapter when I discussed the toothless voice. Instead, vococentrism means thinking of voice as a stable representation of one's subjectivity, one in which the body (able or disabled, vocal or mute, sexed or gendered) is of secondary concern.

Ocaña reflects that idea in how it portrays Ocaña herself as a vocal body. An integral part of the film is its alternation between Ocaña's public drag performances and what I would call the "at-home" Ocaña, in which he speaks directly to the camera about his experiences being queer, Andalusian, and poor. In these scenes, Ocaña's narrating voice is often heard as an unassailable reflection of identity. In a typical example of this kind of hearing, Teresa Vilarós has written that Pons's representation of Ocaña re-creates him visually as a "nueva maja goyesca" (a new Goyan *maja*): "Reflejando el espejo a un Ocaña vestido, este desnuda, sin embargo, su alma" (The mirror reflects a dressed Ocaña, even so, his soul is naked).[20] Her observation on the nakedness of Ocaña's soul is based in the presumption that, unlike the performances he makes with his cross-dressed body, his voice—meaning not just his way of speaking but his narrative about himself during the interview—communicates to the listener a transparent representation of his identity. This idea of voice as representation persists in the popular imagination, and even in some academic circles, notwithstanding the fact that, as Michel Chion has pointed out, such notions confuse the sonic materiality of voice with the signifying function of speech. In film, accompanying that understanding of voice as identity is a concomitant belief in the realism of the documentary form. As Chion has written, "We are often given to believe . . . that the body and voice cohere. . . . The voice, smell, and sight of 'the other': the idea is firmly established that all these form a whole, that the [subject] needs only to reconstitute it by calling on his 'reality principle.' But in truth, what we have here is an entirely *structural operation* (related to the structuring of the subject in language) of grafting the non-localized voice onto a particular body that is assigned symbolically to the voice as its source."[21] What Chion signals here is a tendency to hear into voice a representative understanding of sound, one that can be pegged to the speaker as an enunciation that reflects their body.

Hearings of Ocaña that perceive an opposition between his "true" voice at home and her voice as drag performance perpetuate this assumption. Yet this kind of hearing is also present in literary and cultural analyses that suppose even the most deconstructionist approaches to language. Time and again philosophers have presumed that subjects hear their own voices as representations of their selves. As he seeks to deconstruct this notion, for example, even Jacques Derrida hears voice as an interiorized subjective identity that, according to his reading of Western philosophy, is always already bound to writing as part of a larger system of signification from which voice as phone cannot escape. As he writes in *Of Grammatology*, "The system of 'hearing (understanding)-oneself-speak' through the phonic substance—which presents itself as the nonexterior, nonmundane, therefore nonempirical or noncontingent signifier—has . . . produced the idea of the world, the idea of world-origin, that arise from the difference between the worldly and the non-worldly, the outside and the inside, ideality and nonideality, universal and nonuniversal, transcendental and empirical, etc."[22] But while Derrida presumes to demonstrate that writing, as the opposition to what he calls *phonocentrism*, is always already present as a trace in that presumption of breath and speech as internal voice, he does not move beyond this understanding of voice as subjectivity in order to comprehend its material sonority. This makes sense in his larger project of questioning the metaphysics of presence that underlies the predominant epistemological construction of language's relationship to an exterior world. But it does not really question the idea that the "essence of the phone would be immediately proximate to that which within 'thought' as logos relates to 'meaning.'"[23] Though he invokes the phonograph as evidence of how the voice can escape its speaker's presence, he uses that merely to assert that phonetic writing as the "medium of the great metaphysical, scientific, technical, and economic adventure of the West, is limited in space and time and limits itself even as it is in the process of imposing its laws upon the cultural areas that had escaped it."[24] He does not question the relationship between thought and voice as a subjective experience, he merely suggests that voice is informed by writing as system.

Interpretations of Ocaña that hear him as split between an authentic, accented person who can narrate himself as Andalusian and queer, and the cross-dressing Ocaña—who challenges such assumptions by performing women's voices—continue to hear in voice this binary division between logocentric and phonocentric concepts of language. In this approach, logocentric meaning is the emplacing narrative that represents the subject, while sound is an embodied performance that can be displaced from that subjectivity.

Still, accent plays a large part in how that vococentric subject has been mediatized in recent decades. For his part, Mladen Dolar has argued that there is always a binary struggle between normative and accented language: "A heavy accent suddenly makes us aware of the material support of the voice which we tend immediately to discard. It appears as a distraction, or even an obstacle, to the smooth flow of signifiers and to the hermeneutics of understanding."[25] Nevertheless, while I agree with his argument that the materiality of the voice maps to the sensation that, as I argued in the Introduction, can be dismissed as making a speaker unintelligible, I disagree when he writes that "the regional accent can easily be dealt with, it can be described and codified. After all, it is a norm which differs from the ruling norm."[26] For, as *Ocaña* (as well as the novels by Marsé, Vasallo, and Roig that I interpret below) shows, in situations like Barcelona's where multiple languages are not only the norm but the daily reality of speech, immigration and accent complicate such binary understandings of language as a norm.

Pons's suggestion that *Ocaña* was "una película catalana hablada en andaluz" (a Catalan film spoken in Andaluz) responds specifically to the political and cultural concerns around language in Catalonia that reached new levels of discursive power after Franco's death. Beginning in the 1960s, but especially following the Transition, many artists and intellectuals pushed for new literature and film to be made in Catalan, recuperating a process of national identity construction that had begun in the nineteenth century and abruptly stopped when Franco won the Civil War. The decree in the 1978 Constitution that granted Catalonia and the Basque Country autonomy (a move that left other autonomous regions like Andalusia yearning for the same) was further inspiration to create a Catalan soundscape across a number of regions and cultural spheres. Yet there was also a divide between a conservative nationalist perspective that viewed Catalanism as a pure identity that would be symbolized by a normative and correct use of the Catalan language, and a more inclusive understanding of Catalanness as multicultural, not to mention regionally inflected. Some proponents of restoring Catalan culture pushed the concept of the Països Catalans, which would draw under the rubric of Catalan not just the language of Catalonia, but also that of Valencia, the Balearic Islands, and the eastern parts of Aragon.[27] The debates that ensued drew attention to the fact that the idealized Catalan language instruction that was being proposed in those areas privileged a particular form of Catalan, mainly spoken in Barcelona, as the norm.

Another, longer debate concerned whether or not Catalan identity could include those, like Ocaña, who lived in Catalonia but who did not speak Catalan.[28] Beginning in the 1920s, but especially after the Civil War, people from southern Spain, often called *murcians* or *xarnegos*—pejorative terms whose geographical differentiations often echo centuries-past conflicts between Arab/Islamic and European/Christian communities—had immigrated to Barcelona looking for work. These *altres catalans* (other Catalans), as Francesc Candel called them in 1964, were distinguished from middle-class Catalans not just by their socioeconomic level but by their southern-accented Spanish.[29] The tension between these groups and the Catalan bourgeoisie initially emerged from the fact that they were seen by certain intellectuals, like Josep Antoni Vandellòs i Solà, as contaminating the "patrimoni espiritual del nostre poble" (spiritual patrimony of our people), which was linked to the Catalan language.[30] The new arrivals also angered some working-class Catalans, since at times, without their knowledge, they were hired to come to Barcelona so they could serve as strikebreakers. In general, they were associated with lower wage jobs and the Confederación Nacional de Trabajo (CGT; National Confederation of Labor) or Federación Anarquista Ibérica (FAI; Iberian Anarchist Federation), workers' unions that were founded in 1910 in Barcelona. Many "immigrants" associated with these unions also helped forge the *ateneus*, local neighborhood groups that sought to educate workers about their situation. Over time, anti-immigrant prejudices were exacerbated by the continuing arrival of southerners and their installation in *barraques*, shantytowns on the outskirts of the city, forming communities in places like Sant Andreu or Montjuïc.

As Fernàndez has explored, the immigrant experience has been largely left out of Catalan cultural production, as well as political narratives about the nation's roots, despite the fact that only 25 percent of those residing in Catalonia around the year 2000 could claim ancestry located there in 1900.[31] In part, this is because of their language, which was both Spanish and yet not the Spanish of Madrid. As Kathryn Woolard succinctly put it: "The Castilian or Spanish political adversary of Catalonia is often acknowledged to be different from the Castilian immigrant in Catalonia, but both are symbolized by the same language and referred to by the same term."[32] If the sound of Andalusian can be defined by its *seseo* and *yeísmo*, among other sonic markers, the pejorative *xarnego* is also defined by the use of Castilian itself, looked down upon for the language's association with the Spanish central government, with which Catalonia has had tensions since the early twentieth century; the sound of Castilian in any form can be heard as the sound of imposition over Catalonia.[33] So *el andaluz* has a double valence, as both markedly southern and thus derided, and yet too Castilian to fit in a Catalan soundscape.

With its focus on *el andaluz*, then, Pons's film draws attention to the material sound of voice that, as Michel Chion has explored, can be understood to "thingify difference," since it highlights not just how accent differentiates Ocaña's speech from the Castilian norm, but also how his vocal presence in Barcelona challenges Catalan notions of linguistic and cultural identity.[34] By focusing on Ocaña's language, the film hears Barcelona as a space of immigrant mixing. Rarely is a Catalan word heard in the film, and the protagonist is the only speaker. When he speaks, Ocaña *cecea* (pronounces the letters <s>, <z>, and <c> before <e> or <i> as [θ], or lisps), elides intervocalic /d/, aspirates final /s/, and assimilates /l/ and /r/, all of which are characteristic of Andalusian pronunciation (though not necessarily exclusive to it). Still, unlike the stereotype of the *xarnego*, Ocaña is not portrayed as an *obrero* (worker), and his accent is not particularly strong—not as strong, for example, as that of the children portrayed in *Los jóvenes del barrio*, from 1982, part of the *cine quinqui* movement that focused on the migrants who lived in slums on the outskirts of town, in the *barraques*.[35] Instead, Ocaña's *hablar en andaluz* (speaking in Andalusian) seems more to highlight the presence of linguistic communities of difference not just within Catalonia but within the Castilian language. *El andaluz*, in this reading, marks within Catalonia a dialect of Spanish that is contrasted with the kind of Castilian heard there as a sign of an overly controlling central state. The film also destabilizes the performance of *el andaluz* associated with Franco by resignifying the music, poetry, and theater that helped produce the idea of Andalusia in Spain, a topic I explore in depth later.

In other words, by asserting a relationship between queer performances of sexuality and the sound of "immigrant" voice, the film draws attention to the role echoic memory plays in producing the aural imaginaries of Catalonia, Andalusia, and Spain and their reliance on an emplacement of voice. Rooted in vococentric subjectivity, these echoic memories ignore the daily presence and movement of the body in local spaces. As I explained in the Introduction, echoic memories are the feedback loops between stereotypes of sound and how the ear processes the meaning of sound in the moment it impacts the ear. Aural imaginaries of place are important to the production of these echoic memories. By emplaced voice, then, I mean an aural understanding of subjectivity that converts the sound of voice into identity by hearing it as the origin and stable, "authentic" representation of a person. Moreover, emplacement maps the supposed authenticity of voice to place, be it geographical, sensorial, political, or metaphorical. Emplaced voice is the aural counterpart to the "spirit of the nation" that so often emerges in Western states after the nineteenth century and has continually haunted Spain despite its multilingualism; it is the aural foundation for the imagined, singular "mother tongue" that is supposed to unite body to language to place. It is for that reason that, although they are from the same modern geopolitical entity, Spain, Andalusian Spaniards are heard as "immigrants" to Barcelona.

It is my contention here that those who hear an emplaced identity in the physical pronunciation of language hear the traveler's or migrant's accent as a sign that their body is out of place because their voice *sounds* another place in the moment of its articulation. This conflict, in fact, represents the modern version of the colonial sleight of ear that has informed what Juan Pablo González Rodríguez has called the *locus auditivo* (locus of audition) or *puntos de escucha* (points of listening) that reflect a listener's epistemological standpoint.[36] Although his focus is on Latin American music, González Rodríguez's term is useful because it echoes the postcolonial understanding of subjectivity as voice encapsulated in the term *locus of enunciation*, while also dialoguing with decolonial conceptualizations of Western epistemology.[37] The locus of enunciation is important for recognizing the colonial ideologies that have produced myriad conflicts between North and South, center and periphery, capitalism and dispossession; it is foundational for those who hear the populations of the Global South as having the right to speak up against the imposition of globalized capitalism and neocolonialism.[38] This place does not need to be an actual place, as multiple postcolonial theorists have discussed; it is more an imaginary of place than a territory. The *locus auditivo* similarly draws on the metaphor of place to articulate more succinctly the epistemology from which one hears. I am suggesting that when it comes to accent, geographical place often *is* the locus that listeners imagine as the origin of the voice doing the enunciating, whether the speaker hears themselves as speaking from that place or not. It does not matter what a subject says about where they are coming from, since, as an epistemological structure tied to the cultural politics of echoic memory, the *locus auditivo* means that listeners hear voices that are different through ideological constructs such as nationality and other forms of exclusionary identity based on an idea of origin.

This conceptualization of accent as the perceived emplacement of voice, like the sound of voice as a representation of subjectivity, thus leaves out the material presence of the body in the moment of speech, even though it is the movement of the lips and tongue that produces the accent the ear hears in the first place. At stake is an epistemology of place that hears a geographical residue into the sound of the speaking voice. This is the other side of the aurality implicated in what Steven Feld called *acoustemology*: a sonic way of knowing. As he writes, "Hearing and voice are connected by auditory feedback and by physical resonance, the immediate experience of one's presence through the echo-chamber of the chest and head, the reverberant sensation of sound, particularly one's own voice. By bringing a durative, motional world of time and space simultaneously to front and back, top and bottom, and left and right, an alignment suffuses the entire fixed or moving body. This is why hearing and voicing link the felt sensations of sound and balance to those of physical and emotional presence."[39] For Feld, acoustemology implies that place and aurality are both implicated in how subjects know through what they hear. The term I use here, *emplaced voice*, is meant to reflect the imposition made on a speaker when their accent is heard as signifying that they do not (and cannot) speak or hear from the place in which their body is present. The trouble with this way of hearing voice is particularly evident when one's aural frame of reference—that is to say, one's echoic memory—imposes an idea of place on the sound of accent even when the place being imagined is unrelated to the actual provenance of the speaker. This is the case when a person from Barcelona hears migrant speakers in their city as *moros* (North African) or *murcians* (from Murcia), or in the United States, when someone hears Latinx immigrants as "Mexicans," even when they are not from those places.

But this way of hearing another place into voice, and performing a sleight of ear in the process, was also the normative way of hearing national identity under Franco, as the sound of *el andaluz* was mediatized as a sign of Spanishness. In the 1920s, an exoticized image inaugurated by British and North American Romantic poets a century earlier was recuperated by Federico García Lorca and the composer Manuel Falla in their attempts to collect folkloric representations of the Andalusian spirit. As Joshua S. Walden has written, the idea of authenticity in music was fundamental to the theory of sound of European Modernism, which in the first decades of the twentieth century sought to blend ethnographic scholarship and nationalist ideology with sound.⁴⁰ This idea bled into musical compositions like Falla's that intended to capture folk music traditions as representative of marginal communities. This musical understanding also resonated with stage traditions in Spain. The Álvarez Quintero brothers, for instance, whose work Ocaña evokes in the film, wrote numerous stage plays that sought to transpose to writing the Andalusian accent as a folkloric sound to be captured and preserved: "¿Zí, verdá? . . . ¡Qué cozitas pazan a ezas horas, mi arma!" (Yes, right? . . . What little things happen at these hours, my love!), exclaims Ildefonso in the 1907 *La zancadilla: entremés*. Refugio replies, "Gueño, Irdefonso, pos no me engañe usté: ¿era usté, por casualidá, er que cantaba y er que tocaba?" (Well, Ildefonso, so don't trick me: were you, by chance, the one who was singing and playing?) This kind of speech, as critics have commented, seems to simultaneously capture and idealize a romantic notion of Andalusia, where there are setbacks but not conflicts, drama but not hunger, optimism but not ignorance.⁴¹ The accent produced on the stage elevates the Andalusian sound to a moral category, just as Lorca and Falla sought to elevate its sound to a kind of spiritual realm outside the daily reality of rural life. During the Second Republic, filmmakers like Francisco Elias and León Artola perpetuated this romanticized fiction of Andalusia in their works.

But perhaps most recent in Spaniards' minds when Pons directed *Ocaña* was the cinematic tradition of the Francoist *españolada*. As Juan-Fabián Delgado has pointed out, Franco's focus on Andalusia as a symbol of Spanishness had a huge impact on film production during his dictatorship. Cifesa (La Compañía Industrial de Film Español; the Industrial Company of Spanish Film) produced a number of "folkloric" films that sought to "alegrar las caras de los sufridos españoles" (put a smile on the face of the suffering Spaniards) by "institucionalizando la andaluzada" (institutionalizing the *andaluzada*).⁴² This Andalusia was not geographically limited, but rather pervaded Spain, entwined in an imaginary that highlighted Madrid as Spain's center, with the sound of Andalusian voice and music playing a major role. Manuel Vázquez Montalbán writes in *Crónica sentimental de España*, for example, that even flamenco, the romantically quintessential sound of the south, under Franco became the sound of Madrid: "La épica madrileña no evoluciona en las canciones. La propia Carmen Sevilla, flamenca en el candelero . . . [pregona] entre taconeo y taconeo: ¡Ay, qué bonito es Madrid!" (The Madrid epic does not evolve in the songs. Carmen Sevilla herself, *flamenca* in the spotlight, [proclaims] between stamps: Ay, how beautiful is Madrid!)[43] This expansive hearing of identity and origins aided in the production of a broad, political concept of Spanishness as race, where race is *not* the folkloric authenticity García Lorca sought to capture, nor a skin-tone distinct from whiteness, but idealized participation in the nation-state. At the same time, there is a subtle racism in the association of "immigrants" to Catalonia—many of whom actually come from Castile—with Andalusia. As Eva Woods-Peiró has pointed out, there is a long history of contemporary Spain denying there is any problem with race in the country, and yet Andalusia has long represented a threat to the nation's ability to finally achieve modernity: Andalusia's "proximity to and historical ties with Africa increased its potential to contaminate and, at its most threatening, to obliterate the boundary between Spain and Africa."[44] This sentiment has an uneasy relationship with the rhetoric of similarity that, as I mentioned in the Introduction, historically justified colonial "civilizing" projects in Africa; but the association becomes a way for Catalonia to distinguish itself as more modern than the rest of Spain.

The portrayal of Ocaña's voice questions this aural imaginary of emplaced voice, and it also plays into a longer history of cinematic attempts to use sound to undo the Francoist locus of audition. Pons made *Ocaña* during a time in which experimental film took many forms in Spain, largely thanks to the Encuentros de Pamplona in 1972, a festival seeking to bring art and music together, which was dominated by the figure of John Cage, perhaps the most influential Western sound artist of the twentieth century.[45] As is by now well known, his *4′ 33″* revolutionized the music world by delinking transcription from music as an isolated art form and connecting it to a more holistic listening experience of the body in space. At the same time, Cage's use of everyday objects to make music reflected the importance of found art in the sonic realm. These themes, as well as the film theory of Jean Mitry (which became a link between that period and the *vanguardistas* of the early twentieth century), were present in the films influenced by the Encuentros. One particularly influential idea was that of film as a performative art, "del arte como un *acto literal* realizado en un tiempo presente y efímero" (of art as a *literal act* carried out in an ephemeral and present time), which combined with an *intermedialidad* (intermediality) that saw artists and actors work together across genres.[46]

Several Catalan films prior to *Ocaña* portrayed the shift from state control of the media to a more liberated Spain by using this experimental approach to depict popular culture, specifically music, as a playful challenge to the regime. The films' activism often took the postmodern forms of parody, pastiche, or excess rather than direct confrontation. Basilio Martín Patino's 1971 *Canciones para después de una guerra*, for example, used the counterpoint model of montage, imposing popular songs of the postwar period onto still frame and cinematic images of daily life in Spain. The at times satirical commentary produced by the discord between song and image led the government to prohibit screenings of the film. In 1976, Antoni Padrós directed *Shirley Temple Story*, which he called a work of "musical terrorism," and which had its theoretical grounding in the work of the Situationists. Similar in tone to Terenci Moix's short stories from the same period, the movie poked fun at Franco and his opposition by using spoofs of Hollywood actors to parody them.

But while these films operated under a postmodern, avant-garde assumption that viewed genre as the medium to be destabilized, *Ocaña*'s focus is on the body and the voice not as reappropriation, but as presence. Thus, it engages in dialogue best, perhaps, with Francesc Bellmunt's 1976 *La nova cançó*, which is more concerned with illustrating the existence of a Catalan musical movement than the other films I have discussed. The documentary sought to highlight the political power of a newly articulated Catalan music in Francoist Spain, by filming singers from the group Els Setze Jutges like Maria Bonet del Mar and Lluís Llach as they performed at Canet de Mar in a sort of Catalan version of Woodstock. Epitomized by singers such as Raimon, Joan Manuel Serrat, and Lluís Llach, the *nova cançó* movement was influenced by French *auteur* performers like Georges Brassens and shared the revolutionary politics of Latin American performers such as Mercedes Sosa and Violeta Parra.[47] Ocaña's performance at Canet de Mar, included in Pons's film, responds to Bellmunt's film by showcasing a voice that was left out of the construct of the *nova cançó*: Ocaña's naked, sexualized body voicing a Spanish in *andaluz* is an immigrant occupation of the stage on which a new sound of Catalan identity is performed.

In what follows, I attempt to tease out the movement of ears and voices that emerge in the mix of immigrant accent and gender in Ocaña's voice as, in his own words, he "breaks things." What we can glimpse is that in a Barcelona that is on the verge of democracy, discovering its identity through language, rupture is not just a question of gender roles, transmediality, or camp. Ocaña also breaks the spatial relationship between place and sound by asserting the presence of her voiced body as mobile and temporary, yet fundamentally at home locally, even when an emplaced notion of voice would hear him otherwise.

Two places are central to the sound and space of the film: Barcelona and Andalusia. Each one is complicated by how it is portrayed aurally. The city space of Barcelona is portrayed as open through the use of long shots and traveling shots, and if Ocaña's voice is heard during those scenes, it is almost always while singing or performing on stage. These outside performances occupy large spaces and present his voice as a performance. When Ocaña speaks, though, he is at home, which is portrayed as closed and labyrinthine, cluttered with his art. He is mainly framed in close-up, and the camera visually asserts a relationship between interiority and proximity. The only sounds besides his voice are those of birds chirping and the occasional moped speeding past his window. Several scenes open with what appears to be a direct shot of Ocaña speaking, only to have the camera reveal that the Ocaña the viewer has seen is in fact a reflection of himself in the mirror on the wall. Metanarratively commenting on the disconnect between representation and reality, this cinematographic move draws attention to the camera itself. The effect, as Michael Chanan has written of documentary generally, is a sense of destabilization: "What the camera captures is the subject's self-presentation in the situation that confronts them, which includes the presence of the camera. The situation inevitably produces a filmic version of the Heisenberg principle, which says not that the presence of the observer changes the behavior of the observed, but that it introduces uncertainty."[48] Notwithstanding these visual interventions, Ocaña's accented narrating voice remains unchanged, except in terms of its differentiation from his public performing voice. Ocaña's public drag performances take place mostly in the physical spaces of Barcelona—Les Rambles, the Plaça Reial, the bars, the streets of the old city—but they consistently engage with an echoic memory of Andalusia. The double sound of Ocaña's voice as private and public, queer and Andalusian, allows the film to question Francoist normativity.

At the same time, recuperating an "Andalusian" voice in a city whose Catalan soundscape was only just emerging makes a fairly strong statement against Catalan linguistic normativity and nativism as well. *Ocaña* participates in a camp deconstruction of Franco's ideological depictions of Andalusianness through aural and visual challenges to these emplaced notions of authenticity, but in the process, it rethinks Barcelona's space as one of migrants and difference. In so doing, I argue that it produces what I am calling *vocal chords*, which rethink the emplaced voice that underlies both national identity and accent, entwining them both with a queer spatial sensibility. Steven Connor has conceptualized voice as being of and beyond the body, a *vocalic space* that floats in an aural nowhere, between the mouth that enunciates it and the ear that receives it.[49] In this understanding, voice is breath without signification until the ear inscribes it into an echoic memory of meaning.

Drawing on this idea, Freya Jarman-Ivens has argued that voice is inherently queer, because its sound destabilizes signifiers. She invokes a spatial understanding of voice to communicate this idea: voice's "bodily origin and destination, and its operations across borders and through borders, and its traversal of the space between bodies, collectively give the voice a physical location in two bodies and no body at all, and its meaning arguably arises in all three locations too."[50] The idea that voice exists simultaneously in three locations at once—the speaking body, the listening body, and no body at all—creates a layered structure of understanding of the voice that I would argue we can think of as a vocal *chord*, a voice that is a melding of several notes made in different temporalities and places at the same time. For Jarman-Ivens, voice is therefore a "slippery beast" that cannot be pinned down because it is always already potentially queer.[51] As Douglas Kahn has illustrated, moreover, when we think about the material sound of voice, there is a difference between its interior reception by a speaker and that of listeners outside it because the presence of the body comes into play in a way theories of subjectivity have often ignored: "At the same time that the speaker hears the voice full with the immediacy of the body, others will hear the speaker's voice infused with a lesser distribution of body because it will be a voice heard without bone conduction: a deboned voice. Where bones once stood, there will be only the air within which the voice's vibrations dissipate. Thus, the presence produced by the voice will always entail a degree of delusion because of a difference in the texture of the sound: the speaker hears one voice, others hear it deboned."[52] The voice with which Derrida is concerned, the voice which has dominated Western approximations of subjectivity, and can be dismantled through a postmodern ludic play of accent, is what Kahn here calls the *deboned voice*, because it leaves out the sound of voice as an aural phenomenon that is materially produced by the speaker in a place in which they can hear themselves: the deboned voice is the voice of narrative as emplacement, which contrasts to the felt sense of voice the speaker has of themselves.

The deboned voice (the foundation of vococentric subjectivity), then, is a bodiless voice addressed to an earless listener: even as the sound of accent is tied to an imaginary of place, it is a place that, like Connor's breath, floats outside its embodiment in a specific subject's mouth, lips, and tongue. At the same time it originates prior to, and continues beyond, the physical impact of the accent on the listener's ear, folding into an imaginary that I have called echoic memory. But the vocal chord is that sound of voice located in several places at once. It draws our attention not only to the no-place of voice as construct, but also to the physical location of voice in the speaker's body *and* in the listener's ears as a phenomenon that is at once present, past, and a projection into the future.

One of the film's key scenes, for example, occurs when Ocaña performs Federico García Lorca. Lorca and Falla drew attention specifically to the *cante jondo*, or *deep song* of southern Spain, a particular folk music which they saw as an almost-magical, primitive art that preceded *flamenco*. For Falla, the *cante jondo* revealed a cultural essence linked to Spanish identity; for Lorca it seemed to invoke a deeper understanding of selfhood related to love and death.[53] Notably, as Charles Hirschkind has shown, Falla's conception of that Spanish identity was grounded in a recognition of the Arab influence in southern Spain, which was a corrective to his Catalan mentor's assertion that this sound had its roots in the Byzantine liturgy.[54] Musical *vandalism* (Andalucism) as a shared *fondo sonoro* (sonorous foundation) therefore links Andalusia as al-Andalus to both the Middle East and Europe, in the process sonically overturning the temporal break between the Islamic world and the Christian one marked by the date of the Muslim and Jewish expulsion from Spain in 1492.[55] In this way, *flamenco* "becomes the site for a reconfiguring of the affective and epistemic dispositions from which the virtual territories of Andalucismo can be charted."[56] By performing Lorca in drag, Ocaña *breaks things* by transcending gendered, national, and social norms, creating a vocal chord as an event that moves beyond its geographic and hitherto publicly imagined bounds.

The scene portraying Ocaña dressed as a flamenco *cantaora* (singer) walking alone in a cemetery suggests as much. Twenty-seven minutes into the documentary, the camera pans down from the sky onto a long shot of the cemetery, where birds are singing and Ocaña walks silently past the tombs wearing a *mantilla* (lace shawl), combs, and a colorful patterned dress. Once she is in range of a close-up by the camera, Ocaña begins to hum and then eventually sing the lyrics of a popular *zorongo*, which García Lorca had incorporated into his 1930 play, *La zapatera prodigiosa*.[57] In 1931, Lorca had recorded an album of *canciones populares españolas* (popular Spanish songs), which also included a version of the song, albeit with different verses, sung by Encarnación López Júlvez, known as La Argentinita. Ocaña retains the words from the play but refigures the fast-paced *zorongo* into a mournful *soleá*, thus changing an upbeat tune into a *cante jondo* (deep song).[58] She also begins by humming and singing without words, then ends the performance by ad-libbing lines about "Federico" himself, sung in the same *soleá* style:

> García Lorca, gitano,
> moreno de verde luna
> ¿dónde está tu cuerpo santo, que no tuvo sepultura?
> Se olvidaron de tu cuerpo
> pero en la primavera te están poniendo seda.
> Ayyyy ayyyyy.
>
> (García Lorca, gypsy,
> green moon dark-haired man
> where is your sacred body, which was not buried?
> They forgot your body
> but in the spring they are putting silk on you.
> Ayyyy ayyyyy.)[59]

Of particular importance in this performance are the *cante jondo* "ays" with which, as Nelson Orringer has pointed out, a *cantaor* usually starts his song; in García Lorca's poetry, the *ayyys* communicate an emotion informed by music and representing tragic ideas such as lost love, unfulfilled desire, and fatality.[60] Here, they carry with them a mourning for the loss of García Lorca's body—which Ocaña figures as dark-haired and Roma—even as she redresses her own body to perform the voice of the poet as *cantaora*. In drag, Ocaña restages an already staged understanding of Andalusian culture. Yet by recuperating in a Barcelona cemetery the voice of a poet killed by the regime, he does not merely displace an aurally emplaced understanding of voice to a new place, as the *españoladas* focused on Madrid had done. For, the *ayyys* echo the selfhood García Lorca heard as evoked by Andalusian culture, and the Arab Spain Falla heard into the *cante jondo*, but they also are moments in which Ocaña acoustically takes over the place of the cemetery with an embodied vocality that is not her own and yet also *is* hers because he produces it with his lungs, diaphragm, and vocal cords.

Ocaña's vocal cords, then, are also vocal chords, linguistic instruments that sound an echoic memory of Andalusia that is marked by its simultaneous emplacement both within and without his body. Even as Ocaña embodies the passion of this past, then, she also entwines Lorca's missing dead body with the metaphoric "spring silk" the regime has put over Lorca's image and ideas. The vocal chord directs attention to the material presence of Ocaña's body in the cemetery, enacting a sort of habeas corpus that emplaces an acoustic sense of voice in a physical space even when Lorca's body cannot be there. The vocal chord contrasts not just with the distant Arab-European past of the *flamenco* sound but with the violence of the Franco regime, invoking what Dylan Robbins calls a "wider historical framework linking the sounded utterances of the individual to a more remote and openly violent process," the coloniality of the audible.[61] The scene highlights early on in the Transition the need for what in later years would be referred to as historical memory by focusing on voice as an acoustic occupation of space linked to life and death.

A different vocal chord, tied to the complexities of echoic memory, occurs in an earlier scene, when Ocaña is shown walking down the street, as audioviewers hear the song "Francisco Alegre," a classic *copla* performed by Juanita Reina and included in the 1947 film *La Lola se va a los puertos*. Ocaña is poorly made up in excessively pale face powder, and she is wearing a woman's dress, hat, and shoes, and carrying a fan. In the film, the singer's performance of the song is overlaid on images of bullfighters riding horses and posters announcing bullfights, images that reinforce the stereotypical ideas of masculinity perpetuated throughout Franco's reign. In Pons's film, the camera stays with Ocaña as she pushes a stroller down Les Rambles, the center of Barcelona, fans himself, and flirts with an old man who seems unaware that Ocaña is not a cisgender woman. Eventually, she lifts her dress to reveal her penis and moons a group of onlookers. Even without the music, it is clear that Ocaña is poking fun at gender roles associated with motherhood, femininity, and normative sexuality; that certain spectators are unaware of this draws attention to the rupture between Francoist understandings of Spain and the sexual reality of the Transition, in which Ocaña's sexed body is at the center of the scene. When Pons imposes "Francisco Alegre," complete with record scratches and other acoustic indicators of age, over the scene, he emphasizes that there is a spatial and temporal displacement in Ocaña's transportation of the Andalusian imaginary to Barcelona. This imaginary cannot be mapped to the Francoist cultural production of Andalusia as Spain because the film recognizes the Francoist aurality as anachronistic. With this gesture, Pons questions the way in which a post-Franco ear retrospectively emplaces Andalusian and gendered signs of identity onto a body, even as he reaffirms that ear as the basis from which resistance can take place. By portraying Andalusia otherwise through Ocaña's cross-dressed body, the film harnesses a locus of audition that hears the traditional sound of Andalusia in the accented voice of the *copla*, which cannot be conflated with the image of a trans* woman performing the iconic sounds of Andalusia on Barcelona's city streets. The lack of synchronization between the 1970s documentary footage and the old popular tune suggests the ludic nature of the event, and also draws attention to how the documentary must resort to artistic practices in order to capture for viewers the new experience of accented everyday life in the city.

By offering the audioviewer a different sonic environment from the one in which Ocaña moves through the street, moreover, the film emphasizes not just the camera's intervention in the scene, but how Andalusia, in drag and dragged into another city, is not moved by the body that carries the accent, but by the ear that hears it that way. The song imposed on Ocaña's body is accented for him by another voice, that of the actress Juanita Reina. On one hand, the conflation of Juanita Reina's song with Ocaña's performance on Les Rambles has a unique effect on the song itself. As Sílvia Martínez has affirmed, the *copla* is largely defined by the fact that it tells a story: unlike the *cante flamenco*, which allows the performer to ad-lib and create lyrics, the *copla* is composed—"the only untouchable and irreplaceable elements are the voice and the storyline. Without a plot, there is no *copla*."[62] In *Ocaña*, however, the storyline of "Francisco Alegre," which tells of a woman worrying about her lover, is changed by its postproduction application to footage in which the body it seems to narrate is both accented similarly and yet gendered differently. Lines such as "La gente dise: vivan los hombres" (People say: long live men) and "Yo estoy resando por é / Con la boquita serrá" (I am praying for him / With my mouth shut), or "Dame tu risa, mujé, / Que soy torero andalú" (Give me your laugh, woman, / I am an Andalusian bullfighter) no longer suggest a dialogue between a *torero* (bullfighter) who is the epitome of masculinity and a woman who silently must put on a happy face—notions that would conform with the values portrayed by the *andaluzada*. Instead, overlaid on the image of Ocaña on the street, the acousmatic verses sung by Juanita Reina become an ironic statement on gender roles and their inapplicability to current society, which is embodied otherwise.[63] At the same time, they capture the notion that, despite a Francoist interpretation of the *copla* as Andalusian folklore, it was also a way in which those who were left out of that narrative—LGBTQI+ people, those whose Republican relatives had been killed, and others dispossessed by the regime—could sing about their pain in a covert way.[64]

By imposing the voice of Juanita Reina over Ocaña, the scene also calls attention to the absence of the body in these hearings of accented voice, rethinking a locus of audition that hears voice and accent as emplaced. Both Ocaña and Juanita Reina are deprived of body in voice, although in opposing ways that together signify this queer crossing between the locus of audition (the place from which one listens) and the locus of enunciation (the place from which one speaks). Juanita Reina's body is present only in an echoic memory of Franco's Andalusia, a memory that hears the timbre of her voice and her lenition of the intervocalic /d/ in "serrá" (*cerrada*, closed), or *seseo* in "resando" (*rezando*, praying), as gender and accent. At the same time, Ocaña's unvoiced body in drag is deaf to the song being imposed on it, but enacts the visual counterpart to the narrative this sound of Andalusia constructs. Together, Juanita Reina's acousmatic, bodiless voice and an Ocaña who is without an ear, deaf to the accented voice imposed on him and muted by the sound of another, combine the three locations of accent Jarman-Ivens highlights as queer into a vocal chord. The vocal chord describes, but also masks, the space between an aural imaginary of accent, the material sound of the voice, and the body that speaks it. For Michel Chion, as soon as a cinematic viewer can trace a voice to the mouth from which it originates, a deacousmatization takes place that he likens to a striptease, as the omnipotent power and aura of a bodiless voice deflates into the embodied voice of a person.[65] Here, the inverse occurs, as the imposed voice redresses the body in a way that strips Ocaña's naked body of its power by converting it into a visual sign rather than a presence.

Even though by dressing as a woman he destabilizes traditional norms of gender, religion, class, and more, Ocaña is not carrying out a drag show as a performance of a particular gender. Instead, perhaps ahead of his time, he affirms his disidentification with gender by stating simply that "yo soy una persona" (I am a person).[66] Just as when she gave García Lorca's dead body a place through his vocal occupation of the cemetery, when Ocaña lifts his dress to show his penis, he simply claims a space for a nonheteronormative body in Barcelona. In this sense, although he speaks of himself in opposition to the idea of being *travesti* (a cross-dresser), he is expressing the differently accented notion of *travestí* that emerged in the late 1970s, and which Puig, in the quote above, had articulated as the most radical of subversions even though it was not necessarily related to the need to transition: It was the flexibility and exaggeration of gender norms that was "militant." In direct and open language strikingly similar to today's, Bibí Anderson, Angie von Pritt, and others articulated in the 1978 book *El libro de los travestis* that this kind of performative drag was different from what they considered at the time to be "transsexual." As they explained, the difference was in the body: "El transexual, que vive, actúa y piensa como señora, solo tiene un defecto físico, el pene. Una vez operada, no es transexual, es una señora" (The transsexual who lives, acts, and thinks like a woman, has only one physical defect, the penis. Once she is operated on, she is not a transsexual, she is a woman).[67] Ocaña, who at times affirmed his own misogyny and flashes his penis with pride, was not interested in transitioning, nor in merely performing drag, but in occupying more than one identity. Simultaneously voiced as accented by another (Juanita Reina) and heard as silent because his locally articulated voice is co-opted by the soundtrack, the presence of Ocaña's body as simply a sexual body in Barcelona is what takes center stage.

Nazario Luque's 2016 *La vida cotidiana del dibujante underground* affirms as much when he argues that Ocaña's dress is not a drag element, but rather a piece of clothing that is as interchangeable as any other, just as Ocaña's queer sexuality is as mobile as his accented voice moving through the streets. Nazario, along with his and Ocaña's friend Camilo Cordero, also appear in Pons's film walking down Les Rambles. They and several others shared not just an aesthetic, but a robust sexual life among friends that involved drugs, art, and performance. As Nazario points out, Pons missed this quotidian aspect of Ocaña's presence in his portrayal of the artist as split between an interior experience and exterior performance, converting Ocaña into a sign of resistance rather than a body living it.

As Nazario recalls, in a moment prior to Pons's film, the group Video-Nou captured a more representative scene than any included in Pons's documentary. They had decided to film a video at home in which "Ocaña cantaría, recitaría y se revolcaría por los suelos convulsa en un ataque de sobreactuación" (Ocaña would sing, recite, and roll around the floor convulsed by an attack of overacting).[68] Yet the moment that works in the film is not the performance, but instead, the impromptu gathering that takes place in Pepichek's bed (who is sick with a cold) after the filming is done. Friends like Nazario, Maite, and Onliyú join Camilo and Ocaña as they fall into an orgy around the convalescent Pepichek. The scene, according to Nazario, draws attention to all that is missing in Pons's film: "La presencia de Ocaña, vestido aún con su traje de 'actuar', saltando por encima de todos, jaleando, animando a los participantes, masturbándose de pie sobre la cama, llegaba a un clímax que su director Ventura Pons, no llegaría a alcanzar en ningún momento de la película que rodaría con él meses más tarde" (The presence of Ocaña, still dressed in his "acting" outfit, jumping on top of everyone, goading, animating the participants, masturbating while standing on the bed, reached a climax that the director Ventura Pons would never reach in a single moment of the film he made with him [Ocaña] months later).[69] Placing "actuar" in quotes signals that the dress is not simply a show piece, but part of daily life: Nazario himself throws on a dress Ocaña had brought to change into, and Maite puts on a dress Nazario had worn previously; perhaps, Nazario asks teasingly in his narrative, it was the same one he was wearing when he was thrown in jail? For Nazario, and the counterculture more broadly, this is not about performance; it is merely a day in their everyday lives. What Nazario also calls Ocaña's "presence" as a sexual being, at home, wearing a dress but proudly displaying his penis, is the material body that is missing from the division between public and private, performance and narrative, that Pons's film seeks to elaborate. In this sense, Pons's film debones Ocaña because it leaves out any explicit depiction of sexuality that, according to Nazario, was so central to his life. The deboned voice is not just a voice devoid of the sound of the body, it is a voice devoid of the daily presence of the body that produces a more geographically and temporally encompassing vocal chord than ideology can. Ocaña allows us to hear this difference between a metaphysics of presence and an acoustic presence that is embodied. This sound goes beyond what Roland Barthes considered the grain of the voice, uniting narrative and accent with a sexed body whose presence in space is a temporary occupation that queers aural emplacement of the body in gender. Discipline, as Michel Foucault understands it, comes later, when the soundtrack is imposed over the experience.

Captured by the film, Ocaña's occupation of public spaces also materially changes the soundscape of the neighborhood he traverses, albeit only momentarily. Perhaps the most striking example of this effect of the body in space occurs in a scene filmed in the back streets of a neighborhood in the *ciutat vella* (old city) of Barcelona. After Ocaña narrates the religious rites of his native town to the camera, Pons incorporates a lengthy scene in which Ocaña helps stage a full-blown religious procession down a side street in the Born neighborhood. He, as well as all the couples walking with the altar through the streets, are dressed in women's clothing. Ocaña and his friend Camilo are on a balcony. Ocaña, as the *cantaora*, is dressed in a *mantilla*, with combs in her hair and overdone makeup. As the religious procession of drums and horns makes its way down the street, Ocaña and Camilo carry on a witty conversation about their time in Catholic school when they were young girls, appropriating a common narrative while re-embodying it. Ocaña goes on to sing a mournful *saeta*, full of long runs, to the crowd below. Rather than recording the events from a distance, Pons's camerawork participates in the performance, as it films Ocaña from below, thus highlighting the dominance of the *saetera* over the crowd and emplacing the audioviewer in the scene.[70]

In the Juanita Reina scenes from Les Rambles, the audioviewer observes the joke of Ocaña's camp occupations of the city from a distance. Here, viewing and listening to a rendition of an Andalusian religious procession associated with Easter, in Barcelona, in August, she hears it up close as she listens in on the gossipy conversation between Ocaña and Camilo. Pamela Robertson states that "camp is a reading/viewing practice which, by definition, is not available to all readers; for there to be a genuinely camp spectator, there must be another hypothetical spectator who views the object 'normally.'"[71] Audioviewers who witness this camp spectacle are therefore part of the aural imaginary that ties the *saeta* to traditional religious practices in Andalusia, but they may also participate in the film's subversion of these norms and aural reappropriation of space by being present "inside" the joke. Hearing from two *puntos de escucha* (points of listening), from inside and outside—both from the performance as it resounds in the street and from the intimate talk of Camilo and Ocaña—creates two interconnected aural imaginaries, private and public, local and distant, at odds but overlapping with each other. The doubling of Ocaña's voice is also, then, the doubling of the ear of the audioviewer, who may now hear Ocaña from two loci of audition, that of an emplaced ear and that of a material ear that hears Ocaña's religious Andalusian singing body in a nonstandard time and place. Like the Barcelona audience who has had to navigate a diglossic—or bilingual—Catalan/Castilian soundscape and identity for forty years, Pons's viewer is given the opportunity to do the same, moving between two imaginaries of Spain, Barcelona and Andalusia, neither of which is portrayed according to stereotypical cultural assumptions of them.

As a film, *Ocaña, retrat intermitent* can only present us with a two-dimensional portrayal of Ocaña's body moving through space. But thinking through the presence of Ocaña's body with(in) his voice challenges the construct of the deboned voice that would associate the sound of Andalusian song with foreignness, because it forces the listener to recognize the material presence of the accented body in the sound of voice that occupies the city, resonating in the ear and shaping the acoustic space of the city from moment to moment. Recognizing the simultaneity of his accent, his narrative voice, his sex, and his performance allows us to think about Ocaña not just as a vococentric subject, but, as she wishes, as a person.

THE VOCAL CHORDS OF ACCENT (ECHOES)

If Ocaña's sounding of vocal chords as a part of personhood seems to celebrate the coetaneousness of material difference, in resistance to a structural aurality rooted in the nation-state, in the years following the production of Pons's film such chords exploded and multiplied. Even as the Catalan establishment took charge of the city, eventually, immigration from around the world changed Barcelona from a bilingual to a polyglot space. As Fernàndez has pointed out, one of the main discourses around Catalanness during and after the transition to democracy was the possibility of its reproduction, which, in turn, produced several literary conceptions that consider the role of the woman's body in shaping a future Catalonia.[72] As I will now briefly show, gender and accent echo each other throughout the aural imaginary of Barcelona in the years following the Transition. Nevertheless, presence and personhood gradually fade from the discourses around the sounds of immigrant accent. With the increasingly global state of the media and the influx of tourists, these discourses eventually come to rest the ear on stereotype rather than lived experience.

We may consider first Montserrat Roig's 1982 L'òpera quotidiana, in which the sounds produced by an immigrant Andalusian mouth soon become a point of conflict, as they come to reflect the changing norms of women's rights and the changing shape of Catalanism. The novel narrates how the protagonist, Horaci Duc, "normalizes" his xarnega wife Maria by teaching her, like Pygmalion, how to speak Catalan, reshaping her body by reshaping the way her mouth moves when she speaks. Initially, Maria's accent is irritating, as, for example, she pronounces engegar (turn on) as "enchegar," and commits other pronunciation errors that make Duc angry: "apitxava les jotes, i, la nostra vocal neutra, la feia massa oberta. Jo, això, no ho suporto. No suporto que la llengua grinyoli" (aspirated her <j>'s and pronounced the neutral vowels too openly. I can't stand that. I can't stand the tongue squeaking).[73] Her xava accent, common among speakers whose Castilian interferes with their Catalan pronunciation, grates on Duc's ear; moreover, the double meaning of "llengua" (language and tongue) in the quote above suggests that Maria's mispronunciations are a fault of her physical body.[74] But at the same time, what he then calls her oïda d'àngel (angel's ear) allows her to pick up the "sweet and soft" sounds of Catalan, which Duc hears as sacred, and which eventually she seems to pronounce (29). Thus by the time he marries Maria he considers her "una autèntica catalana" (an authentic Catalan) through which he has helped reestablish a Catalan soundscape and subjectivity in the city: "aquest és el meu trofeo, la meva petita obra perquè Catalunya torni a ser allò que era. He convertit una xarnega analfabeta en una catalana instruïda" (this is my trophy, my little work so that Catalonia will go back to being what it was. I have converted an illiterate *xarnega* into an educated Catalan) (62). Yet when this *ángel del hogar* (angel in the home) fully aligns herself with the *catalanista* movement, her resistance to being dominated by a husband who has taught her to speak but then kept her at home sounds, again, the accent that grates on Duc's ear:[75]

> estic farta d'estar-me tot el dia a casa, . . . m'avorreixo, m'avorreixo, m'avorreixo, ho va dir tres vegades, i totes tres vegades va convertir la *x* en una *ch*, com se es vengés de mi en tornar al català pixat d'abans, el so em va grinyolar a l'oïda, i no sé què em va fer patir més, si que em digués que s'avorria o bé com ho va dir, com si li faltés oli a les consonants.
>
> (I'm fed up with being at home all day, . . . I get bored, I get bored, I get bored, she said it three times, and every time she turned the *x* into a *ch*, as though she were taking revenge on me by returning to the nasty Catalan from before, the sound grated at my ear, and I don't know what made me suffer more, that she said she was bored or how she said it, as though her consonants needed to be oiled.) (66)

Duc, in other words, now hears the harshness of her pronunciation as a sign of a normalization that has gone too far: His wife feels comfortable using the Catalan language to sound off about her needs. In a Caliban-type refiguring of the narrative of instruction, the fact that Duc has converted his *xarnega* wife into a woman who *feels* Catalan and can express herself through the language turns out to be a threat to the hierarchical comfort—itself entwined with the misogyny of *franquismo*—that Duc had felt when he was her instructor. More than just sound, gender and performance are at stake in the vengeance Duc now hears in Maria's voice: "La Maria ja no era la meva Maria, sinó que s'havia convertit en una nova dona, com una roca" (Maria was no longer my Maria, rather she had converted into a new woman, like a rock) (64). This sound of her empowerment finally turns to resentment. Duc had witnessed in person Francesc Macià declare the

Catalan Republic; his father went off to fight and never returned. As a result, he learned at a young age that, as a Catalan "ens tocarà callar, però per dintre hem de cridar ben fort perquè no ens robin les paraules" (we had to be silent, but inside we must scream very loudly so they don't steal our words) (60). Hearing his wife, one of the *altres catalans*, speak defiantly thus strikes at a construct of voice that hears "pure" Catalan as a one-time silence that is now repressed, again, by the noisiness of the accented voices who are unafraid to speak.

Still, as with Ocaña, it is not just the performativity of speech that matters, but the presence of a woman's body with a changed voice that demands recognition. Unlike in the colonial notion of the toothless voice I interrogated in the last chapter, after the Transition the presence of vocal bodies threatened the ideological erasure of violence that had produced vocal order in the past. As the older character Patricia Miralpeix puts it when speaking to Horaci Duc, his problem is not that he aligns with the Catalanist movement. Instead, wrapped up in an imaginary of masculinity and poetic Catalanism that is out of touch with the new Barcelona, "vostè té por d'oblidar. Que té por de ser d'una altra manera. . . . així, no sabrà veure mai la realitat" (you are afraid to forget. You are afraid to be another way. . . . that way, you will never know how to see reality) (161). The reality of post-Transition Barcelona is that the Catalan ear forged seventy years prior cannot coexist with the social changes that interrupt the nostalgic aural imaginary of Catalan identity as somehow pure. This reality is ugly, especially for women, which the novel illustrates not just with the sounds of accent striking Duc as irritating, but with physical illness: Patricia Miralpeix constantly has diarrhea, and when Mari Cruz, Horaci's new love interest, learns that Horaci is obsessed with her: "Vaig anar al water i vaig vomitar, com si les tripes se me n'anessin per la boca i, amb ells, tots els records" (I went to the bathroom and I vomited, as though my intestines were coming out of my mouth and, with them, all the memories) (175). Most importantly, all the characters confront in different ways the uncomfortable reality that the imaginaries of culture associated with the past cannot survive in the transition period. The old *senyora* Altafulla—a sort of Catalan Miss Havisham—for whom Mari Cruz works spends her time imagining nights at the opera from years past. Horaci Duc imagines a spoken Catalan that returns to the prewar Catalan heyday. Mari Cruz dreams of once again feeling the love she thought she was receiving when she was in fact being molested as a girl. But these fictions are replaced with bodies that are suffering and lost among a city that has moved on: the novel ends with *senyora* Altafulla surrounded by trash, listening to *La Traviata* for the ump-teenth time. Duc ends up at an old folks' home lamenting that he is a murderer because he participated in a violent September 11 protest. And Mari Cruz, the *xarnega* who had dreamt of love, ends up drugged out, trolling the streets outside the Cafè de l'Òpera with a transvestite friend. The daily opera invoked by the novel's title is not the sound of high culture, or even the desired sound of Catalan: The daily opera is the lived experience and societal presence of bodies in pain.

The deboned ear has caused this discomfort because it has produced an aural imaginary of a smooth sound of Catalan in conflict with the "grating" noise of multiple accents that make up the everyday soundscape. In the book, during the early days following Franco's death, these sounds affect the Catalan establishment, personified by Duc, most, because he must confront the fact that hearing accented immigrants speak improperly means hearing himself as incapable of instructing. Moreover, there is a lack of willingness to fully include immigrant voices in the dominant Catalan sound: When Mari Cruz asks Duc, who constantly corrects her accent, to teach her to speak Catalan properly, he refuses to do so, telling her that she simply needs to listen harder. The colonial ear through which Horaci Duc thought he could convert the voice of another into what sounds like a "civilized" voice while still not giving her the tools she needs to fully take ownership of it is the same ear that has caused him to repress his screams of frustration, and which has made him hear himself as the victim of an immigrant voice whose accented consonants sound like shrieking. The result is not only a Catalan ideal striving to reassert itself but an undermining of the Catalanist project based on anti-immigrant sentiment that instead fills the streets with dispossessed, seemingly monstrous bodies that threaten the stability of the Catalan state. As I show in Chapter 4, those who feel this exclusion at times use the same sonic script elaborated here to resist inscription into the new, neoliberal model of the city.

The question of dispossession comes up in these post-Transition texts often, and it does so in a way that emplaces accent not just in place, but in class, skin-tone, and gender as factors that stymie an imagined, peaceful national aurality. The next two texts I briefly interrogate illustrate how this accented presence becomes carefully circumscribed by linguistic norms that keep *xarnegos* and other *altres catalans* in their place. Maria Barbal's *Carrer Bolívia* (1999), for example, follows the story of Lina, a woman from Linares who moves to the outskirts of Barcelona with her husband. He is often away, involved in workers' rights movements and protests, and he eventually leaves her for another woman involved in the cause. Like Roig's *L'òpera quotidiana*, the text repurposes the tropes of womanhood celebrated by Franco by opposing Lina's motherhood to her neighbor Sierrita's prostitution.

And here, the reproductive capacity of the female immigrant body emerges not through her children, but through her decision to speak Catalan, even though it is not her native language, and even though she speaks with an accent. Lina and her friend Sierrita, also from Andalusia, decide to speak only in Catalan because Sierrita's father had told her, speaking in Castilian in a text otherwise mainly written in Catalan, "'Si hablas catalán, no te llamarán charnega'" (If you speak Catalan, they won't call you *charnega*).[76] As Lina, narrating, explains, the father "era més analfabet que ella . . . però no pas ignorant. Havia entès que parlar el català no deixava indiferents els del país, aquesta terra afortunada on hi havia feina i es podia viure. Que per això havia vingut, mira tu" (was more illiterate than she, but not at all ignorant. He had understood that speaking Catalan did not leave those from the country indifferent, this fortunate land where there was work and one could live. For that reason he had come, you see) (113). Although she is poor and a prostitute, Sierrita also claims that "la majoria del que tinc, per no dir tot, ho dec a saber català" (the majority, if not all, of what I have, I owe to knowing Catalan) (113).[77] Speaking the local language means being able to pass in ways that being heard as "speaking in Andalusian" does not permit.

Still, in Barbal's novel, Lina affirms, "rèiem molt, perquè al Besòs a penes se sentia el català, mira tu, com si no fóssim a Barcelona, i la gent ens mirava estranyada, quan ens sentien, i com amb un respecte" (we laughed a lot, because in Besòs you barely heard Catalan, you see, as though we weren't in Barcelona, and the people looked at us strangely when they heard us, and even with respect) (114). Here, the emplacement of accent and language works both ways: the fact that Catalan is not widely heard in their poor immigrant neighborhood means the space is not heard as Barcelona from the outside, while the immigrants who hear the women speaking the language look to them with respect, unable to recognize, as Lina does, that her friend speaks Catalan *amb gràcia andalusa*, that is, with an Andalusian accent. Through language, Lina can recognize the failure of her fellow migrants' ears, and as a result Lina and her friend are able to carry out an aural drag show for their community, one that empowers them because, like Ocaña, they can occupy two loci of audition at once. Indeed, Barbal privileges a bilingual immigrant ear as superior to a monolingual one, since a bilingual ear hears two places at once and thus makes possible a distinction between good and bad immigrants, those who linguistically make an effort to assimilate and those who seem merely to invade the city space. Importantly, though, Barbal does not give the women power in any other realm that would allow them to prosper: their laughter is not evidence that the system that emplaces their voices has been overturned; on the contrary, it is the laughter of women who know that their accented voices are the butt of their own self-deprecating joke.

This emplaced hearing of voice continues even when Barbal attempts to dismantle the racial stereotype of Andalusianness by incorporating into the story depictions of English blond and blue-eyed migrants to Andalusia (who, the novel attests in a reaffirmation of the stereotype, stand out because they are not dark). Just as with the "respect" the women receive for speaking Catalan, the aurally emplaced understanding of identity merges with class and national distinctions that inform how locals hear an English engineer who has immigrated to the region and lives there with his daughter: "Li deien, somrient, si havia pensat que, per més que ell li parlés l'anglès, la nena ja no seria anglesa, una velada invitació a tornar-se a casar amb alguna senyora del seu país o retornar-hi" (They asked him, smiling, if he had considered the fact that, no matter how much he spoke to her in English, the girl would never be English, a veiled invitation to remarry with a woman from his own country, or go back there) (65). His daughter, meanwhile, "parlava andalús amb accent anglès" (spoke Andalusian with an English accent) (65). In this novel, the tangled association between language, accent, and origin manifests in accents that occur because of the kinds of movement that migration engenders. But in this case, the English immigrant family are those with control and wealth. The subtle talk of accents as a suggestion to leave is, then, a recognition of language's colonizing force in Spain and a rejection of the globalized opening of Spain to foreign investment that was part of its move to join Europe.

Perhaps the best-known portrayal of accented language in Barcelona, however, appears in Juan Marsé's 1990 novel, *El amante bilingüe*. Marsé is the quintessential literary figure who, despite being from and writing almost exclusively about Catalonia, has come to be known for being excluded from the Catalan literary pantheon for his decision to write in Castilian; he writes from a similar position in the Catalanist debate as the one Pons took when he filmed his documentary. In *El amante bilingüe*, Marsé goes beyond the occasional code switch between Castilian and Catalan, as occurs in his other texts, and instead uses an orthographic portrayal of sound to explicitly parody the attempts by the Normalització Lingüística (Linguistic Normalization) program to regulate the Catalan language. Seeking to complicate, but also largely reaffirming, the tropes that understand Catalanists to be elite, blond bourgeois fanatics about language, and immigrants to be street children and dark-haired *xarnegos*, the novel inscribes the question of language and immigration in the language debates of the 1980s.[78] The book begins the night after Franco dies, when the main character, Juan Marés, discovers his wife Norma (a not-so-veiled reference to the Normalization program) has been cheating on him with a *limpiabotas charnego* (*charnego* shoeshine). Through a complicated flashback structure, the text incorporates scenes of his childhood as well as his present-day attempts to reunite with Norma. He takes on the persona of a childhood friend, the *elegante murciano* (elegant Murcian) Faneca, who describes himself as "un charneguito amigo tuyo que un día se fue del barrio en busca de fortuna y nunca llegó a nada" (your little *charnego* friend who one day left the neighborhood looking for his fortune and never became anything).[79]

Like Ocaña, Marés turns his body into a visual and aural playground. However, in an inverse of Ocaña's portrayal, Marés portrays culture as a costume that can be put on and taken off at will, even as gender itself is never questioned. At one point, a Molotov cocktail thrown by Catalan nationalists burns his face off, and so, attempting to look darker and more exotic, he draws on eyebrows and wears toupees and an eyepatch. He also finds a neighbor's contact lens on the floor and decides to change his eye color from brown to blue. Privately, he hopes to win back his Catalanist wife, at least for a one-night stand, by performing the *limpiabotas* role. Publicly, he busks on the street, playing the part of the impoverished musician, and carrying a sign that identifies him either as a "pedigüeño charnego sin trabajo ofreciendo en Catalunya un triste espectáculo tecermundista" (*charnego* mooch without work offering Catalonia a sad third-world spectacle) or a "fill natural de Pau Casals" (natural son of [musician] Pau Casals) (22). Focused on the spectacle of identity, he occupies both the despicable immigrant stereotype and the nationalist Catalan role as needed. And in my reading, at least, both of these are, in fact, the "natural child" of Pau Casals, if we consider Casals to be, in the novel, a representation of the sleight of ear that creates the Catalan(ist) voice in opposition to the *xarnego*-accented sound of Castilian Spanish. Casals was a world-famous Catalan cellist who was exiled from Spain due to his Republican beliefs. In 1971, when accepting the United Nations Peace Medal, he not only played his famous piece "El Cant dels Ocells" (Song of the Seagulls), he made a speech titled "I Am a

Catalan." In it, he expressed pride for being from a nation that had the first parliament in the world, prior even to that of England; his claim became a part of the Catalan nationalist claim. Here, Marsé seems to be playing with the two extremes of Catalan identity. The same sleight of ear that produces a perceived sound of Catalanism in Pau Casals's music also produces the opposition of that spiritual sound as the interrupting voice of the immigrant worker. Still, thoughts of the reproduction of Catalanism are never present here; the reference to Casals is presented to the reader as a joke at the expense of nationalists' claims of injustice and suffering. At most, Marés and Norma want sex, and the Catalan project in the text is a sign of elitism rather than an effective movement for change.

Just as Marés refigures his body through a variety of disguises, he refigures language by hearing it not as the ideological site of building the nation and the family, but as a meaningless acoustic experience that only reflects the locus of audition of the listener. In this (de)construction of accent, the body in movement that is so celebrated in Ocaña is nonexistent. Instead, the emplaced voice as a stereotype *replaces* the body. The text is full of spellings intended to represent the *xarnego* accent, heavily dependent on replacing <z> with <s>, leaving off final <r>s, and switching <l>s and <r>s, with *xarnego* characters speaking in ways that echo the Quintero brothers' attempts at *costumbrismo*: "Una hiztoria mu bonita, zí, zeñó" (A very nice story, yes, sir) (17); "No se fíe uzté de las apariencias. Su mujé le quiere a uzté" (Don't focus on the apperances. Your wife loves you) (18); "Yo no sé, zeñora, yo zolamente soy un mandao. . . . Pues musha grasia, no la molesto más" (I don't know, ma'am, I am only a messenger. . . . Thank you, I won't bother you anymore) (75).

Yet, his speech is by no means an accurate depiction of a "real" southern accent. Marsé is not striving for mimesis, but rather the contrary: a deconstruction of all linguistic and societal norms through sonic performance of an aural imaginary of *el andaluz* that hears it simply as other, whether that be an other to the norms of the Real Academia Española or the Catalan norms of pronunciation that circulated in the 1980s under the auspices of the Generalitat. By exaggerating the sound of the accent and the Catalan norm it supposedly threatens, Marsé allows the reader to hear the absurdity of the imagined deviance of Andalusian speech. In a series of calls that may be a reference to the "Digui, digui" (Hello, hello) television campaign used to teach residents Catalan in the 1980s, for example, he calls his ex-wife Norma at her work as a linguistic advisor and proceeds to carry out a sort of aural striptease by asking her to name various articles of clothing in

Catalan, proceeding from a winter coat to a blouse to bras and garter belts. In the conversation, he "disfraz[a] la voz con una ronquera abyecta y un suave acento del sur" (disguises his voice with an abject gruffness and a soft southern accent) (26). The carnivalesque nature of this acousmetric strip-tease is explicitly alluded to in the epigraph to Part One of the text, a quote by Antonio Machado that states, "Lo esencial carnavelesco no es ponerse careta, sino quitarse la cara" (The carnivalesque essence is not to put on a mask, but to take off one's face) (7). There is no voice other than the deboned voice, no identity without the ear on the other end of the line receiving and validating its presence.

The satire of the book thus rests on the notion that the immigrant accent is enveloped in an aural imaginary of geographical alterity that echoes the imagined local normalization that would purify the Catalan soundscape. Affirming the slippage between phonocentric and logocentric constructs of speech, the novel ends with the protagonist, who has suffered a crisis of identity, on the street portraying himself as the self-proclaimed *torero enmascarado* (masked bullfighter), playing accordion for tourists outside the Sagrada Família. Previously, he has directed his performance at his ex-wife Norma, and the linguist Jordi Valls Verdú—a seeming reference to the well-known sociolinguist Francesc Vallverdú. He is wearing a bullfighter's sequined *traje de luces* (suit of lights) and the traditional Catalan hat, the *barretina*, and speaking a gibberish consisting of Catalan and Andalusian-accented Castilian:

Pué mirizté, en pimé ugá me'n fotu de menda yaluego de to y de toos i així finson vostè vulgui poque nosotros lo mataore catalane volem toro catalane, digo, que menda s'integra en la Gran Encisera hata onde le dejan y hago con mi jeta lo que buenamente puedo, ora con la barretina ora con la montera, o zea que a mí me guta el mestizaje, zeñó, la barrexa y el combinao, en fin, s'acabat la explicació i el bròquil, echusté una mo-neíta, joé, no sigui tan garrapo ni tan roñica, una pezetita, cony, azí me guta, rumbozo, vaya uzté con Dio y passiu-ho bé, senyor.

(So look, in the first place, I laugh at myself, then at everything and everybody and as far as you want because we Catalan matadors want Catalan bulls, I mean, I truly integrate myself into the Great Enchant-ress [Barcelona] wherever they let me and I do with my face whatever I can, now with the *barretina* now with the *montera* [bullfighter's hat], that is, what I like is *mestizaje* [miscegenation], Sir, the mix and the combination, anyway, that's it the explanation and the nonsense are over [I've gone too far], throw a coin in, fuck, don't be so tightfisted or such a miser, just a little *peseta* [coin], holy shit, that's how I like it, generous, go with God and have a good time, Sir.) (220)

My translation does not do justice to the constant mixing between accents and languages that takes place in the text.[80] The fact that in one moment in this diatribe Marés says *zeñó*, meant to represent accented, Andalusian Spanish, and at the end merely the Catalan *senyor* on the one hand seems to signal the character's bilingualism. Yet an indeterminate accent interrupts both languages: there are moments in which it is unclear if his accented interference is with Castilian or Catalan, as with the use of the word *ora*. This could be Castilian *ora* (now) (itself not necessarily a typical indicator of Andalusian accent, though it could be meant to suggest a quicker pacing of speech, a charge often levied by non-Andalusians listening to those from the south), or a Castilianization of the Catalan word *ara* (now) with the "o" replacing the first "a." The accented, constantly code-switching voice—which is fully legible only to a bilingual reader and listener—draws attention to the relationship between language as identity and voice as performance in a Barcelona that is both polyglot and superficial.

But again, however he speaks, a monstrous character who does not conform to either the ideal or the abject remains present in the center of the city, and specifically a city that has turned its sights to entertaining the world. The sonic portrayal of linguistic *mestizaje*, as the character calls it, alongside his sartorial deconstruction of the visual signs of national identity (both Castilian and Catalan, with the *montera* and the *barretina*), sounds its disconcerting vocal chord in front of the Sagrada Família. This site is one of the most visible signs of Catalan culture and, also, its main tourist attraction. The cathedral simultaneously represents the modern golden age of Catalan culture, which took place between the *Renaixença* and before the Civil War, a public avowal of Catalan resistance and determination, as the project is taken up with fervor after Franco, a future-oriented neoliberal fantasy, and a polyglot space wherein, paradoxically, in order to promote Catalan success, the very soundscape meant to reflect it must be diluted in order to appeal to foreign visitors. In this final scene, Marsé captures both the everyday reality of a Catalan soundscape that can never be linguistically purified, despite the desire of some to shed Catalan of any Castilian "barbarisms," and the fact that, in the lead-up to the Olympics, Barcelona is being gradually sold to a foreign audience who does not understand the complexity of the culture with which they engage. After all, the *mestizaje* to which Marés/Faneca refers is an accented, multilingual, code-switching language that bothers the ears of both Catalan purists like Norma and the sociolinguist Valls Verdú for its sonic impurities, and the tourists, who hear only an unintelligible, poor beggar taking up space on the sidewalk, not a person living in his city.

In that sense, Marés's performance of accent foreshadows the polyphonic soundscape of Barcelona that Brigitte Vasallo represents in *PornoBurka* (2014), a similarly ludic, even more graphic depiction of what she calls *la Barcelona pequeña* (the little Barcelona), by which she refers specifically to the recently gentrified neighborhood now known as El Raval.[81] As she depicts it, this space is occupied by what she calls in an acerbic, tongue-in-cheek tone *lo multiculti* (the multicultural) (78). In that group she includes—among others—*xarnegos*, Pakistanis, Indians, Nigerians, Galicians, Romanians, Roma, folks from across the LGBTQI+ spectrum, feminists and those who can identify as "lesbianas-feministas-de-verdad®" (real-feminist-lesbians®), and a pseudo-Argentine (45). Daily, they confront not just the expectations of a society obsessed with heteronormativity, image, and glamor, but the constant incursion of tourists and uptight *mossos d'esquadra* (police) who are *normalizados lingüísticamente* (linguistically normalized) and ready to jump at any moment at the possible threat to Western notions of civility that are required to keep tourists coming to town (155). The engaging, irreverent narrative revolves around a single moment that throws El Raval into chaos: the sight of a burka on the Rambla. Carrying out a similar, though far more wide-ranging portrayal of accent to that which Juan Marsé created in *El amante bilingüe*, Vasallo's text uses similar strategies of destabilization not only to deconstruct, but simultaneously, as Marina Garcés has suggested, to search for a presence that is also a kind of truth (171). The polyphonic tale takes various characters' points of view, but what emerges on almost every page is a listening ear that is attuned to sound as a culturally inflected phenomenon. Accent in this text does not merely appropriate or play with the emplaced voice, it abolishes it as a legitimate way of hearing by giving insight into the interiority of the bodies that from without appear one way, but within are struggling with how they are emplaced by others.

Unlike the silence of Ocaña's walk or the self-directed interiority of Marés's ear, from the first pages of *PornoBurka*, El Raval is described as a "descomunal caos sonoro" (huge sonorous chaos) filled with children screaming, neighbors praying to Krishna, and *rancheras* playing at all hours

(14).[82] This noisiness, though, is part of the "truth" of the novel, which satirizes the effects modernization, tourism, and immigration have had on the neighborhood. On the one hand, "El barrio del Raval barcelonés sigue amaneciendo como cuando se llamaba Barrio Chino: meado y con resaca" (The Barcelona Raval neighborhood continues waking up in the same way as when it was considered Chinatown: pissed on and hung over) (21). But on the other hand, "llegó Europa" (Europe arrived) and with it, all that was *la vida* (life)—sex, specifically *puterío* (whoring), pain, violence, misery, and the hardness of the neighborhood—was converted into a theme park, an attraction for visitors (22). Within this acoustically rich environment, the silent visual of the burka throws into chaos the epistemology of the emplaced ear that has pervaded the various ideological constructs that sustain the different communities that live there. Seeing the burka but being unable to hear a voice from within it has multiple effects for the people of El Raval: from the fear of terrorism by the *mossos d'esquadra* (Catalan police force) who are focused on protecting tourists, to the fear of the Pakistanis who live in the area that they will be seen as terrorists, to the fear of the limits of feminism by the main character Lo. The narrator, who is not so much omniscient as she is all-hearing, reveals only to the reader the farcical reality of what is underneath the burka: a male private detective, who is likely a riff on Vázquez Montalbán's Pepe Carvalho, and who has been hired by Lo's overbearing, old-fashioned parents to find her. With their focus on the appearance of the burka and what it might represent for each of their various interests in El Raval, not one of the characters bothers to ask who is inside it. Instead, they react only to its silent visual presence. This includes Lo, who has a whole conversation with the person underneath, offering the presumably oppressed woman a speech on feminism as a way toward liberation. Lo does not realize that the person listening is the man who wants to reinsert her into the life she abhors: the normative family structure.

In other words, the question of reproduction as it relates to gender is front and center, but entwined now with a sense of dispossession: Lo blames the nuclear family structure for gentrifying the neighborhood, bringing with it "hordas de personas perfectamente sanas y en edad reproductora, guapos incluso" (hordes of perfectly healthy people of child-bearing age, good-looking even) (13) who interrupt the *mierdez* (shittiness) of the place with the sounds of children screaming. But the wider question is the superficiality of the emplaced way of seeing and hearing culture that assumes only certain performances of gender have a place in society, or that those less desirable ones should be relegated to certain geographic zones, like the old red-light district of the Barri Xino. Lo, "charnega, diseñadora y camarera (o viceversa)" (*charnega*, designer and waitress [or vice versa]), is the main example of the limits of emplacement—after all, she is Galician, so considering herself "charnega" is itself a reappropriation of the idea of migrant origin (13). Similarly, once known as Conchita, and later Cookie, she has chosen the neuter Lo to refer to herself, as she struggles to reconcile her feminism with her ever-changing sexuality. She has tried to identify as lesbian, trans, and eventually a gay man, but none of these fit, as she explains in long narratives that echo the confessional style but also concern themselves less with her sense of self than with her interest in effectively naming who she is. Finally, after trying on a number of terms, she asks, "¿Pero qué soy cuándo me follo a otros? Un hombre. Decidí que yo era un hombre" (But what am I when I fuck others? A man. I decided that I was a man) (18). Specifically, she decides that she and her partner, Buenaonda, "un exiliado presuntamente argentino bien plantado y bohemio" (a supposedly Argentine, handsome and Bohemian exile) are now gay men (22). In the end, Lo's concern with naming their relationship changes the balance of power, such that her partner gets in touch with his maternal instinct and, because of the dialectical nature of their relationship, forces her to become the breadwinner, a role she did not want, all while "follábamos como siempre: su alma de mujer y mi alma de hombre, su bio-cuerpo de hombre y mi bio-cuerpo de mujer, los cuatro heteros " (we fucked like always: his woman's soul and my man's soul, his bio-body of a man and my bio-body of a woman, the four heteros) (18–19). Despite Lo's focus on defining their gender roles, in this quote the reader learns that it is not so much that Buenaonda wishes to take on the role of woman's soul in a male "bio-body," but rather that he is faking his entire identity and is frustrated with Cookie's "historias de si soy hembra o soy merluza . . . de si soy bi o tú eres tri" (stories of if I'm a woman or I'm a fish . . . if I'm bi or you're tri) (23). Although we first hear him speaking with all the familiar Argentine slang, "Dejalo, boluda . . . Vení acá, es muy pronto para tanto quilombo" (Leave it, *boluda* [silly]. . . . Come here, it's too early for such *quilombo* [ruckus]) (13), it turns out that he is not Argentine at all, but rather a guy from Cuenca "sin más habilidades que liar buenos porros, caer bien a la gente y tocar un poco la guitarra" (with no skills other than rolling good joints, being liked by people, and playing a bit of guitar) who discovers that in Barcelona, land of democracy, you can be whatever you want: "descubrí mi auténtica vocación: yo quería ser argentino" (I discovered my true vocation: I wanted to be Argentine) (24).

This satirical riff on the Western capitalist idea of the self-made man is thus taken to its most ironic extreme through the character's superficial use of the sound of voice. Exaggerating the role the sound of voice plays in representing identity, Buenaonda's fake Argentine accent can become a life's mission. The absurdity of this goal—not just to embody, but to live the stereotypes associated with the sound of accented immigrant voice—draws attention to the way a nationalizing geography of the ear continues to hold sway over society, now as an emptied-out ideology, despite the radical changes in cultural composition that have taken place in the city.

Moreover, the notion that accent can be mapped onto one's locus of enunciation is also firmly destabilized in the text. Lo's parents are themselves immigrants to Barcelona, having come from Galicia and speaking in a mix of Galician and Castilian, and yet still associating "normalcy" with the Spanish state: Lo's father, Jacinto, meeting with the Catalan-speaking private detective he has hired to find his daughter, thinks, in a mix of Galician and Castilian, "¿De qué carallo ríese y por qué fala catalán si aínda estamos en España?" (What the hell is he laughing about and why is he speaking Catalan if we're still in Spain?) (48). Her mother, meanwhile, explains, "Yo me vine a Cataluña que es España y al llegar me encontré que no entendía ni el dialecto que falan aquí. . . . Cuando llegaron los moros pues la cosa siguió como es normal. Ellos querían que el barrio fuese Marruecos. Nosotros les decíamos que en España había que vivir como españoles. Lo normal" (I came to Catalonia, which is Spain, and upon arriving I found that I didn't understand at all the dialect they speak here. . . . When the Moors arrived, well, things continued on as normal. They wanted their neighborhood to be Morocco. We told them that in Spain you have to live like Spaniards. The normal thing) (49). The debate over Catalonia's place in Spain is evident here, but by blending Galician and Castilian in these internal monologues and outward speech, the text draws attention to the difficulty of emplacing identity through accent, or even language. The paradox of a bilingual person living in Catalonia but rejecting its right to language, or even affirming a monolithic state's right to order, pokes fun at both the simplistic idea of a singular sound of linguistic identity as well as Catalonia's presumption of its own bilingual status as somehow unique within Spain.

In some ways, these code-switchings and questionings of a linguistic standard for culture recall Juan Goytisolo's rewriting of Spanish culture in *Señas de identidad* (in fact, he writes the prologue to the novel). But more than that, the text emphasizes the effects global migration and tourism have had on a city not necessarily (or not only) confronted with the legacy of

Arabic on the Spanish language (as in Goytisolo's novel), but rather with the effect multiple languages have on a space that can no longer be heard through an emplaced distinction of local versus foreign, Castilian versus Catalan. Impacted by the global flows of sounds and loci of audition, the ear confuses the boundaries of place on which linguistic identity have been founded.

In the process, the soundscape of the city has been reshaped. Lo's parents, for example, had come from Galicia and moved to the Barri Xino (now El Raval), but they were forced out by the increase in rent and other costs associated with tourism and gentrification. Instead, they moved closer to the Arc de Triomf, to live near Chinese immigrants whom they are able to cast as the new threat to Spanishness: ". . . salimos del Barrio Chino que estaba lleno de moros para meternos en Chinolandia. Con chinos de verdad. . . . Y un buen día todos os carteles do barrio y todas as cousas estaban en chino y en catalán. Como si España no existiese" (We left Chinatown which was full of Moors just to end up in Chinaland. With real Chinese people. . . . And one fine day all the posters in the neighborhood and all the things were in Chinese and Catalan. As though Spain didn't exist) (52). In this commentary, the sound of language becomes the stand-in for racist critique that is factually misinformed due to the echoic memory of the colonial ear, which is now global in its reach:

> Yo lo único que pido es que los chinos devuelvan las enes adonde tienen que estar. Que la confecsió se diga confección, que la importasió, vuelva a ser importación. Que se enteren que Cataluña es España. Es normal que un hombre pida eso. Pero entonces la niña dejó de venir. . . . Que dijo que yo era racista. Ya me dirás cómo puedo ser racista si los negros me dan igual. Pero desde el tiempo de los chinos y las enes no ha vuelta a venir y no sabemos dónde para.

> (The only thing I'm asking is for the Chinese to put the n's back where they belong. That they say dressmaking [confección] instead of dress-makin' [confecsió], that importatio' [importasió] go back to being importation [importación]. That they realize that Catalonia is Spain. It's normal for a man to ask for that. But then the girl stopped coming around. . . . She said I was racist. You tell me how I can be racist if I'm fine with Black people. But since the time of the Chinese and the n's she has not come back and we don't know where she's gone.) (52)[83]

Reflective of recent debates on immigration, here one of the sounds that indicates the linguistic difference between Catalan and Castilian (*confecció* versus *confección*, *importació* versus *importación*), is heard as being a deficient pronunciation of Spanish by the Chinese immigrant community, rather than a sign of a successful integration of these immigrants into a system that has been privileging Catalan over Castilian for several decades now; *confecció* and *importació* are not mispronounced Spanish but a sign of the success of Catalan education programs. Lo's father, who is still hearing language through a local/other binary, rooted in a Castilian aurality, through which even his own native tongue should not be used in the public arena, cannot contemplate a space where Catalan is the norm, and thus does not even recognize the language he is hearing. Moreover, assuming that racism is something that can only be carried out against Black people, he perpetuates a binary construct of race that does not account for the global flows of migrants discriminated against by nativists. The inadequacies, but also persistence, of the colonial ear thus resonate in this scene, sounding at once outdated and humorous, yet also potentially dangerous. Moreover, those who stand to lose the most in this persistence of the colonial ear are the immigrants who are heard as out of place even when they have learned the language, as immigrant accent becomes either a sign of alterity to be exoticized or a sound of complete abjection in need of salvation.

In the end, the book demonstrates how a monolingual ear, grounded in the double aurality of voice enacted in the *Requerimiento* (see Introduction), crosses linguistic lines and dominates contemporary media apparatuses that convert migration into a political spectacle. In one of the central scenes in the book, a Pakistani grocer named Lahore is approached by a brigade of integrationists, accompanied by TV3 cameras, who hope to demonstrate the (im)possibility of immigrants becoming Spanish or Catalan because they are (un)able to master the languages. A *mosso d'esquadra* (police officer under Catalan authority) and a *guardia civil* (police officer under Spanish authority) take turns asking Lahore to perform linguistic competence as a sign of integration. While he has rehearsed a line of each language that satisfies both groups at first, soon someone in the audience yells out, "Un moment! . . . Aquest moro parla català amb accent xarnego!" (Wait a minute! . . . That Moor speaks Catalan with a xarnego accent!) (101). What follows is a series of language tests. First the Catalans have a go:

—Feu-li la prova de la doble ela! Grita una filóloga desde el fondo.

El Mosso d'Esquadra se acerca a Lahore. . . . Coloca la lengua en una cuidada posición y pronuncia, alargando las eles: "Lluís Llach, Lluís Llach."

—Yuís Yac, Yuís Yac, responde Lahore.

—No, no, més lateral, menys postalveolar i més velar lateralitzada—instruye el Mosso.—Lluís Llach!

(—Give him the double L test! shouts a philologist from the back.

The *Mosso d'Esquadra* approaches Lahore. . . . He places his tongue in a careful position and pronounces, lengthening his <l>s: "Lluís Llach, Lluís Llach."

—Yuís Yac, Yuís Yac, responds Lahore.

—No, no, more to the side, less postalveolar and more lateralized velar—instructs the cop. Lluís Llach!) (101–2)

Despite the call-and-response structure, in which Lahore is expected to take instruction on the movement of his tongue in order to repeat back the proper sounds of the language, his mouth and tongue cannot conform to the norm; because of that, he cannot pronounce the name of the *nova cançó* singer whose name is meant to represent Catalan sonic culture as well as a perfect Catalan pronunciation. Later, the *guardia civil* takes a turn:

—A ver. Pronuncia: Ehpaña!

—Ssssbania . . .

—Ehpaña!

—Ssssbania? Murmua Lahore.

El público niega con la cabeza: indudable acento moro.

(—Let's see. Pronounce: Ehpaña!

—Ssssbania . . .

—Ehpaña!

—Ssssbania? Murmurs Lahore.

The public shakes its head in disappointment: indisputable Moorish accent.) (102)

The Pakistani immigrant caught between these two language norms is thus cast out by both groups due to the accent that marks him as out of place—either as *xarnego* or as "Moorish," neither of which accurately reflects where he is from, and excluding him from any language at all: "Este jodido inmigrante no habla ningún idioma" (This fucking immigrant doesn't speak any language) (102). Whatever their linguistic aurality, the public sleight of ear hears Lahore as voiceless and earless, because their ear assumes that his inability to reproduce the normative sounds of a local language means he is incapable of both listening and speaking. As I show in Chapter 4, a similar way of hearing has affected the way Moroccans and other non-Romance speakers, particularly those associated with Islam, are heard in the news media as only being capable of nonvocal sound.

To conclude, we may consider how the increasingly globalized sound of Barcelona depicted in Vasallo's book portrays Lahore's accent as perhaps the most threatening sound of the city. This is not because his Spanish or his Catalan are imperfect, but because, in yet another vocal chord, his accent denotes Spain's need to adapt to a global reality in which English is more threatening than any other linguistic sound because it signals not just tourism and immigration but a global geography of the ear in which both languages are trumped by English. When Lahore fails his public language hazing, then, the "integrationist brigade" concludes, "¡Al final tendremos que aprender nosotros inglés para pedirles documentación, joder!" (For fuck's sake, we're going to end up having to learn English just to ask for their documentation!) (102). Indeed, the polyglot nature of contemporary Barcelona comes into play in the work's address to a trilingual reader, whose ability to sound out English contrasts with the characters in the story. Thus, ©Jor-dee, whose copyrighted name refers to his desire to brand himself using a common Catalan name (Jordi) and build a company based on his image, determines due to a mishearing of language that Lahore must be a terrorist. He is shocked Lahore has enough money for an iPhone and asks him about it. Lahore responds: "Yes, yes. . . . Really iPhone. I bought it" (142). Although they have had sex and hung out a bit, ©Jor-dee shows he shares the sleight of ear with the policemen and linguists above, because he is shocked that Lahore can speak at all. Moreover, whereas a reader who knows English reads "I bought it," ©Jor-dee, still trapped in a binary aurality informed only by Catalan and Spanish, hears something different:

—A ver. ¿Qué has dicho? Aiboutit, aiboutit. . . . ¿Qué idioma es ese, dime? ¿Es pakistaní? Debe ser un dialecto de alguna tribu, supongo. Aiboutit. . . . Suena muy bonito, ¿sabes? . . .

—I bought it—repite Lahore—. I bought the iPhone. . . .

©Jor-dee está maravillado. Maravillado y muy decepcionado. O sea, Lahore habla, pero no entiende. ¿De qué coño le sirve entonces que este tarado no sea mudo?

("Let's see. What did you say? Aiboutit, aiboutit. . . . What language is that, tell me? Is it Pakistani? It must be some tribal dialect, I guess. Aiboutit. . . . It sounds really nice, you know?"

"I bought it," repeats Lahore. "I bought the iPhone. . . ."

©Jor-dee is amazed. Amazed and very disappointed. That is, Lahore speaks, but he doesn't understand. What the fuck good does it do him, then, that the fool isn't mute?) (142–43)

©Jor-dee, thinking at first that the language can only be a "tribal dialect," doesn't believe his ears when he realizes Lahore speaks English because it implies a more nuanced, everyday aurality of a global, trilingual city than he expects, and because it gives Lahore, an immigrant, an advantage ©Jor-dee cannot possibly conceive of: "—¡Coño! ¡Inglés! ¿Inglés?—El artista no da crédito. ¿Pero cómo es posible? ¿Dónde has aprendido tú el inglés, por dios?" (Fuck! English! English? The artist can't believe it. But how is that possible? Where have you learned English, for god's sake?) (143). Then, his assumption of Lahore's lack of ability to learn any language, and the fact that he speaks English, the contemporary language of empire in a global world, turns to suspicion: "¿Tú no te juntarás con gente rara, verdad, de esa de la tuya?" (You're not going to get involved with weird people, right, those of your kind?) (143)

Lahore's English voice is thus emplaced in a geography of the ear that associates language not only with national identities, but with echoic assumptions about class that suppose that immigrants have predetermined stations, deviations from which can only be explained as suspicious. Such suspicions are the result of a truncated ear with a complete lack of historical understanding of the colonial condition (the result of which is a Pakistani immigrant who speaks English, if not Spanish or Catalan). When they finally use the iPhone itself to translate, the mishearing bleeds into a diatribe by ©Jor-dee on the place of immigrants in Spain. Earlier, ©Jor-dee had understood himself to be open minded, forcing oral sex on Lahore as a means of (in his mind) liberating him from what he perceived as his overly conservative Muslim values. Yet when faced with the immigrant's ability to speak English and purchase an iPhone, he quickly goes down a rabbit hole of stereotyped and malicious expressions about Lahore's right to be in Spain, supposing that Lahore must be connected to al-Qaeda. Linking immigration and tourism to terrorism, he explains (via the iPhone translator) that al-Qaeda is only good for big business invested in travel, like RyanAir, EasyJet, and Vodafone. Grabbing Lahore by the neck (the violent conquest that is the other side of fetishizing and then having sex with him), he "escupe tantos gritos que ni él mismo se oye" (spits out so many screams that he doesn't even hear himself) (145). Among them are:

¡o hacéis mucho ruido o sois una amenaza silenciosa! . . .
¡iros a vuestro país!
¡esta es nuestra casa!
¡vuestro nuestra!
¡norte sur!
¡arriba abajo!
¡hombre mujer!
¡barça madrid! . . .
Ensuciáis nuestra ciudad,
Saqueáis nuestra riqueza,
. . . ¡pedís que os hagamos mezquitas! (146)

(You all either make a lot of noise or are a silent threat! . . .
Go back to your country!
This is our house!
Yours ours!
Up down!
North South!
Man woman!
Barça Madrid! . . .
You dirty our city!
You take our wealth!
. . . You ask us to build mosques for you!) (146)

These ludicrous binaries—derived from a stereotyped "civilization/barbarism" opposition—reinforce the notion presented through Jacinto's hearing of Chinese people speaking and writing Catalan that a monolingual or even bilingual ear continues to inform the aural imaginary in Barcelona, despite the fact that migration has complicated the emplaced origins of vocal sound to such a degree that the old modes of hearing one's place through voice no longer work. Within the polyglot space, an expectation that immigrants will be silent and unspeaking, and thus remain anonymous, has replaced any desire by the listener to expand his geography of the ear, with ©Jor-dee concluding that Lahore is "un moro cualquiera como todos los demás" (just another Moor like the rest of them) (148).

Although from a historico-cultural and aesthetic perspective *PornoBurka* is in many ways different from the early exploration of immigration to Barcelona seen in Ventura Pons's film, it nonetheless dialogues with Ocaña's idea of "breaking things." It does so, moreover, by invoking the emplacement that emerges in accent, which overlooks the very person it claims to represent, the "truth" that is so elusive and yet *is* real because the body in space is real. At stake, as I explore in the next chapter, is the physical space not of the city as a whole, but as composed of local neighborhoods, and how residents hear the voices that occupy them as they confront a globalized flow of investors, migrants, and tourists whose voices and music change the soundscape of Barcelona in even the smallest of ways, every day.

IMMIGRANTS AND ACCENTS

One person's voice does not an accent make: Voice is an ephemeral corporeal production of breathy sound emerging from a speaker's body, but in the moment it is sounded and heard as accented, it forms part of a vocal chord that reverberates in the bodies that are present and in the aural imaginaries of emplacement that engage those bodies through the listening ear. The boneless emplacement that constitutes the Western conception of voice as subjectivity echoes the imaginary inscription of national culture in place. But as migration and travel become more the norm rather than the exception, new ways of hearing identity are going to be needed if we are to avoid perpetuating—perhaps even worsening—the colonial sleight of ear that has allowed this aural imaginary to flourish throughout not just the Global South, but old centers of power, in the early part of the twenty-first century.

Fran Tonkiss has suggested that accent and polyglot sound challenge the monolinguistic acoustic order imposed by architecture, tradition, language, and historical memory because, as unfamiliar sounds, they demand attention in a space that is otherwise heard as silent—silent not because it is not voiced, but because silence presumes linguistic sameness: "Speaking the same language is always a first requirement of 'assimilation,' but the city as polyglot soundscape is a space in which differences remain audible and translations incomplete."[84] For Tonkiss, the sounding of otherness through accent challenges the fictions of individualization, freedom, and autonomy that remain wrapped in monolithic understandings of voice, since subjects "glance at sounds in the city, we don't gaze."[85] However, this kind of walking-the-city approach to sound, which resonates with Michel de Certeau's visual understanding of the city, presumes a kind of *flânerie* that is inconsistent with the daily interactions with shopkeepers, tourists, and other inhabitants that are common to life in Barcelona. In a critique of Barthes, Tonkiss asserts that "For all his insistence that it should be seen as a kind of 'writing' or 'text,' Barthes' city kept bursting into speech in ways that go beyond the mute language of architectural symbols."[86] We may be caught off guard by the sound of an audible immigrant voice we do not recognize when we are not directly interacting with those around us, but in those interactions our attempts to make accent mean lead to linguistic sound shaping our spaces both representationally and geographically.

As immigrants in Barcelona, those who "speak in Andalusian," the Pakistani shopkeepers, the Moroccan women, and others, are present in the space of the city as listening, speaking bodies, despite the colonial ear that extracts their bodies and their presence from the aural imaginary and reduces them to an idea of foreignness or out-of-placeness. Discussing whether immigrants should be heard as Catalans, or *els altres catalans*, or (il)legal (im)migrants, or if they can or cannot form part of the national culture, is in the presence of their materiality irrelevant. They are here, and to ignore them, to hear their voices as mute and their ears as deaf, in need of punishment, exclusion, instruction, or policing, is to hear voice as an emplaced concept that mistakes an imaginary *of* place for placement, ideology for body, alterity for presence. This is the difference between an aural imaginary *of* immigrants and accents, an emplacement always produced at a temporal distance from accented bodies and listening ears; and immigrants and accents occupying a place in a territory or society. Although I now turn away from immigrant accents to focus on radio and music, it is that acoustic occupation of city space, and the potential for protest and resistance it implies, that is the topic of my next chapter.

3 Radiophonic Restlessness

In the last chapter, I demonstrated how immigrant accent calls into question the concept of emplaced voice by identifying the *vocal chords* accented bodies produce when they occupy different places in the city, however transitorily. My focus there was on how the presence of an acoustic body that sounds multiple places at once confronted the geographical ear that associates normativity with place. In this chapter, I wish to consider a different kind of movement than those created by the accented voices of immigration. Here I am interested in the underground sounds that cropped up in the 1980s through illegal radio stations, punk music, and experimental sound culture, which Jaime Gonzalo has argued formed part of the *ciudad secreta* (secret city) that was born in the years between Franco's death and the institutionalization of Spain's new democratic system. As Gonzalo argues, this "secret" sound has been left out of the narrative about the city because "en la incipiente Barcelona de diseño cualquier disidencia de esa dogma, cualquier reacción, resultaba severamente reprimida por los gustos populares, también llamados modas" (in the incipient designer Barcelona, any dissidence from that dogma, any reaction [to it] was severely repressed by popular tastes, also known as fashions).[1] In some ways a continuation of the counterculture I explored in the last chapter, and in other ways a shift in political priorities, this period of acoustic experimentation just after the turn to democracy did not fit into the Catalanist program of the period and therefore has been largely forgotten until recently even by Barcelona's own population. It was nonetheless "una vanguardia no menos poliédrica que condenada a la impopularidad generaría, no obstante, su propio circuito en los márgenes de lo minoritario" (a no less multifaceted avant-garde that, although it was condemned to unpopularity, generated its own circuit in the minoritarian margins).[2] For my purposes, it reflects a restless movement in and around sound that Xavier Mercadé, citing the fanzine *Pravda*, has referred to as the *meneo barcelonés* (Barcelona sway), a term he contrasts to Madrid's well-known and "self-exultant" Movida.[3] As he explains, while the Movida was understood as a sudden rupture in a previously staid Madrid, Barcelona's movement in the 1980s was a continuation of something that had begun in the '60s with magazines like *Star, Vibraciones,* and *Disco Express,* progressive rock and the anti-Franco *nova cançó* movement, and punk: "Era una lucha ya asimilada y consciente" (It was a conscious and already incorporated fight).[4] At the same time, the *meneo* "no entendía de tribus urbanas" (did not have urban tribes) but instead included everything from funk to surrealist pop, rockabilly, and street punk, to the early skinhead movement.[5] Local neighborhood relationships were especially important to how underground sound circulated in the city. I focus here on the neighborhood of Gràcia, which has its own aural history of resistance within Barcelona that echoed through the punk and sound art that emerged in the 1980s; to consider the *meneo* as solely a post-Franco or aesthetic phenomenon is to overlook the echoic memories of dispossession it captures, which are largely linked to intensely local spaces around the city, and have been since even before the dictatorship. With echoic memories of anarchist thought and the working-class lived tradition of the neighborhood *ateneus llibertaris* (libertarian atheneums), combined with an ear turned to music outside Spain, the restless and overlapping geographies of sound that traversed through free radio stations, punk concerts at local nightclubs, and self-recorded experimental sound art were heard nightly in different areas of the city.[6] Improvisation and a DIY ethic were key to these sonic productions, as were an ideal of utopia that Víctor Nubla, Julià Guillamon, and Pau Riba described through the work of rock historian Greil Marcus, whose *Lipstick Traces* conceived the history of the twentieth century as defined by "movements of total negation (Dada, situationism, punk)."[7] As they explain, they were drawn to Marcus's idea of *alter music* as "the search for utopia: progressive utopia, the utopia of associative creation, technological utopia and also the cry of negation inherent in free music and punk, when utopia showed itself to be unviable."[8] For them, at stake was "the pulse of revolt against the system, even when the system has been embodied in the music."[9] This pulse included not just experimental music, but in the 1990s, the *polipoesía* movement, which incorporated sound poetry, action poetry, and other experimental continuations of bruitism into a network of ephemeral sound happenings performed around the streets and bars of Barcelona. The aurality at play thus embraced an overlap between aesthetic forms, as well as a political position that associated sonic experimentation with direct action, as Nubla once put it: "A partir de Dadá, cualquier acto poético es terrorista" (After Dada, every poetic act is a terrorist act).[10] Nevertheless, as the publication of these artists' experimental sound theory work *Alter músiques natives* from 1995 illustrates, this acoustic terrorism had its limits: After all, the book was published by the Generalitat de Catalunya, part of the very institutional system that sought to elevate Barcelona from its perceived marginal position within Europe.

I therefore hear these musical and radiophonic experimentations as microresistances that are short-lived, often unheard outside the circumstances of their immediate performance, but whose repeated resonances transform public and private spaces over time, creating a constantly changing network of contingent sonic encounters. These encounters participate in what aesthetic theorist Nicolas Bourriaud describes as "tiny revolutions in the common urban and semi-urban life," the creation of split-second "micro-communit[ies]" that ebb and flow with the sounds they produce.[11] Bourriaud claims that these kinds of microencounters rework the Situationist understanding of daily revolution by exceeding art, since what is at stake now is no longer art as a closed system but rather the relationality of form, which moves in the interstices between multiple realms of lived experience, not just the aesthetic. Indeed, at times the aesthetic overlaps with the political in unexpected ways. Thinking of the multiple sound projects I study here this way, they illustrate what Brandon LaBelle calls a restless acoustics: a daily sound that emerges in the interstices of the public life of the city to create a "distributed agency," a "dwelling *in* difference," that "support[s] an event of *association*, or a space of radical sharing, or a more than one."[12] This notion of sound as radical sharing extends the radiophonic from how it relates specifically to broadcasting to the contingent relationships that obtain around it.

Specifically, I listen in to this *meneo* through the fragmented associations that radiate out from the longest-running free radio station in Barcelona, Ràdio PICA (Promoción Independiente Coordinación Artística-Alternativa [Independent Artistic-Alternative Promotion Coordination]), founded by Salvador Picarol in 1981. Inspired by the free radio movement in Italy and based in a local understanding of working-class realities, from its inception, the station sought to open up the city's soundscape to other ways of hearing local and global culture, with a focus on noncommercial, free access to information. Essentially, Ràdio PICA extended the print circulation of counterinformation through fanzines and other homemade media to the airwaves, sonically occupying radio with information and sound that would otherwise be inaccessible to the public. In this sense, its struggle was against the modernity/rationality aspect of coloniality, which Aníbal Quijano presents as the idea of "society as a closed structure articulated in a hierarchic order with functional relations between its parts . . . and a rationality consisting in the subjection of every part to that unique total logic."[13] The underground, as Ràdio PICA and those in their orbit conceived it, was an attempt to make audible all that did not fit in the closed structure that came to close in on Barcelona's democratic, neoliberal, linguistic development controlled by the government. Musically, it became the center of punk culture in the city, both for the access it gave to local and international music unavailable elsewhere in Spain, and for its connection to antisystem protests and an overall culture of resistance. At the same time, experimental sound artist Víctor Nubla was producing an underworld of sound, theorized through avant-garde ideas and informed by literature, that would slowly acquire a small-scale globalizing presence within the city, as it brought artists and musical experimentation from around the world into the soundscape of the city. The founder of the sound art and performance space Gràcia Territori Sonor (GTS), Nubla is perhaps best known for starting the experimental band Macromassa with Juan Crek in 1976, when they produced the first self-published record in Spain, *Darlia Microtónica*. Attuned to the technological possibilities of sound in a new era of recording mobility, the album was recorded at one of the primary places of underground congregation, the Sala Màgic nightclub. From that point on, Nubla collaborated with over four hundred artists and recorded over a hundred albums. He participated on Ràdio PICA in programs like *La Oreja Abierta* (1982–84), *Monstruo sin Cola* (1993–97), and *Territori Sonor* (1997–99), but he also created an entire acoustic universe out of recorded albums, live performances, and writings grouped under the larger notion of the Submundo Pérez (Pérez Underworld); his experimentation involved mixing the musical with the literary, the oral, and the performative in order to question any systematization of the arts.[14]

Present in the restless acoustic movement exhibited by the acoustic struggle that echoes through Picarol and Nubla is the fractal geography of a local ear, the underside to national or global aural imaginaries of sound and language. As Wai Chee Dimock argues, "Fractals is the geometry of the irregular and the microscopic, what gets lost in the big picture."[15] Although for Dimock fractals reflect deep time, within Barcelona they also share a microtemporality (what Nubla and Macromassa call the *microtonic*) that reshapes the sights and sounds of the city by pushing, from within, at the limits of a generalized neoliberal (and neocolonial) ear. As Nubla, Guillamón, and Riba wrote, artists such as they should not "forget that many of the musical initiatives we are dealing with move in circuits of a very limited scale, with very small runs and practically without access to the media."[16] Still, as Dimock argues, the fractal is not limited to the local: "It is a geometry of infinity, of what keeps spinning out, endless spirals," the pocks and marks that create an irregular "family of shapes" in what otherwise appears to be a smooth fabric.[17] In this case, the fabric in play is the Barcelona soundscape, whose scale could never be geographically circumscribed by the city or neighborhood alone. The radio program "Microtopies," for example, which Nubla broadcast on Ràdio PICA, sounded "miniatures sonores de música i geografia" (sonic miniatures of music and geography) that were "micrologías poéticas, paisajes moleculares, monólogos anfibios. . . . indicios de patologías sonoras de diversa índole en general" (poetic micrologies, molecular landscapes, amphibious monologues. . . . indexes of sonic pathologies of different types in general). They performed this microtonic broadcast by bringing together recordings by artists from Mexico, Argentina, the United Kingdom, France, Italy, and Spain.[18] Because they are pockmarked, irregular, and often dismissed, if not completely unnoticed, these loosely related movements among the underground *ciutat secreta* represent for those who heard them a lived experience that is out of sync with the dominant national(ist) narratives, aligned with the Normalizatció (Linguistic Normalization) movement and the building of Barcelona into a cosmopolitan city.

The art practices associated with free radio and sound experimentation thus share a *locus auditivo* that I hear as *radiophonic*, which I do not use solely to refer to sound that is broadcast through the radio. The term for acoustic broadcasting over the airwaves and a geometric radius are homonyms in Castilian (they are *ràdio* and *radi* in Catalan, respectively). I draw on these overlaps to illustrate a fractal constellation of moments, practices, and projects that hear Barcelona as possessing an intensely local acoustic geography that also turns the ear outward in differing scales. In this sense, the radiophonic also reflects a space-making philosophy that draws from the perspective the Radios Libres del Estado Español (Free Radios of the Spanish State) described in 1983 as "a whole practice of communication based on a radical confrontation against all kinds of social relations of domination."[19] And it is a form of artistic production that, as

Félix Guattari has written, is an "experimenting with a new type of direct democracy. . . . Direct speech, living speech, full of confidence, but also hesitation, contradiction, indeed even nonsense."[20] In the *meneo barcelonés*, this free democratic speech is imbued with an aural imaginary of revolution and resistance, such that the radiophonically restless ear imagines the sounds of the underground spaces of the metro, the sewer, the marginalized *Ciutat satèl·lit* (satellite city) on the outskirts of Barcelona, or even just the act of hanging out at clubs and bars as daily engagements with broad antisystem thought.

As I detail below, the microtonic sounds produced in the *meneo* often echo into and through the neighborhood of Gràcia, home to Víctor Nubla and Salvador Picarol, a space in which both aural resistance and the experimental sound art movement are still quite strong. El Raval has long been the semiofficial site of resistance in Barcelona, a place that in the 1920s was designated by some as a "nursery for revolutionaries," but Gràcia was a separate town until it was annexed by the city in the nineteenth century, with a history of workers' associations, revolt against Spain, and an identity wrapped up in its annual *festa major* (annual festival).[21] The town developed rather unevenly as a result of the construction of textile factories in the early to mid-nineteenth century, which shifted it from a *vila* grounded in agriculture into a town inscribed within a larger Barcelona industry scene. Because of this change, as Chris Ealham notes, in Gràcia, the working class mixed with the middle and upper classes well into the twentieth century, a phenomenon that had decreased in the rest of the city. Even though there was never a complete geographical isolation of the classes, in general workers in other parts of Barcelona became relegated to certain neighborhoods—especially after the development of *barraques* and *cases barates* (cheap houses)—while the bourgeoisie moved into newer areas.[22] Lafarga i Oriol has described the paradoxical situatedness of Gràcia's architecture, as a working-class, residential neighborhood that butts up against the orderly, rational, luxurious planned community of the Eixample: "Es com si dues persones es parlessin en idiomes diferents. Estaven una enfront de l'altra, però no sabien què dir-se" (It is as though two people were speaking different languages. They were in front of each other, but they did not know what to say to each other).[23] Gràcia's workers also have a history of defining themselves as *menestrals* (craftspeople) more akin to artisans than workers, who often have "un pequeño negocio para los clientes que son los vecinos de la calle y del barrio, de las personas que quieren orden, bienestar, progreso, solidaridad, colaboración, que en las asociaciones de todo tipo se encuentran con sus similares para disfrutar de la vida, con un sentimiento catalanista y progresista" (a small business for the clients who are their neighbors, on the street and in the neighborhood; people who want order, comfort, progress, solidarity, collaboration, who find similar folks in associations of all different kinds in order to enjoy life, with a progressive and Catalanist sentiment).[24] It is from within this local context that both Picarol and Nubla engage in their sonic interventions in the city and, thus, as I show in the last section, they harness an echoic memory of the local that has long been at odds with a top-down Catalanist project.

In its nightly alegal broadcast of music and live concerts that were considered by the mainstream to be dirty, unsavory, or simply out of the norm, Ràdio PICA and its radiophonic resonances carried out a subtle resistance: Like both its anarchist forefathers and *els altres catalans* whose descendants were often drawn to it, Ràdio PICA *picava* (itched), making itself a constant, if small, thorn in the side of the establishment. The *meneo* it produced throughout Gràcia and beyond is not an interruption, then, but a continual sounding of discord with the norm that draws attention to the superficiality of the neocolonial smoothness created by institutionalized economic and political mechanisms of democracy. Interrogating these different examples of aural subterraneity allows us to hear the radio and music of the 1980s and 1990s as part of a longer continuum of aural dispossession in Barcelona that has forced alternative media underground. Below I focus on the moments in which this aural underground collides with institutionalized radio programming, the music industry, and official definitions of linguistic identity in the city.

RADIOPHONIC BARCELONA

To understand the microtonics of Picarol's Ràdio PICA and Nubla's construction of Gràcia as what Nubla considers a *territori sonor* (sonic territory), it is necessary to address the cultural history of post-Transition radio in the city. As Julià Guillamon has explored, in the period I examined in the last chapter, the counterculture exemplified by the performances of Ocaña, the writings of Biel Mesquida and Quim Monzó, and the magazines *Ajoblanco* and *Star* represented a *ciutat interrompuda* (interrupted city), defined in part by the fact that the new, democratic city was a space "sense memòria ni referents" (without memory or referents).[25] Still, by the early 1980s, in the new Barcelona, acoustic authority was linked to commercial intelligibility, the smooth and orderly modes of radio transmission that were framed according to economic and legal notions of possession.[26] Previously the central Spanish state had controlled information through overt acts of censorship. Tight controls on media under the dictatorship meant that only eight stations occupied the FM dial in Barcelona when Franco died, and there were only two television stations.[27] In other words, state radio out of Madrid was the most widely followed radio in Spain. As a result, by the time of the *golpe* (coup; here, the final attempt at Francoist restoration) on February 23, 1981, of the 89 percent of Spain's population who were following the events, those listening live on radio heard them transmitted through Cadena SER (Sociedad Española de Radiodifusión [Spanish Broadcasting Company]).[28] Music production was similarly tied to (the production of) mainstream tastes coming out of Madrid, even as music industry executives in Spain tried to define a sound that would be a radical break from *los dinosaurios* (the dinosaurs).[29] The industry was still overly focused on Spain and Spanishness, even as a new genre of music, punk, was exploding in England, the United States, and elsewhere. Beginning in the early 1980s, though, the Barcelona government, directly allied with global capitalism, sought to control the circulation of information through regulatory practices that determined who could hold radio broadcast licenses; it also allied media production with specific linguistic policies. Law 10/1983 created the Corporació Catalana de Ràdio i Televisió (Catalan Corporation of Radio and Television), which, under the auspices of the Catalan government, has regulated all television and radio operations in the Generalitat de Catalunya ever since, although its name was changed in 2007 to the Corporació Catalana de Mitjans Audiovisuals (Catalan Audiovisual Media Corporation). Its main objective, according to the law, was to "contribuir de forma decisiva a la normalització lingüística i cultural de la ràdio i la televisió de Catalunya, així com tenir un paper cabdal en l'impuls i el desenvolupament de la indústria audiovisual catalana" (contribute in a decisive way to the linguistic and cultural normalization of radio and television in Catalonia, as well as playing a key role in the promotion and development of the Catalan audiovisual industry).[30] A simplified history of radio in Barcelona, then, would point to RAC 105 (Ràdio Associació de Catalunya [Radio Association of Catalonia]; founded 1982) and Catalunya Ràdio (founded 1983) as being the first post-Transition-era autonomous stations to broadcast in Catalan after Franco's death, since those were the first official stations run by the new democratic government.[31] For the Catalan establishment, the ability to control the radio after forty years of censorship was a watershed moment because it made official Catalan available to a wider public after being heard as largely unintelligible by the Franco regime.[32]

In response to that history, Jordi Costa has suggested, in short order a *gusto socialdemócrata* (social democrat taste) linked to the new democratic government adopted and financially supported much of the counterculture, converting what had been a movement of resistance into mainstream popular taste. They did so, moreover, as a sort of *arqueología sentimental* (sentimental archaeology) that denuded the counterculture of its very life: "El contracultural muerto—o pretérito—se convertirá, así, en una figura homologable con el bárbaro expuesto en la vitrina del museo de una realidad modélica" (The dead—or past tense—counterculture person will thus be converted into a figure who is homologous to the realistic model of the barbarian on display in a museum case).[33] For Donald McNeill, invoking both the *cronista* Maruja Torres and the urban historian Thomas Bender, the result is Barcelona as "The City Lite," a pseudospace geared toward gentrification and tourism.[34] This superficial acceptance, and even promotion, of the values of the counterculture or the marginalized spaces of the city, as an instance of a barbarous past to be remembered but sterilized—meaning, beautified, marketed, and sold—over time inscribed the noisy fun and spontaneity of the counterculture as a presence into the economic logic of the neocolonial ear, now turned inward and focused on producing Spain's eventual arrival at modernity. Or, as Marc Caellas puts it acerbically in his *crónica* (chronicle) *Carcelona*, art confronted a capitalist-governmentalist mechanism of control that seemed nearly inescapable:

> La cultura en Carcelona se convirtió en un departamento del Ayuntamiento cuya principal misión era detectar cualquier atisbo de rebeldía para, una vez localizado, seducirlo y comprarlo asumiéndolo como propio, conviritendo a los agitadores, a los innovadores, a los provocadores en pseudofuncionarios anestesiados con un buen sueldo a fin de mes.

> (Culture in Carcelona became a department of the City Council whose main mission was to detect any hint of rebellion and, once located, seduce it and buy it, assuming it as its own, turning the agitators, the innovators, the provocateurs into pseudoofficials anesthetized by a good salary at the end of the month.)[35]

The post-Transition sense of aural resistance I am studying here, in fact, came about largely because, as punk culture arrived on the scene in the early 1980s, whatever perceived openness there might have been, exemplified by Ocaña's accented, trans* occupation of Les Rambles and bars of Barcelona with a celebration of heterogeneity (see Chapter 2), was soon threatened by economic and political policies that regulated "civility" in the name of progress.[36] Rebuilding the city as a center for tourism and development aligned with the country's path to entering the European Community (the predecessor to the European Union), which took place in 1986. This also entailed Spain lending support to causes like the US-backed contras in Nicaragua, even as Barcelona was demolishing the *barraques* (shanty-towns), often occupied by workers coming to Barcelona from other parts of Spain, in order to rebuild the city for the 1992 Olympics.[37] The Ajuntament sought to put the city directly in touch with a globalizing investment project, in a sense going beyond the Spanish state as it rebranded the city for foreign tourism, eventually supporting projects like the "Barcelona, posa't guapa" (Dress up, Barcelona!) campaign. Social geographer David Harvey has referred to global speculative practices linked to economic projects such as those that Barcelona embraced and into which the media were inscribed as *the new imperialism*, since they forge transnational relationships between private interests and government entities that perpetuate colonial geographies and practices within local neighborhoods.[38] For Harvey, these globalized models of economic development privilege the creation of new economic centers but also create knots of dispossession in local places elsewhere. In Barcelona, over time those knots of dispossession appeared locally, not just in areas of the city being gentrified, but in the media apparatuses that dispossessed some of the city's own residents by supporting the projection of a sleeker new idea of the modern Catalan city.

Aurally, this dispossession was not necessarily about which language was used but about how sound art and music obtained access to a listening public. Despite narratives about Catalan being fully erased in the public ear under Franco, there was not a sudden switch from a Castilian to a Catalan soundscape following the Transition. Instead, like the *meneo*, the Transition could be considered part of a longer extension of the Civil War and its political effects; less a transition, as Resina writes, than a "transmission" of existing Francoist values; but also a transmission of a supposition that culture needed to be legitimized by institutions. After communism and anarchism had been all but wiped out, the gradual opening of the public sphere to Catalan under Franco was permitted primarily as a means of consolidating the development of neoliberal institutions that would keep any revolutionary political development at bay.[39] Even before Franco died, the first two broadcasts in Catalan celebrated, respectively, the one hundredth birthday and fiftieth anniversary of the death of the nineteenth-century poet Jacint Verdaguer. And in 1971 Catalan radio and record producer, novelist, and screenwriter Jaume Picas was allowed to produce the television program "En totes direccions."[40] The program featured interviews with film directors, musicians, writers, and artists, all in Catalan, which reestablished the language as a language of information and the present day, rather that something relegated to music or folkloric remembrances of the past. In 1976, Ràdio 4, tied to Radio Nacional de España, began broadcasting in Catalan.

However, as Kathryn Crameri has demonstrated, the choices made about which sounds to privilege were related to the idea that Catalan high culture was the purview of identity, that the lettered production of culture could be distinguished from the transcultural and bilingual realities of everyday life lived dually in Catalan and Castilian: "The basic social knowledge of the Catalan people is mixed with that of the rest of Spain. . . . 'High culture', on the other hand, is much more easily defined separately, normally through the basic distinction of the language used. Works of written 'high culture' in Catalan could not easily be claimed as part of a homogenous Hispanic tradition. It is therefore a secure marker of identity . . . safeguarding 'high culture' becomes synonymous with safeguarding the national identity itself."[41] Even before Franco's death, then, debates around how best to create and produce Catalan literature as the height of culture were powerful, with banks and government agencies participating in the publishing industries as they attempted to produce what they considered to be a "mature" literary tradition that would put Catalonia on par with other European nations and distinguish it from Castilian Spain. From the 1960s on, there was a translation drive in Catalonia that aimed to physically and linguistically broaden the Catalan offerings on bookshelves in order to help Catalan culture "mature" by normalizing itself. But, as Crameri points out, another way to read the drive for translation is that the Catalan culture was lacking in some way and needed translation to fill the gaps.[42] Local literary and artistic movements like the *Gauche Divine* or the Barcelona School of Film, firmly anti-Franco, were also mainly tied to the Catalan intellectual elite, and there were spatial associations between them and certain areas of the city that reinforced a sense of the place of culture as more or less bourgeois; the members of the *Gauche Divine*, for instance, were known for frequenting the disco club Bocaccio. The *nova cançó* movement, as well, whose music was known across Spain as a protest against Franco, was also inscribed in the Catalan bourgeois imaginary. Their diminishment in the public consciousness over time led to political debates in the 1990s around which music was supported by the government-regulated radio, which was not, and why.[43]

Not all Catalan radio production, and certainly not all popular Catalan sound, either before or after the dictatorship, was necessarily "high culture," even when it was part of the establishment, though. Two months before Franco's death, Joaquim Maria Puyal began narrating soccer games in Catalan on Ràdio Barcelona. And newer writers like Quim Monzó regenerated the language even as they engaged with a ludic counterculture resistance associated with comics and writers in the United States like Robert Crumb, Donald Barthelme, and John Barth. Nevertheless, recognizing the need for popular culture to also be part of the Catalan mediascape, in the mid-to-late 1980s, the Generalitat also began to promote Catalan rock and radio stations, such as Flaix FM, which was specifically associated with the popular television personality Mikimoto (Miquel Calçada), who left television to join efforts to beef up Catalonia's radio presence.

By the late 1980s, the Generalitat was supporting the so-called *boom del rock català* (boom of Catalan rock), whose emphasis was on producing rock in the Catalan language. Bands like Sopa de Cabra, Sau, Els Pets, and Sangtraït reached such popularity that in 1991 they performed Catalan-language hits in the largest concert to date at the Palau de Sant Jordi at Montjuïc, which had only a few years before been home to migrant shantytowns.[44] The Organizing Committee of the Barcelona Olympic Games (COOB '92) also commissioned a special song, "Barcelona," to promote Spain's emergence as the new democratic hope for the world. The song featured global pop star Freddie Mercury performing with the operatic soprano Montserrat Caballé, who had graduated from Barcelona's Liceu Theater, the musical pride of the city. The lyrics were an odd mix of English and Castilian Spanish that combined dreaming and seeing a beautiful woman with pleas for God's will and the refrain "Barcelona"; the grand swells of strings, orchestral drums, and overall majesty of Caballé's runs combining with the unique vibrato voice of one of the hottest global pop stars of the day sounded an aural affirmation that Barcelona had finally arrived on the global stage.

What distinguished Ràdio PICA, then, which has been broadcasting in Catalan and Castilian since March 9, 1981, two weeks after the unsuccessful 23-F coup attempt, and beginning more than a decade before the Olympics arrived in town, is that it operated outside any official institution, bringing with it music and sound art that had not been validated by the state or the neoliberal establishment, be it from Madrid or Barcelona.[45] At the time Ràdio PICA began, music by the Sex Pistols, Joy Division, GRB, the Ramones, and the like could only be obtained through purchases made in Andorra or via overseas trips taken by local young people; friends sold mix tapes to each other, setting up an informal, underground circuit of distribution, or they relied on the underground store Pirata's, which did not have a storefront but would sell to buyers who called them first to set up appointments.[46] Even if there had been music stores devoted to the sound, most punks probably would not have shopped there, given their resistance to mainstream fashion and capitalist production. Instead, Picarol sold records that he had brought over from France or Andorra in a flea market stall.[47] Locally, punk and hard-core bands like Shit S.A., Último Resorte, L'Odi Social, Subterranean Kids, Monstruación, GRB, Kaoss Urbano, Anti/Dogmatikss, and others, who played at clubs like Zeleste, Mágic, or El Kafe Volter, were often excluded from the recording industry, so they could only be heard live.[48]

Ràdio PICA provided the first on-air outlet for this music in the city, publicizing and broadcasting local concerts on air and playing records from outside of Spain that were unavailable for purchase locally. Broadcasting only after midnight, first on 91.8 FM, and later on 96.6 FM, the station provided free access to music, mainly experimental, punk, industrial, and hardcore, as well as experimental literature, poetry, and theater, and counterinformation about protests and other events related to the local music scene, opening a portion of the city's ears to other places and experimental sound through music. In just one example of how imports happened, one of the early DJs on Ràdio PICA, Trashmike, brought rockabilly to the city after visiting New York, where he first heard the Stray Cats. As Jordi Llansamà has said, Ràdio PICA thus produced "la instauración de un sonido y una actitud" (the establishment of a sound and an attitude) that reflected the post-Transition sense of disillusionment and captured the defiant "háztelo tú mismo" (do-it-yourself) approach that defined the punk movement in Barcelona.[49] As the Barcelona-born rocker Loquillo would later put it in his memoir, "En BCN, solo Ràdio PICA parece plantarle cara al aburrido panorama de la ciudad: una radio libre escorada hacia una realidad que diverge de la mía. . . . Mezcla de punk, futurismo, anarquismo y no sé qué más. . . . Todo en Radio PICA es transgresor y combativo" (In BCN [Barcelona], only Ràdio PICA seems to stand up to the boring panorama of the city: a free radio leaning toward a reality that diverges from mine. . . . Mixture of punk, futurism, anarchism and I don't know what else. . . . Everything on Ràdio PICA is transgressive and combative).[50] (It is important to note, however, that Loquillo himself was seen by the underground as a sellout, since he was fully embraced by the Spanish music industry, representing for many a superficial aesthetic of the underground rather than a legitimate revolutionary force.)[51] At a time when *El País* was decrying the pirating of music as a "cancer" killing the industry, Ràdio PICA's commitment to the unrestricted circulation of alternative media was heard as harmful to Spain's future as a modern democratic society, despite its supposed movement away from censorship after Franco's death.[52] Acoustically, free radio's irreverence was evident not just in its embrace of punk, hardcore, or experimental sound art, but in its very existence on the air, its occupation of the radio dial despite the government's insistence that it stay silent.

That being the case, in a way we can hear Ràdio PICA sharing a history of clandestinity with the alternative media that, prior to the rise of the Generalitat, had similarly filtered into Spain's airwaves despite Franco's attempts to control them. From 1981 until today—with a four-year break when it was shut down by the government in 1987—the self-described underground radio broadcaster has operated outside the regulatory limits of the state, carrying out what Picarol has called a *guerra de guerrillas sónica* (sonic guerrilla warfare) in support of the free circulation of information and art.[53] As Joni D., who broadcast a show in 1985 called "I Don't Care" (named after a Ramones song) claimed, broadcasting was a political statement: "Era un programa de punk polític i tant el títol (No m'importa) com el grup de la sintonia eren una contradicció amb el contingut del programa. Aquestes contradiccions buscades formaven part del moviment punk" (It was a program of punk politics, and both the title (I Don't Care) and the radio station group were a contradiction with the show's content. These sought-after contradictions were part of the punk movement).[54] Part of that contradiction, itself a resistance to the Catalanist cause, was the mixed use of Spanish and Catalan, and even English, during the broadcast, which reflected the daily speech of the city; Fèlix Villagrasa has written that "Ràdio PICA, a part de La Campana de Gràcia, ha estat, conjuntament amb Ràdio 4, durant molts anys, l'emissora que més hores en català ha estat radiant, fins al boom d'emissores d'FM aparegudes durant la dècada dels noranta" (Ràdio PICA, apart from La Campana de Gràcia, was, together with Ràdio 4, the station that for many years broadcast the most hours in Catalan, until the boom of FM radio stations appeared during the nineties).[55] In addition to occupying the airwaves with unrestricted, "barbarous" mixing of everyday Catalan and Spanish, later, the radio also became entwined with the *okupes*, or squatters, who literally occupied private properties with a punk aesthetic and rebelliousness. The mainstream rejection of this association between punks and politics is evident in a 1984 article, in which *El País* referred to the squatters as a "dolorosa caricatura de nuestra peripecia económica cotidiana. De ahí la repulsa que inspiran" (painful caricature of our daily economic vicissitudes. Which explains the repulsion they inspire).[56] This repulsion would characterize much of the official reaction to punk culture, even as it drove many bands to embrace noise as a way of being heard.

As part of their rebellion, in the early days—and in contradistinction to the United Nations International Telecommunication Union's *Acuerdo* and *Plan de Frecuencias sobre Radiodifusión* signed onto by Spain in 1978, which required radio stations to provide coverage from 8 a.m. until midnight—Ràdio PICA broadcast only after midnight. It did so, moreover, from a secret location, moving from apartment terrace to apartment terrace in an attempt to escape detection.[57] At first it broadcast live, with programs like "Caprichos matinales" covering the early morning hours. But by 1984 individuals were editing their shows at home on cassette and either posting them by mail or leaving them at designated drop points, like a bookstore in Gràcia, for Picarol to pick up and broadcast later. Eventually, broadcasting with a mere one hundred watts from the top of Tibidabo, a hill overlooking Barcelona, the station reached beyond the center of Barcelona to a radius of about thirty kilometers, from Baix Llobregat to Mataró, including the *Ciutats satèl·lits* on the outskirts of the city, places associated with a large influx of southern migrants. The broadcast radius thus contributed to the daily production of an aural community distinct from either the uniform, national(ist) or commercial ones provided by state-sanctioned radio stations, or from the internal sense of mental and geographical separation that differentiated working-class communities from those who lived in Barcelona's center.[58] With Ràdio PICA, to quote Caleb Kelly's observations of sound generally, underground sound "turned corners," uniting spaces through the airwaves, particularly at night, that might otherwise have remained either separated or mediatically homogenized.[59] In 2000, seeking an even broader reach, the station went online (the first in Catalonia to do so). In recent years, its web presence has included links to Democracy Now, Kaosenlared, Lahaine.org, and, until the program lost financing in 2018, Más Voces, a Madrid-based free, community radio association program that consolidated news and had an online presence across Europe, Latin America, and the United States. For forty years Ràdio PICA has been funded only via donations, and it has refused to run ads. Nevertheless, in 2016, after three decades seeking a license from the Generalitat,

Picarol definitively left the FM dial after he was hit personally with a sixty-thousand-euro fine for interfering with Ràdio Barcelona, of Cadena SER, a legally operated frequency. At the time of this writing, the program is still available live online twenty-four hours a day, and it has a presence on Facebook, Instagram, Twitter, and Spotify, although the number of listeners today is down substantially.[60]

From its highly mobile subterranean space, Ràdio PICA has engaged an echoic memory of radiophonic resistance that had been present in Spain since before Franco's death, but whose spontaneity contrasts with any notion of organized political movement.[61] For years REI (Radio España Independiente) had provided counterinformation to that controlled by the Franco regime. Founded by Dolores Ibárruri, better known as la Pasionaria, the station was popularly called La Pirenaica and was thought to have been broadcast from just over the border. In reality, REI broadcast over 100,000 programs, from 1955 until 1976, mainly from Bucharest.[62] Much as the later free radio stations would do, REI depended on networks of volunteer reporters who provided socialist, alternative information to that put out by the regime.[63] Following conventional broadcast formats, programming included the popular *Correo de la Pirenaica*, in which listeners could write in and have their letters read on the air, and women's programming such as *Página de la Mujer* and *Charlas Femeninas* that reflected women's sense of repression by Francoist norms.

In the early post-Franco period, however, free radio defined itself by opposing the Transition's insistence on capitalist ownership as a means of advancing the nation, equating freedom not with socialism (which, officially in power under the Spanish Socialist Workers' Party, antisystem proponents like Picarol thought to be a farce), but with an understanding of the airwaves as conduits for independent, noncommercial circulation of music, media, and information. The first free radio station in Catalunya, Radio Maduixa, had been founded as part of the *Ones Lliures* (Free Waves) movement in Granollers in 1979; stations subsequently appeared throughout Spain, in Iruña, Madrid, Bilbao, Terrassa, and Barcelona, among other places. In the years following the founding of Ràdio PICA, other free radio stations in Barcelona, such as Ràdio Linea 4 (founded 1982; 103.9 FM), Radio Bronka (founded 1987; 104.5 FM), and Contrabanda (founded 1991; 91.4 FM), emerged.[64] In many cases, both for practical reasons and to embrace the DIY culture that defined punk, free radio broadcasters built their own equipment, following guides published in alternative magazines. Many of the broadcasts followed Italian models and advocated on behalf of causes that had been censored under Franco, including those associated with homosexuality, feminism, ecology, antiauthoritarianism, and civil rights.[65]

The relationship between censorship and neoliberalism was at the top of the agenda. The "Manifiesto de Villaverde," which was put out in 1983 by Radios Libres del Estado Español, and was still present on the website of Ràdio Linea 4 at the time of this writing, explains this connection, which was particularly strong in a country coming out of forty years of dictatorship:

> En una sociedad cuya realidad está altamente centralizada e informatizada, donde los medios privados y públicos de comunicación son poder y están al servicio del poder, las RADIOS LIBRES surgen ante la necesidad y el derecho de toda persona individual o colectiva a expresar libremente sus opiniones y criticar y ofrecer alternativas en todo aquello que le afecta directa e indirectamente. . . . Las radios libres pretendemos potenciar toda una práctica de comunicación basada en un enfrentamiento radical contra todo tipo de relación social de dominación y, por tanto, apostamos por una forma de vida alternativa a la actual.

> (In a society whose reality is highly centralized and computerized, where the private and public media are power and are at the service of power, FREE RADIO arises from the need and the right of every individual or collective person to freely express their opinions and criticize and offer alternatives in everything that affects them directly and indirectly. . . . Free radios aim to promote a whole practice of communication based on a radical confrontation against all kinds of social relations of domination and, therefore, back an alternative way of life to the current one.)[66]

This dedication to a noncommercial form of broadcasting that also resisted the consolidation of information as a source of power distinguished Ràdio PICA and its cohort from the ideological goals of the REI, or even the pirate stations that had begun to emerge in Europe, specifically off the coast of Spain, in the 1960s.[67] Spain's unique geographical position in Europe made it the contentious site of one of the best-known commercial pirate radio stations in Europe, Radio Mi Amigo, a sister station to Radio Caroline. The pirate station initially broadcast in Dutch and English from a ship circulating in the North Sea, but in 1964 it settled in Platja d'Aro and was supported illicitly by the Ministry of Tourism in Girona, a city north of Barcelona, catering to Belgian and Dutch tourists. Tensions arose between this local government ministry, which was hoping to draw attention to the town internationally in order to attract more business, and central government officials. The latter thought they could potentially use their power against Radio Mi Amigo to force other countries to quash REI, the anti-Franco broadcast which was also based outside of Spain. At the same time, beginning in 1959, Franco allowed the United States to occupy the Platja de Pals, north of Barcelona, with the anticommunist Radio Liberty, directed back at the USSR.

Notwithstanding these geopolitical tensions, listeners often confused free radio stations with these foreign pirate radio stations. The difference is subtle and ideological. Most free radio broadcasters reject the term *pirate* since it reaffirms the legality of commercial radio stations, while suggesting that free radio is illegal.[68] Pirate stations, such as Radio Mi Amigo, often had commercial ends. Salvador Picarol, the founder of Ràdio PICA, has therefore repeatedly insisted that the station he manages is *alegal* because it exists outside the binary of legal or illegal, and is therefore truly antisystem. By the 1990s, he took issue with the notion of *free radio* itself, preferring instead to be called "independent and noncommercial": "El concepte ràdio lliure ja no defineix bé el fenomen perquè, en el fons, de lliures, més o menys, en són totes les ràdios que no estiguin excessivament sotmeses a un patrocinador o a una administració pública. . . . Definir-se com a independents i no comercials era afinar molt més la delimitació del que era una ràdio lliure i la primera emissora a definir-se així va ser Ràdio PICA" (The concept of free radio no longer properly describes the phenomenon because, basically, free radios, more or less, are all radios that are not excessively subject to a sponsor or a public administration. . . . Defining themselves as independent and noncommercial was more in tune with the limits of what a free radio station was, and the first station to define itself in this way was Ràdio PICA).[69] Because the location of Ràdio PICA was unknown, however, rumors began to circulate in the early 1980s that, like Radio Mi Amigo or Radio Caroline, Ràdio PICA also broadcast from a ship in international waters, a story that—as part of his Dadaesque restlessness—Picarol has never denied, although he will at times admit it is untrue. Adopting this "pirate" status as a joke, at one point, Picarol staged a photo shoot of himself broadcasting on a boat, using visual imagery to turn oral misinformation into a public rebuke of the authority of conventional sources of information and system-based thought.[70]

Although, like Ràdio 4, Ràdio PICA broadcast in Catalan, sonically it was out of lockstep with the drive for Catalan national identity because, like most of the artists I discuss below, Ràdio PICA communicated in the mix of Castilian and Catalan that was common of everyday speech at the time.[71] In some cases, as in Joni D.'s program, this bilingualism spilled into trilingualism, as he inserted English phrases and words into the conversation.[72] Moreover, the extreme independence of Salvador Picarol lent itself to friction with the establishment, as well as other free radio stations, in particular Radio Contrabanda. Unlike the commercial radio stations that developed in the 1980s, Ràdio PICA was considered a collective effort, common of free radio models of the time. Yet Picarol was largely the person coordinating the programming, and he has been the only person present over the course of the entire forty years. A marginal counterculture figure, before founding Ràdio PICA, Picarol had worked at Ràdio 4 and participated in the early *Ones Lliures* that opened up free radio in Barcelona by working at La Campana de Gràcia, another early free radio station. He had also acted at Club Helena with La Esquella theater group and performed as a mime. A graphic designer, he sold posters at markets and outside the Fontana metro station and was arrested early in the 1970s for disseminating "subversive" propaganda that denounced the murder of Salvador Allende and Víctor Jara in Chile.[73] Already walking a fine line between law and art, Picarol was instrumental in delivering a copy of Nazario Luque and the Farriol brothers' subversive comic, *El Rrollo enmascarado*, to a publisher after they were arrested.[74] He even advertised his "mafia" persona when he published a piece in the write-in section—titled, aptly, *cloaca* (sewer)—of the counterculture staple *Ajoblanco*, which included tongue-in-cheek ads for everything from sex to apartments to jobs:

> La mafia "Picarol" desea contratar a nuevos miembros para su organización. Indispensable dominar la guerrilla urbana y algo la rural, para importante campaña de revolución cultural o contracultural. Imprescindible estar inforciado al menos en alguno de los siguientes medios de difusión: radio, televisión, prensa y revistas especializadas de: arte, teatro, cine, cómics, fotografía, literatura, y música. Sueldo según aptitudes, se aceptan personas de ambos sexos. Escribir a: "Picarol" Apartado—9.242 de Barcelona. Nota: este anuncio no es pasote, es tan verdad como que el Llobregat está contaminado.

> (The "Picarol" mafia wants to hire new members for its organization. It is essential to master the urban guerrilla and to a lesser extent the rural one, for an important campaign of cultural or countercultural

revolution. It is essential to be informed in at least one of the following means of diffusion: radio, television, press and specialized magazines of: art, theater, cinema, comics, photography, literature, and music. Salary according to skills, people of both sexes are accepted. Write to: "Picarol" Apartado—9.242 from Barcelona. Note: This ad is not a joke [going over the line/*un pasote*], it is as true as that the Llobregat river is polluted.)[75]

In another posting, he put out a call that poked fun at the vagrancy that so ruffled the feathers of the establishment by connecting it to the democracy that did not yet exist:

"Picarol" quiere formar una asociación. Se llama "EX-BARISTAS A PARTIR DE LAS 1130 DE LA MADRUGADA." Tendría carácter nacional y consistiría en formar por toda la geografía hispana locales privados en los que los asociados pudieran hablar de fútbol, tomar vino entre otras cosas, comer tapas y jugar al millón. Según las últimas encuestas realizadas en las capitales de provincia existen ya más de 25.000 electores, lo cual convierte esta asociación antes de existir en la de más porvenir del país. Los interesados se pueden poner en contacto con la oficina electora a través del apartado 9.242 de Barcelona.

("Picarol" wants to form an association. It is called "EX-BARISTAS AFTER 11:30 IN THE MORNING." It would have a national character and would consist of forming private venues throughout the Hispanic geography in which members could talk about soccer, drink wine, among other things, eat tapas, and play the lottery. According to the latest surveys carried out in the provincial capitals, there are already more than twenty-five thousand voters, which makes this association, before it even exists, the one with the most future in the country. Those interested can contact the electoral office through section 9.242 in Barcelona.)[76]

With casual ties to Ocaña, the counterculture comics of Nazario and journals like *Ajoblanco*, then, Ràdio PICA and its illicit sounding of the punk movement and the *meneo barcelonés* traces a legacy of aural resistance back to before Franco's death, even though it seemed to burst on the scene out of nowhere in the early 1980s when it "contaminated" the dial with punk, electronica, and hardcore music.[77]

Narratives like the one by punk fan and one-time Ràdio PICA radio host Joni D., *¡Que pagui Pujol! Una crònica punk de Barcelona dels 80*, as well as Jordi Llansamá's oral history of punk, *Harto de todo*, and Xavier Mercadé's photographic album *Odio obedecer* (all of whose titles invoke the names of songs by punk bands L'Odi Social or the Subterranean Kids), have chronicled the role Ràdio PICA played in producing the underground sound scene through its DIY culture and devotion to the free circulation of art and information. The vast quantity of information contained there still needs to be investigated. These narratives associate Ràdio PICA not just with bands like Joy Division, Sentido Común, GRB, L'Odi Social, and others the station was the first to broadcast in Barcelona, but with the larger culture of resistance that included punks who hung around the Plaça Reial with mohawks and pins, the youth that squatted houses in protest of military conscription, and the crowds that participated in rowdy concerts and antiskinhead brawls that spilled into the streets late at night. The antisystem attitude embraced by punk was also buoyed by smaller fanzines and independent magazines like *Subtítulo*, NDF, *Drama del Horror*, and the antimilitarist *La puça i el general*, among many others, that did the daily work of transmitting the punk ethos beyond the sound of the radio. Early magazines like *Star* and *Disparos* provided photographs of the London underground and hippies, Black Panthers, and other images of the United States' counterculture. And self-produced fanzines published information on new bands, anarchist counterideology, illustrations of punk style, ribald images, and song lyrics. But like the *meneo*, generally, the fanzines also aligned themselves with specific and distinct ideologies, as in the case of NDF, founded in 1983. Its title

quiere decir algo así como "Niños Drogados por Frank [Sinatra]," y es una declaración de principios, estamos contra la droga y contra la gente que se enriquece gracias a los infelices que mueren drogados, contra los cerdos mafiosos como Frank Sinatra, adorado mito de nuestras madres. También declaramos ya de principio que somos anarquistas, pacifistas, ecologistas, anti-policiales, anti-nukleares, anti-racistas, etc . . .

(means something like "Children Drugged by Frank [Sinatra]," and it is a declaration of principles, we are against drugs and against people who get rich thanks to the unfortunate who die drugged out, against gangster pigs like Frank Sinatra, adored myth of our mothers. We also declare from the beginning that we are anarchists, pacifists, environmentalists, antipolice, antinuclear, antiracists, etc.)[78]

NDF, as well as other fanzines such as *Subtítulo* or *El general i la puça* thus articulated the counterinformation ideology that undergirded not only the "repulsive" sound of punk, but what Stephen Vilaseca has referred to as the *bodily thinking* of the squatters' movement through which they would eventually bring these ideas into lived spaces.[79]

The ludic nature of these texts—and many others tied to the *meneo*—are themselves an attempt by those involved to thumb their noses at anyone who stakes any kind of serious claim. As Guillamon has written, "Es vivia una sensació de fi del món. Si la contracultura havia basat les seves reivindicacions en la protesta, els que van venir després no van voler aprendre el llenguatge de la negociació ni van voler legitimar el poder amb les seves reclamacions" (There was a sense of the end of the world. If the counterculture had based its revindications in protest, those that came after did not want to learn the language of negotiation, nor did they want to legitimate power with their claims).[80] In this sense, the groups eschewed even the language of debate, focusing instead on absurdity to articulate disgust with the system. A particularly resonant example and justification of this notion can be found in an early *Ajoblanco* article from 1978, penned by someone supposedly named Timothy Schmurz. Almost certainly a pseudonym, the name recalls Boris Vian's 1959 *Les Bâtisseurs d'empire ou le Schmürz*, an anticolonial critique of empire written during the height of the decolonization period in Africa. Vian's play revolves around an abject creature occupying the house of a bourgeois family, which he names a Schmurz, whose body is ever present, overlooked yet grotesquely haunting.[81] The *Ajoblanco* article appropriates the unknown creature in order to associate its seeming incivility with the body, anarchy, sound, and revolution:

> SCHMURZ es la lengua sacada
> el pedo sonoro y feliz
> el moco sacado a tiempo
> LA REVOLUCIÓN.
> Schmurz es Anarquía.
>
> (SCHMURZ is a tongue sticking out
> a happy and noisy fart
> a booger taken out just in time
> the REVOLUTION
> Schmurz is Anarchy.)[82]

Taking up the disgusting as a rallying cry, punk culture, and that of the underground generally, resisted the civilization/barbarism dichotomy of the Catalan establishment not by rejecting the frame, but by occupying the barbarous side of it. Moreover, the call for revolution as anarchy, associated here with boogers and farts, takes a new, absurdist turn in the avant-gardist tradition, while also maintaining a throughline of Barcelona culture that goes all the way back to the anarchist Civil War icon, Buenaventura Durruti, to whom the first issue of the magazine, in 1976, was dedicated.

Musically, Barcelona punk culture adopted barre chords, heavy drums, and discordant, screeching voices, while lyrically it defined the city as immersed in a culture of shit. Thus, like the counterculture adoption of the Schmurz, it embraced an impure, decaying, unpolished sound that was out of sync with the larger urban project of civility the government had in store for the people of Catalonia, and which would go live in 1985 under the slogan "Barcelona, posa't guapa." One of the first local punk bands was called Basura (garbage), for example, and intentionally irreverent lyrics called attention to sewers, rot, excrement, and filth in order to irritate the ear of a "civil" society, which had been informed not only by Franco's insistence on polite normativity and on the censors' enforcement of morality associated with purity, but on the Catalan establishment's long-standing valorization of high culture.

In the case of the urban space specifically, the proud history of architecture and spatial organization as signs of national identity are taken on in one of the first hard rock songs performed in the city, La Banda Trapera del Rio's "Ciutat podrida" (Rotten City), sung in Catalan.[83] The song is perhaps one of the clearer reconsiderations of the city space, since it provokes its listeners into celebrating the less desirable aspects of the city, which emerge at night and are attractive to the youth because in their fire and rot, they allow freedom that is otherwise not available:

Ciutat podrida. . . .
Ens portes la nit i la por
ara que ets adormida
els carrers són plens de foc.
Vull sortir d'aquest infern
on els crits
dels perduts s'obliden,
on és presoner
L'esclat del vent

i la llibertat no camina.
Ciutat podrida. . . .
. . . Aquest és el moment
en el que ha mort la vida.
No m'importa el ponent.
Puc caminar sense guia.

(Rotten city . . .
you bring us night and fear
now that you are asleep
the streets are full of fire
I want to get out of this hell
where the cries of the lost are forgotten
where the gust of wind is prisoner
and freedom does not walk.
Rotten city. . . .
. . . This is the moment
that life is dead
I don't care about the sunrise
I can walk without a guide.)

In addition to figuring the city as a rotten hell because it forgets those who are lost, the song suggests that one can only walk freely before sunrise, when life has died and experimentation "without a guide" is possible.

However, another of their songs, "Venid a las Cloacas" (Come to the Sewers), is in Castilian. A fast-paced piece with several guitar riffs, it calls on those in Sant Ildefons, where the band is from, to come join the rats in the sewers because they are already living in shit:

Creéis que estamos salvados, pero estáis en un rincón
En un rincón de mierda, de control y represión.

(You think we're saved, but you're in a corner
In a corner of shit, control, and repression.)

The idea of repression seems to refer to the built environment of the *ciutat satèl·lit* (satellite city) of Sant Ildefons, one of several that ring Barcelona proper and are defined by a combination of factors that place them, economically and culturally, somewhere in between the *barraques* and Gràcia. Sant Ildefons was a rural area until the 1950s, when several factories moved to the area, attracting large numbers of working-class migrants from the south of Spain. Originally, they built shantytowns akin to those in Somorrostro or Bogatell, but the large numbers of new arrivals, combined with housing shortages, led the Comissió d'Urbanisme de Barcelona to propose demolishing the improvised housing and constructing huge blocks of housing known as *polígons* (polygons). With most of the residents working outside the area—and without adequate transportation into the center of the city—the control and repression that define the "rincón de mierda" (corner of shit) seem to refer as much to the isolation of the place as to the imposition of huge homogenous structures over the land. Together, the song suggests, these uniform apartment blocks are worse than running through the sewers to freedom, where the escaped can see firsthand the "heces que vierte la Sociedad" (feces Society dumps out). The song ends with the sound of the singer doing a raspberry that, like the anarchist SCHMURZ, sounds both like a fart that sonically accompanies the production of sewage, as well as a sign of the absurdity of ignoring these realities. Its theme thus resonates with songs like the Basque ska band Kortatu's "Mierda de ciudad," which also complained that Saturday night was "always the same city shit," or the Madrid fanzine *Kaka de Luxe*, through which the eponymous band promoted itself, adopting shit as a symbol. The fact that La Banda Trapera del Río's song is in Spanish, while the previous one about Barcelona is in Catalan, highlights the bilingualism of local punk music, and draws the ear to the daily existence of a linguistic soundscape heard as "impure" by those who had instituted linguistic norms for the pronunciation and perfection of the Catalan soundscape. It also broadens the geography of a centralist Barcelona ear out to the regions of the city occupied primarily by working-class migrants, and into a listening ear spanning all of Spain.

For their part, the band Shit S.A., who did not record albums, but performed at festivals and clubs like Zeleste, sang songs such as "Cagando y vomitando (Todo a la vez)" (Shitting and Vomiting [All at the Same Time]), with screaming vocals, hardcore drumming, and guitar riffs whose speed increases as the metaphoric purging of the singer reaches a climax. The intentionally "vulgar" references to the human sounds of defecation and vomiting in these songs pierce the rhetoric of civility associated with bourgeois values, and they also echo the absurdist avant-garde experimentation that circulated among the fanzines and comics associated with the underground. Similar imaginaries that produced alternative mattering maps of Barcelona were also evident in the punk Ràdio PICA circulated nightly over the radio.

There were political concerns filtered into these intentionally disgusting sounds as well. L'Odi Social's "Odio obedecer," for instance, begins with a fast-paced drumroll and guitar riff on the Marines' Hymn that evokes a frenzied drive for conformity, a gesture toward the mandatory conscription into the army that punks were among the first to question. Combined with the critiques of religion, monarchy, and fatherland, the lines

> Yo no tengo honor, no tengo bandera
> Ni estoy orgulloso, de mi madre patria
> . . . No quiero estar ciego, ni creer en dios

> (I have no honor, I don't have a flag
> Nor am I proud of my motherland
> . . . I don't want to be blind, nor believe in God)

knock all major institutions off their pedestals and into a cesspool of conformity evoked by the rushing speed of the song.[84]

We can also hear the punk movement and Ràdio PICA's underground use of cynical humor and irreverent hardcore sound as performing against the seriousness of both popular Spanish rock and the normative Catalan musical culture that had previously been associated with the *nova cançó* inaugurated by Raimon, Lluis Llach, María del Mar Bonet, and the rest of the Setze Jutges only a little more than a decade earlier. The *nova cançó* had challenged Franco's repressive policies with an acoustic sound, usually mapped to Catalan lyrics, that reverberated with the *chanson française* (French song) by Georges Brassens or Jacques Brel; the Latin American political *cantautores* Silvio Rodriguez, Víctor Jara, or Violeta Parra; and even the North American folk music of Woody Guthrie and Bob Dylan. Songs like Lluís Llach's solemn "L'estaca," about a stake in the ground that could fall if everyone pulled together (an obvious anti-Franco reference in its call to unity in the face of something seemingly unmovable), or Raimon's "Ahir (diguem no)," which implored listeners to say no to hunger, bloodshed, and fear, associated national identity with a serious negative affect of sadness or anger.[85] Raimon's "Jo vinc d'un silenci" (I Come from a Silence) claims to participate in a silence that is "antic i molt llarg" (ancient and very long) and which is not an intermittent appropriation of small spaces, but a grand, solemn anonymity that represents a "lluita sorda i constant" (constant deaf struggle). Although Jaume Ayats and Maria Salicrú-Maltas suggest that such lyrics express a "collective euphoria before a democratic future," the pacing and tone of the song is more of a solemn anthem than a rousing celebration.[86] In Bellmunt's 1976 film *Canet Rock*, meanwhile, founder of the Zeleste club Víctor Jou would say, "La música pop española o el rock español está fuertemente condicionado por toda la problemática de colonización anglosajona. Esto es un problema que es general en la música y en cualquier otro arte que se haga de un país con una economía débil" (Spanish pop music or Spanish rock is strongly conditioned by all the problems of Anglo-Saxon colonization. This is a problem that is general in music and in any other art that is made in a country with a weak economy).[87] Jou associated commercial development with British imperialism, presenting Spain as underdeveloped. But the punk groups that would include Último Resorte, L'Odi Social, and the Subterranean Kids just a few years later did not hear themselves as part of a lost Catalan or Spanish empire, or even as victims of colonization, but rather as out of alignment with an internal system that was on board with capitalism and neoliberalism as a strategy. Moreover, sonically they could be easily distinguished from other music from noncentral Spain, like *rock andaluz*, influenced by American GIs who brought records from home, or *rumba vallecana*, born out of the slums of Madrid, which represented "the dark side of *desarrollismo*" (developmentism) yet achieved this sound by experimenting with the existing rhythms of *nacionalflamenquismo* (*copla*, flamenco *palos*, and the like).[88] The international sounds of anarchy found in Joy Division, the Ramones, and the Sex Pistols were not the Barcelona punks' competition, but rather examples of a new geography of the ear informed by outward listening, in which the sound of punk in any language drew an alternate map of Barcelona in an international frame attuned to the antisystem struggle.[89]

The radiophonic *meneo barcelonés* thus sounded resistance through the squawking power chords of the guitars and aggressive double-quick beat of the drums, as well as irreverent punk lyrics that decried all association with the establishment, of whatever language or nationality, from the imported English "I am an anti-Christ / I am an anarchist" of the Sex Pistols to the anti-Catholic "fuego, fuego, fuego" (fire, fire, fire) of Último Resorte's hardcore re-sounding of Barcelona as an ironic and dark "hogar, dulce hogar" (home, sweet home) that has become, according to the song's title, a "cementerio caliente" (hot cemetery). The singer screaming at the listeners, "¡Somos el peligro social!" (We are the social danger!), both parrots the establishment critique of punk and inscribes the listener in this new dangerous soundscape by using the first-person plural to be sung by those listening.

The Almen TNT song "Ya nadie cree en la revolución" (No One Believes in the Revolution Anymore) is now widely associated with the punk resistance, and it expresses punk's acoustic and political break with Catalanism, in the sense that Catalanism is but another example of a capitalist system that has overtaken everything in the city. The recording opens with music in a 1940s style, followed by the sounds of gunshots and police sirens, indicating a shift away from tradition toward the noise, and violence, of the street. Carrying out the collage mixing of sound sources that defined Basilio Martín Patino's musical documentary *Canciones para después de una guerra* (Songs for After a War) to produce what Jo Labanyi calls a "usable past," this montage similarly rethinks everyday sound in order to critique the recording industry's separation of song from noise.[90] Several well-placed tonal shifts by the singer also highlight a critique of conformity that eviscerates Catalonia's previous musical innovation, the *nova cançó*. In the song, a hopeful musician knows aligning with the cause of "the people" will sell records:

> aspira con la guitarra
> a cantante de *nova cançó*
> siempre al lado del poble
> que hoy da mucho millón
>
> (aspires with the guitar
> to be a singer of *nova cançó*
> always with the people
> which today give a lot of money)

The front man pronounces the word *cantante* (singer) in a high-pitched voice that lends it an intentionally out-of-place sound, signaling the perceived superficiality of the *nova cançó* movement when compared to the rougher sound of the other lyrics. At the same time, his overinflected pronunciation of the world *poble* (people) in Catalan, instead of the Castilian Spanish *pueblo*, suggests that the Catalan language is a superficial industry that "da mucho millón" (makes a lot of money). This sonic engagement with Catalanism through the ironic vocal tone is pointed and radical, because the *nova cançó* that had critiqued the oppression of Catalonia by Spain is now only in search of money. Even as his high-pitched, pop-friendly voice rejects the folk-music sound of the *nova cançó* as insincere, the lyrics reject the market politics at stake in Barcelona's total acceptance of Western capitalism as the way of making the future:

Ya nadie cree en la revolución
Están de rebajas en el Corte Inglés

(No one believes in the revolution anymore
There's a sale on at the Corte Inglés)

A critique of everyone, not just Franco, this song shifts the politics of song from a cry for democracy to a cry for recognizing that the democracy is oppressive because it is capitalist. With lyrics and sound mixing, Almen TNT called out the ease with which post-Franco Spain had been held in thrall to dreams of capitalist success, forgetting the meaning of resistance in the same way it had conveniently forgotten the Franco regime's crimes under the *pacto del olvido* (pact of forgetting).[91]

Even if the particular musical sounds of the *meneo barcelonés* were new in the late 1970s and early 1980s, then, the aurality of dispossession, or the imperialism of the colonial ear that produced it, was not: Free radios and punk resistance in Barcelona, informed by libertarian and anarchist ideals, were a continuation of a struggle that, prior to Franco, was constant but unheard by the bourgeoisie until it overflowed into the streets first during the early months of the Civil War, and in the 1980s as a result of the community that free radio made possible. Moreover, since the *meneo* fits into a longer continuum of musical development in Catalonia, it would be difficult to assert that there was a sudden musical differentiation between *nova cançó* and punk rock. The overlaps and subtle transformations that take place over time are evident spatially in the transformation of the club Zeleste from the origin site of the *Ona laietana* into a place for popular punk by Loquillo or Los Rebeldes, and eventually international bands like Radiohead, after the experimentalism of progressive rock and any promise of a Catalan record industry collapsed.[92] But they are also evident aurally in the shift from Pau Riba's 1968 single "Noia de porcellana" (Porcelain Doll), which originally shared the soft, folk-music sound of other *cantautores*, to the version featured in Francesc Bellmunt's celebrated 1975 documentary of the *nova cançó* movement, *Canet Rock*. In the 1975 film, the song includes solo electric guitar jams, a saxophone, and drum riffs, all of which add a somewhat distorted element to the clean sound of the original version. The new version seems to strive for a Jimi Hendrix guitar vibe, while visually invoking a David Bowie-esque play on androgyny that questions the heteronormativity of the poetic voice. The rougher rock sound aligns as well with Pau Riba's self-depiction in the film as *not* being linked to the bourgeoisie that supports the Catalanist trends of the *cantautor* tradition, saying that he sees himself as part of the Catalan culture only because his goal is to destroy it.[93] From a lyrical perspective, the song "Noia de porcellana" can be read allegorically as a critique of the fragility and vacuousness of a national identity that is merely a pretty doll, "freda i inhumana" (cold and inhuman) with "carn translúcida i repel·lent" (translucent and repellent flesh). By 1975, the harder musical tones suggest that this image has already cracked, as the gritty reality of the sound that accompanies the lyrics aurally marks Riba's shift over seven years from the inscription in Catalanist identity of the *nova cançó* to his embrace of the internal fissures that reject it from within.[94]

Ràdio PICA contaminated the regulated radio dial with noise not just because the sounds it broadcast were unfamiliar; it also sounded "vulgar" language and rowdy behavior that traditional notions of decorum—and the penal code relating to *pudor colectivo* (public modesty)—would consider inappropriate, and played nonprofessional or foreign music acts the musical industry in Madrid did not control. Ràdio PICA also looked to avant-garde models for inspiration. One of Picarol's early companions was sound artist Javier Hernando, founder of the experimental, anticommercial group Xeerox and, later, Melodinámika Sensor. Throughout the '80s he broadcast his show *Los silencios de la radio* on Ràdio PICA and even appeared with Picarol on the Spanish television program *La edad de oro* with Paloma Chamorro, a program that helped define and legitimize the Movida Madrileña, which had a broader reach across Spain than the *meneo*, but also reached at times into it.[95] With his experimental *electrónica* and industrial music, Hernando has explained, he was responding to the avant-garde sound concepts Filippo Tommaso Marinetti pioneered, as well as to the synthesis of *radio* and *radioactivity* in the German band Kraftwerk's 1975 album, *Radio Aktivitat*, which suggested that radioactive decay and radio communication were of a piece. By invoking both an aesthetic approach to the "modern" that had defined early avant-garde sound poetry and the new sound of electronic music, this program, like many of Víctor Nubla's sound experiments of the time, put technological innovation at the center of the Ràdio PICA project. Likewise, although grounded in anarchism, the station's dedication to sound independence meant putting the artist's work above any political connotation it might have (for example, the fact that Marinetti's futurism could be heard, in effect, as a protofascist aesthetic). The program played everything from electronica and house to agit-pop and Muzak. Hernando couched his justification for the program in an anecdote that metaphorically figured the radio dial as an empty spectrum to be occupied:

> Cuando comencé a pensar en cómo podía ser el programa que nos habían propuesto, me atraía enormemente una experiencia que pocos años antes había tenido: tras un corte en el suministro de luz acudí a escuchar la radio, tenía el presentimiento que era en toda la ciudad e incluso a más zonas, en efecto al mover el dial no se sintonizaba ninguna emisora hasta que en un punto ya casi al final y de manera muy lejana sonaba "Leader of the Pack" de las Shangri-Las, cuyos rugidos de motocicletas de fondo se mezclaban con el ruido del espectro radiofónico, después de esta señal y hasta el final de nuevo prácticamente silencio. Esta conjunción de música, ruido y silencio me llevó a bautizar el programa como Los Silencios de la Radio.

> (When I began to think about what the program they had proposed to us might be like, I was greatly attracted to an experience I had had a few years earlier: After a power outage I went to listen to the radio, I had a feeling [the power] was out all over the city, and even in more areas beyond it; in fact, when moving the dial no station came on until at one point, almost at the end, in a very distant way you could hear "Leader of the Pack" by the Shangri-Las, whose background motorcycle roars mingled with the noise of the radio spectrum; after this signal, and until the end, again, practically silence. This conjunction of music, noise and silence led me to baptize the program as Los Silencios de la Radio.)[96]

The white noise of the dial in a darkened city evokes the underground's sense of isolation from the celebration of modernity that was being broadcast elsewhere, even as it opened up the possibility of new, transnational acoustic discoveries. By hearing the emptiness of the dial as geographically extended across the city, Hernando also suggests that this emptiness is not a one-time blip, but rather a radial (spatial and radiophonic) experience ripe for appropriation.

More than simply a vehicle for the transmission of a contestatory form of experimental music and punk, Ràdio PICA also provided listeners with direct access to the daily life of a Barcelona in transition, rather than a burnished vision of what it could be. In this way the station seems to have embodied Bertolt Brecht's desire for a radio that would let the listener speak as well as listen, hence establishing community networks, instead of just a one-to-one relationship of transmission and reception.[97] The first *okupa* group, Colectivo Squat de Barcelona, for example, was intertwined with the punk of Ràdio PICA. Joni D.'s show and others did not just participate in and cover the squatting of the first house in 1984; perhaps more importantly, they broadcast about the evictions and detentions carried out by police in response. Eventually the *okupa* conflicts were covered in mainstream news outlets, such as the newspaper *El Noticiero Universal*, who provided a typically conservative headline that laughed at the punk movement: "Cinco squatters ya tienen hogar: el juez los mete en la cárcel" (Five squatters now have a home: The judge throws them in jail). But punk radio, fanzines, and posters served to propel the movement into other protests, like those against Spain joining NATO. Ràdio PICA also joined with anti-imperialist groups with broader global concerns, such as El Salvador's Radio Venceremos, who had an office on Carrer de Pelai, while also hosting concerts in support of the Sandinistas in Nicaragua and against militarization generally, as seen in the 1986 "Manifestación Mili-KKK."[98] Not long after the Transition, Ràdio PICA was among the first to open up telephone lines, cassette recorders, and microphones to everyday speech and the noisy, underground corners of a society that had built its identity around a hierarchical idea of civility and linguistic grammar. This was a country in which only recently the Minister of Information and Tourism, Manuel Fraga, had proclaimed "la calle es mía" (the street is mine) after the police in Vitoria substituted rubber bullets with real ones, killing multiple people during a workers' strike, as Manuel Vázquez Montalbán reminded readers in his 1985 *Crónica sentimental de la Transición*.[99] Literature, theater, and poetry all had a place on Ràdio PICA, even if what is usually remembered are the nighttime broadcasts from bars and concerts, or the call-in show in which those who were jailed in La Modelo prison could write letters or call in to request songs.

In just one instance of this approach that draws attention to the emplaced sound of punk as a practice and not just an aesthetic, we may consider Toni Ignorant, Víctor Nubla, and Salvador Picarol's *Informe semanal*, in which they parodied the Televisión Española (TVE) show *Informe semanal* (originally titled *Semanal informativo*), which was itself a symbol of the Transition's openness to democratic processes and freedom of information. But while the mainstream television program focused on reporting events of national interest, Nubla, Pica rol, and Ignorant visited clubs and transmitted live the punk concerts taking place there. Linking free radio to an aesthetics of found art combined with direct access to information, in these moments, Picarol and his friends adopted an informal role as on-the-ground war reporters, where the battle being fought was not just about the form of art, but about access to it, and the free ability to disseminate it. In the process, they critiqued the openness of media communication even after Franco's death. They also critiqued the Francoist laws still on the books, like the Ley de Peligrosidad y Rehabilitación Social, by flouting them in public. In 1983, for example, the three managed to sneak backstage at a festival hosted by Luis del Olmo (whether it was *flamenco* or *folklórico*, Picarol could not recall), accompanied by some punk women friends. Recently, the first all-female punk band to achieve a national audience, Las Vulpess (from Bilbao), had caused a stir by performing a cover of the Stooges' "I Wanna Be Your Dog"—"Me gusta ser una zorra"—on television. The program led to the Spanish prosecutor's office taking the program and the singers to court for the crime of *escándalo público* (public scandal), a law that obligated people to uphold good manners and modesty in public.[100] Picarol, Nubla, and Ignorant presented themselves backstage at the festival with the women, who pretended to be Las Vulpess. They ended up giving interviews and even appearing on stage.[101]

Taking to the streets with their microphones, these mobile voices of *Informe semanal* can be heard as carrying out a distorted echo of the use of radio pioneered in Spain by Ramón Gómez de la Serna in the 1930s, prior to the dictatorship, only this time one that occurred from outside the establishment. Although Ramón and Picarol came at the medium with very different backgrounds (one a renowned *vanguardista* [avant-garde] writer and orator, the other an unknown radio aficionado, graphic designer, and mime with a penchant for the free circulation of ideas), both Ramón and Salvador Picarol occupied the airwaves with spontaneous talk radio in a short-lived historical moment before new, rigid controls on radio were established in their respective cities (first by Franco, later by the Ajuntament). On November 22, 1929, the Madrid newspaper *El Sol* reported on Ramón's new "micrófono ambulante" [walking microphone], with which he broadcast discussions with people walking through the Puerta del Sol.[102] Prior to that point, Ramón's radio voice played with formal language by employing *greguerías onduladas* spontaneously on air, orally philosophizing by combining the daily with the metaphorical.[103] With a microphone in his office that he could turn on at any time, the orator wielded extreme broadcasting power, able to plug in at any time he wished to broadcast his thoughts. But taking his microphone to the streets, he gave voice to *panaderos* (bakers), street vendors, and passersby. In this sense, both he and Picarol produced a spontaneous aural *flânerie*, transmitting the fragmentary nature of daily life that had previously been published in newspaper *crónicas*. Like any good *cróna*, both seemed to extend the ear of the listener, allowing her to access spaces from afar, and both provided access to fragments of daily life. Yet unlike the distant *flânerie* of the sort described by Charles Baudelaire or Walter Benjamin, this walking of the city transmitted the sound of voices directly to the audience. While Ramón took to the streets, however, Ràdio PICA's traveling microphone used its mobility to let listeners directly into underground sites of resistance to the norm—clubs, concerts, and nightlife in the *cutre* (seedy) parts of town—providing access to the sound of another "real" Barcelona, rather than an idealized one portrayed by the *nova cançó* or the Catalan establishment. In a sense, then, Ràdio PICA and those in its radius were more in line with Marcel Duchamp, who ludically desecrated museum space with his "fountain," than with Ramón.

Víctor Nubla, for example, experimented with sound and the limits of music by theorizing the noise of everyday objects as instances of Jorge Luis Borges's hrönir. As postmodern critic John Barth had written in a piece brought to Barcelona in 1983 through a Quim Monzó translation, Borges's 1940 short story "Tlön, Uqbar, Orbis Tertius" stands in for the kind of changes captured by Marshall McLuhan in his famous dictum "The medium is the message." In literature, that meant the lines between form and experience had become blurred: "The story Tlön, etc., for example, is a real piece of imagined reality in our world, analogous to those Tlönian artifacts called hrönir, which imagine themselves into existence. In short, it's a paradigm of or metaphor for itself, not just the *form* of the story but the *fact* of the story is symbolic; the medium is (part of) the message."[104] Once the line between reality and fiction has been blurred, even the most minor detail in a person's narrative about the world can throw their understanding of reality as a whole into chaos, the story suggests. At the same time, because these details can be so minor, this radical change in realizing there is no difference between the medium and the message may go entirely unnoticed by the larger public.

Within that frame, experimental music could stand in for the underground, with hrönir being duplicates of lost objects that appear elsewhere and thus are both old and new, one and multiple, at the same time, like the punk women in the example above "being" Las Vulpess when they were not. Nubla's performance piece "Aquí no es allá el día de hoy" (1993) is perhaps the most fascinating example of this notion. The work draws on a collection of lost puzzle pieces Nubla had amassed over the years. Photographing and cataloging each one beginning in 1984, in the 1990s he exhibited them to-gether and turned them into a "musical score." Like the music of John Cage and Terry Riley, Nubla's score eschewed traditional composition, since it did not place the music on a staff but instead presented the puzzle pieces, laid out in a particular order on graph paper, as their own graphic repre-sentation of sound, accompanied by a textual description that was more like a script for the performer than a score. These small pieces of simulated landscapes and other images formed a new reality when brought together, even as they represented multiple other realms in time and space and a crossing between sound and image, music and noise. Like the hrönir, they altered the listener's present even as they alluded to a past that is unknown but imagined by the listener, changing the listener's reality in the process. In their own way, then, these hrönir are a representation of echoic mem-ory. They gesture both backward and forward in time, both to the most local space and spaces elsewhere, representing a fractal change: tiny on the surface, but all-encompassing in their significance to an experiential un-derstanding of a place.

Nubla's band Macromassa sought to produce the same idea on albums like the 1989 *Piedra Nombre* through their *Método de Composición Objetiva*, in which they rearranged normative sounds of found objects, which could be both material items converted into instruments—hammers on metal, for instance—and fragments of songs turned into objects through their frag-mentation, and overlaid them on the natural sounds of wind, electronic rhythms, and distorted voices. By breaking apart existing sounds and re-combining them, Macromassa's sound experiments decontextualized not just the sound itself but the entire frame of music at a time when Cata-lan Rock, sponsored by the Generalitat, had a hold on what music was and should be. Similarly, Ràdio PICA's found sounds on its nightly broadcast conceptualized a future that did not mean modernizing the city for con-sumption, but rather engaging in protest to create change. In one case, the cause was related to public toilets that were to be installed in Portal de l'Àngel; several punk bands and their affiliated fanzines plastered posters around the city critiquing the government's spending of funds on public toilets instead of on beer and music. During the protests, Picarol recalls, he sneaked a Walkman into a cop car outside the police station and recorded what was being put out on the police radio to broadcast later.[105] These frag-ments of free radio thus challenged state regulations of media from within their own acoustic space of authority and control by taking the concept of found art directly into the realm of anarchist activism.

The reason for this protest was not just aesthetic: Picarol has felt from the beginning that Ràdio PICA, grounded in Gràcia, has suffered from a precariousness that exposes how his own working-class background has informed his political sensibility. In an interview in July 2019, when I asked if he felt that the Catalan independence movement excluded immigrants, he turned to another kind of exclusion rooted in his own long background speaking Catalan and being from a multigenerational Catalan family:

> Yo me he sentido excluido, siendo mi idioma, ¿por qué? Porque mi familia ha sido trabajadora. . . . No perteneció a la burguesía. . . . Hay una burguesía que es la que está en el gobierno catalán y es la que cerró Radio PICA, y es la que, si pudiera, acabaría o que intenta acabar con Radio PICA y con su voz. A que no exista. Lo ha intentado. O sea, sí que hay una discriminación. No será que lo intentan disimular de manera nazi, como diríamos, aunque hay otros que lo son dentro del independentismo. No es el caso de la CUP (Candidatura d'Unitat Popular), pero hay un sector burgués catalán medio-alto, que lo es, y que son peligrosos. . . . Claro, soy muy de base, soy muy de calle.

> (I have felt excluded, with it being my language, why? Because my family has been a working-class family. . . . My family did not belong to the bourgeoisie. . . . There is a bourgeoisie that is in the Catalan government and closed Ràdio PICA, and they are the ones that, if they could, would end, or are trying to end, Ràdio PICA and its voice. So that it does not exist. They have tried. In other words, there is discrimination. It is not that they are trying to cover things up like the Nazis, you might say, although there are some [Nazis] who are within the independence cause. That is not the case with the CUP [the political party Candidatura d'Unitat Popular (Popular Unity Candidacy)], but there is a middle- to upper-class, bourgeois Catalan sector that is, and they are dangerous. Of course, I am very much of the base, I am from the streets.)

The class differences between Ramón and Picarol thus symbolize how the aural imaginary of established radio broadcasting normalized the authoritative voice of the lettered intellectual over time, while the free-radio transmission of underground nightlife, even while using the same technologies and model of the *micrófono ambulante* (walking microphone), could be heard so differently by its publics, participants, and those who wielded the power to determine what was suitable for the airwaves and what was not.

The acoustic expansion of the airwaves provided by Ràdio PICA also extended into politics in unofficial ways. The *mossos d'esquadra* closed the station down in 1987 and impounded the broadcast equipment, claiming that its emissions were interfering with radio traffic from the Barcelona airport, which was being revamped for the Olympics. Today, Picarol still insists that the accusation of interference was technically impossible, and this seems to be confirmed by the fact that the government returned PICA's embargoed equipment in 1990 after they were unable to prove his wrongdoing. Ràdio PICA returned to the same (unoccupied) frequency on the dial in 1991. Later, it was shown that Radio 3 was the station interfering with the Prat airport.[106] Nevertheless, while he was shut down, supporters took to the streets with graffiti that read, "Si nos cierran la boca hablarán las paredes" (If they close our mouths, the walls will talk).[107] Picarol, meanwhile, stood outside the Ajuntament every day for a year (except, he likes to say, when it rained) with a patch over his mouth that displayed the Catalan flag. Like the expression of punk sound and attitude evident in the DIY rendering of the body with pins, mohawks, and other signs of nonconformity, with its occupation of the physical spaces of the city with talking walls and corporeal performance of how the government muzzled voices, Ràdio PICA transmitted just how deaf the official ear was to other forms of expression. By placing his muzzled body on the steps of the center of government, Picarol drew attention to the limits of voice. Moreover, he repeated this corporeal public performance of the distortion of voice years later. In the first decade of the twenty-first century, he told me in an interview in 2018, he was sued after attempting, with other free radio stations, to jam a frequency run by "un tío de estos ultracatalanistas, ultranacionalistas, que tenía una emisora de música disco" (one of those ultracatalanists, ultranationalists, who had a disco station). Accused by the plaintiff of being a hippie, Picarol claimed in the interview with me that, in the courtroom, he put on a performance that echoed those of the early Ràdio PICA years. In this case, he ludically drew attention to the stereotype that associates punk music and radio resistance with youth:

Bueno, me puse una joroba, me encorvé, lo juro, y fui con un bastón (risas). La abogada cuando me vio se reventaba de risa. Claro, [el demandante] no me había visto nunca. No sabía si era así. Entonces yo le dije: "A ver, usted qué cree, que yo estoy para estas movidas. Además, ya ve la edad que tengo," que además era verdad. Y el tío alucinó. Total que lo perdió.

(Well, I put on a hump, bent over, I swear, and I went with a cane [laughs]. The lawyer burst out laughing when she saw me. Of course, [the plaintiff] had never seen me. He didn't know if I was like that. Then I said to him: "Let's see, what do you think, that I am up for this stuff? Besides, you see how old I am," which was also true. And the guy was gobsmacked. Anyway, he lost.)

Whether or not this narrative is historically accurate, it highlights the role of humor in fighting the system, even when one will likely lose. Although today official policy is that free radio stations are tolerated as long as they are not commercially motivated and do not run ads, in 2016 Ràdio PICA was successfully kicked off the dial for good for "stepping on" the frequency associated with Ràdio Barcelona, of Cadena SER.

One of the limitations of the microtonic, the fractal, or the hrönir, though, is the difficulty of maintaining a distance from systematic or institutional forces in order to preserve the underground aspect of these sounds: Over time and with the accumulation of sonic artifacts posted online, the ephemerality its creators cite as its strength becomes hardened into if not a movement, then an antimovement that defines itself based on what it opposes. Neither Picarol nor Nubla can be defined by outright resistance to the state, despite their stated anarchist postures: Picarol tried several times, but failed, to achieve licensing for his radio station in the '80s, and for a short period of time in the early '00s, in order to stay afloat economically, Ràdio PICA shared a frequency with Radio Gladys Palmera, a commercial station devoted to Latin-Caribbean music, which now has a global online audience in Latin America and Spain.[108] His continued dedication to resistance could, then, also be read as a reaction to exclusion, at the same time as the potential hypocrisy of his current situation has forced him to emphasize the independence of his venture over its libertarian freedom from capitalism and government control.[109] Over the last few years, as the station has moved to online platforms like Spotify just to have an audience, Picarol has been unable to control the appearance of ads before the shows, which undermines the noncommercial stance of the platform, a fact that is lamented in a note on the Ràdio PICA webpage. Similarly, by 1992, the year of the Olympics, Nubla was producing sound art as part of the "ArtFutur92" program sponsored by the Caixa-Forum, the Ajuntament, and other mainstream industry organizations like Canal+. By the time of his death in March 2020, he had achieved a presence at the MACBA (Museum of Contemporary Art of Barcelona), which speaks to his gradual incorporation in, and recognition by, the very institutions that represented neoliberal development for the city. In an interview with Nubla in 2019, in fact, I asked if it was contradictory for him and his organization to take the Ajuntament's support, which Ràdio PICA, for example, has never received. His response encapsulates the structural approach that underlies the ideological concept of the underground: "PICA ha tenido apoyo del Ayuntamiento porque le cerraron la radio el Ayuntamiento. Es otra forma de apoyo. . . . Puedes robar o que te roben. . . . Eres parte de la lucha" (PICA has had the support of the City Council because the City Council closed the radio. It is another form of support. . . . You can rob or be robbed. . . . You are part of the struggle). On one hand, this libertarian perspective on art is informed by anarchist thought and the internationalist spirit of the *ateneus llibertaris* that have such a long history in the city.[110] But it also reflects how the microtonic always shares a radius with the structural forms of power, even if they wish not to be defined by them.

Picarol, for his part, rejects the ideas proposed by Caellas and Costa, that resistance is no longer possible. In an interview with me in 2018, he said, "Es una tontería esto de destruir un anarquismo, la contracultura y la vida y la Rambla" (It is stupid this idea of destroying anarchism, the counterculture and life and the Rambla). Rather, like Mercadé, hearing a restless continuity between the underground of the 1970s, punk in the 1980s, and certain strains of hardcore *okupes* today, Ràdio PICA states openly that its sound has long been unheard by the mainstream, but that it is there in the background making noise all the time:

> Estamos entusiasmad@s que la política, el gran capital y la mass media menosprecie el 'underground'
> aparentemente no existimos, somos invisibles. . . .
> Funcionando subterráneamente de manera diaria y constante durante décadas avanzamos fuera del status de la superficie de un decadente sistema endogámico y corrupto que lo invade todo.

> We are excited that politics, big capital and the media despise the 'underground'
> apparently we do not exist, we are invisible. . . .
> Operating underground daily and constantly for decades we move beyond the status of the surface of a decadent endogamous and corrupt system that invades everything.[111]

As I will show in the next chapter, this continuity as echoic memory repeats in the *okupa* protests that still take place in Gràcia today. In fact, an argument could be made that the nocturnal occupation of space from which Ràdio PICA began its constant underground sound echoes in the music of today's marginalized groups in the city, such as Latino immigrants, who "problematizan su identidad y adscripción nacional" (problematize their identity and national assignation) by dancing in clubs to reggaeton, bachata, and merengue, among other imported sounds.[112] Among other examples, as Íñigo Sánchez Fuarros has pointed out, a *música mestiza* began in the 1990s with the arrival of immigrants from Latin America and elsewhere, renovating beats and rhythms in local music, even as the Ajuntament erased them from public view. In 2004, the same year that the Fórum de les Cultures sought to draw world travelers to the city to explore "cultural diversity, peace, and environmental sustainability," drumming and music-making were banned from the public spaces of the Parc de la Ciutadella.[113] Yet such fractal sounds pop up whether the government wants them to or not.

At the same time, by establishing itself as a station dedicated to combating notions of dispossession tied to the circulation of and access to information, Ràdio PICA has echoed into what would later be called *hacktivism*, as it pertains to issues surrounding internet access and "copyleft" privileges. In one of only a few music videos that are posted on Ràdio PICA's online page at the time of this writing, a video of the song "Acció Directa" (Direct Action), from 2009, by anarchopunk hardcore band Kolumna d'Odi Proletariat (KOP), updates the notion of sonic guerrilla warfare that Ràdio PICA carried out in the 1980s, to emphasize how technology and power are still intertwined. The video portrays a hacktivist group using cell phones and laptops, running through the streets of Barcelona and evading capture by shady *Men in Black*–type characters, as they take down the Borsa de Barcelona (Barcelona Stock Exchange) in the wake of the economic collapse of 2007. It ends with a shot of CNN and other news media announcing the hack, as well as a shot of the Torre de Collserola, built in 1991 on Tibidabo— from which Ràdio PICA had at one time broadcast—in anticipation of the Summer Olympics. By denouncing this change in the Barcelona landscape as a sign of globalized capitalist information and control, the song allows us to recognize cell phones and laptops as the extension of the cassette tapes and mobile radio transmitters with which Ràdio PICA had begun taking on broadcast media more than two decades earlier. Moreover, like its punk predecessors, "Acció Directa" openly affirms its commitment to sonic guerrilla warfare:

No rimo per rimar, omplo les cançons
De denúncia social i de rebel·lió
Mira al voltant, és un món global
Digue'm, que veus?
Més presons per omplir, centres comercials
Tanquen fàbriques, més precarietat
Vagues generals, càrregues policials
I ara, sents la ràbia?
Ràbia, gàbia, màfia
En front, un kop, black block:
Sents trets, saps bé què fer
Cor fort, de nou, puny roig
Acció directa!

(I don't rhyme to rhyme, I fill in the songs
Of social denunciation and rebellion
Look around, it's a global world
Tell me, what do you see?
More prisons to fill, malls
They close factories, more precariousness
General strikes, police charges
And now, do you feel anger?
Anger, cage, mafia
In front, a kop, black block:
You hear shots, you know what to do
Strong heart, again, red fist
Direct action!)

The spatial relationship between the underground and the mechanisms of power that control the city are evident here: Factories, shopping malls, and prisons are the physical sites of abuse and resistance. Just as years earlier Almen TNT had lamented the commercialization that had obscured the revolutionary power of leftists in Spain prior to Franco, here the proliferation of capitalism has not just stymied the revolution, it has created the conditions for it anew. If "Nadie cree en la revolución" was a humorous denunciation, here the call for direct action takes place in a violent setting, with references to black block, red fists, and direct action invoking an international anarchist movement with origins in twentieth-century Europe. Even as it is going up against a global system, its anarchist leanings also echo a very local lived experience of dispossession in Barcelona, expressed in Catalan. When KOP sings,

> No rimo per rimar, omplo les cançons
> De denúncia social i rebel·lió

> (I do not rhyme for the sake of rhyming,
> I fill the songs with social denunciation and rebellion)

they are recuperating a lived experience that is not aligned with Barcelona's mass-media beautification. It sonically echoes, instead, the displacement and resistance against control that took place well before modern communications technology even existed, investing their art with political action that takes up anew the anarchist past of the city. Their use of *omplir* (fill) to "fill" songs with rebellion reoccupies the "fullness" of shopping centers and prisons that define the global today in terms of consumers, taking a harsher approach to the sense of dispossession that Almen TNT had already sung about in 1979. The song thus occupies the aural imaginary of Barcelona with a memory not just of its anarchist roots but of an anarchist sentiment that is still present.

GRÀCIA, TERRITORI SONOR (ECHOES)

Free radio, punk, and the varied projects associated with the *meneo barcelonés* are about freedom of information and experimentation, but they are also about the acoustic and aesthetic occupation of local spaces. As I mentioned above, Ràdio PICA began in the terraces of Gràcia, and Víctor Nubla located his "Centre Místic de l'Univers" (Mystical Center of the Universe) in the Plaça del Raspall, an old plaza known for being a site of Roma gatherings. Their microtonic acoustic occupation of space—both through the airwaves, and through the rethinking of sound as a found object that can be put to use to revolutionize spatial relationships through time—was linked to avant-garde notions of communication that found their stride in the years after Franco's death. I would like to link these two aspects of

Gràcia's music scene in the 1980s and 1990s to a longer aural history of the *vila* (town) that hears sound as resistance.

Of special importance to Gràcia's community identity is the sound of the Marieta clock tower in Vila de Gràcia (at the Plaça Rius i Taulet), which is located across from the town government and which is still the site of local festivals, such as the famous Festes de Gràcia held each August.[114] Since the late nineteenth century, the tower has been associated with resistance to authority, and it is also associated with secularity, since it is the only nonreligious bell tower in Catalonia placed at the center of a plaza.[115] As Alain Corbin has shown, in nineteenth-century France, the sounds of bells "shaped the habitués of a community or, if you will, its culture of the senses. They served to anchor localism, imparting depth to the desire for rootedness and offering the peace of near, well-defined horizons."[116] They also inscribed listeners into an acoustic understanding of local and foreign space and "were supposed to preserve the space of a community from all conceivable threats."[117] In this case, the Marieta is associated with a constant ringing that took place in April 1870, during the Revolta de les Quintes (Bell Revolt). In the 1870s, working-class populations in Gràcia, as well as Sants and San Martí, rebelled against the reinstitution of obligatory military service by Spain. In response, the Spanish government sent troops in to put down the insurrection, resulting in twenty-seven deaths and numerous houses being sacked. The mayor of Gràcia at the time happened to be Josep Fabra, the father of famed grammarian Pompeu Fabra, who would later compose the authoritative grammars and orthography of the Catalan language.[118] The sounds of the bells ringing to warn of the slaughter have since been reclaimed to form part of a longer narrative of Gràcia's independent spirit and willingness to stand up to the government. Throughout most of the twentieth century, in fact, the bell took center stage annually in the posters symbolizing Gràcia's Festa Major.[119] Today, Gràcia's page on the Ajuntament de Barcelona website recalls the importance of the bells ringing as a sign of pride in resistance against the Spanish government:

> Pero también quedó el orgullo, un orgullo que tenía el sonido de la campana "Marieta," la campana de la torre de la plaza de Orient, Rius i Taulet después y Vila de Gràcia ahora. Su toque resquebrajado e infatigable avisó a los *gracienses* de la entrada de las tropas. Le lanzaron cañonazos desde todas las direcciones, pero no la hicieron callar, no la destruyeron, sólo la hirieron y por eso ha quedado como el símbolo de aquella revuelta ciudadana.

(But there was also pride, the pride that the sound of the "Marieta" bell, the bell of the tower of the Plaza de Orient, Rius i Taulet later, and Vila de Gràcia now. Its cracking and indefatigable ringing warned the Gracians of the troops arriving. They fired cannons at her from all directions, but they did not silence her, they did not destroy her; they only wounded her, and that is why she has remained as the symbol of that citizen revolt.)[120]

This echoic memory of resistance as signified by the ringing of the bells resonates in the history that led to the punk and experimental sound art of the *meneo barcelonés*. Just before beginning Ràdio PICA, Picarol, who made his living in and around Gràcia, participated in the founding of another free radio station that was an early example of the broader *Ones Lluires* (Free Airwaves) movement that arose in Barcelona after Franco's death. This station, La Campana de Gràcia (on air 1978–82), discussed local affairs and also invoked the Marieta bell. By occupying the airwaves through an underground apparatus with a reduced broadcast radius, the station repeated the gesture of the nineteenth-century bells, creating the sound of a local community defined by access to its sounds, but also serving as a sort of shadow voice to that of the local government. Like PICA's later expansion of punk rebellion to the satellite cities of Mataró and other marginal communities that did not hear themselves in mainstream media and politics, this sound coexisted alongside the public ringing of bells that formed the official community. It is possible to interpret this as the radio station having taken up a hauntological presence that marked the absence of voice for the people.

Still, hearing the bells solely in relation to Revolta de les Quintes does not fully capture the echoic memory of the Marieta as it changes shape over time. The name of the radio station, La Campana de Gràcia (The Bell of Gràcia), also invokes its eponym, a long-lived Catalan magazine. Informed by Masonic and anticlerical ideals, *La Campana de Gràcia* ran from 1870 to 1934, and was founded in the wake of the Spanish slaughter of those who participated in the revolt. The magazine is known for being the first long-running publication in Catalan, and it also is an example of pro-Republican popular literature that forms part of a longer use of satire as resistance, with its articles being framed from the first issue as *batallades* (battles).[121] In the first few decades after its founding, "a la majoria d'Ateneus populars i a les cases menestrals, *La Campana* sustituïa, literalment, *La Bíblia*" (in the majority of the popular *ateneus* and artisan houses *La Campana* literally substituted the Bible).[122] In line with its association of bells ringing as a sign of local resistance, the first issue of the magazine featured an engraving of the bell tower by Catalan artist Tomás Padró; later that year the drawing of the bell incorporated written sound effects in the space around the bell, as though to indicate the occupation of the city space by its reverberations. Initially subtitled "Setmanari Bilingüe" (Bilingual Weekly), by 1872 the text was being published only in Catalan. The magazine was shuttered several times by the Spanish government, and it often retaliated by publishing an alternative magazine, *L'Esquella de la Torratxa* (The Cowbell of the Turret), whose title was tongue-in-cheek: Having been shut down by the central government in Madrid, the editor, "en vista de la situació, pensà, 'No volen sentir campanades? Doncs sentiran esquellots!'" (given the situation, thought, "They don't want to hear tower bell chimes? Then they'll hear cowbell chimes!").[123] Drawings of a giant cowbell replaced those of the tower in these substitute publications. In 1878, when the sale of *La Campana de Gràcia* was once again prohibited, then-editor López Bernagossi recuperated the sound of the bell ringing in protest, but this time making a public spectacle that was also a sonic disruption of authority: He installed a little store on the Rambla del Centre, in front of the government administration, that rang a bell on Saturdays to announce the new publications. In a place where overt dissension against the Castilian government was not possible, this positioning of the body and occupation of public space with the sound of another bell stood in for all that could not be said, in much the same way that Ràdio PICA's occupation of the dial in the name of access to information occupied a public air space as a symbol of the government's repression.

However, the specificity of that sense of repression shifted toward the beginning of the twentieth century. Gràcia was annexed into the larger city of Barcelona in 1897, and in 1906, with the founding of the right-wing Catalanist political party, the Lliga Regionalista, *La Campana de Gràcia* and its symbolism of resistance to oppression shifted. What was once a popular, worker-oriented, and anti-Catalanist publication favoring the pro-Republican Alejandro Lerroux joined the cause of Solidaritat Catalana, a coalition that brought together parties including Carlists, the Lliga, and others in order to defeat Lerroux's more radical leftism. In this way, the echoic memory of the bells as a symbol of autonomy acquired a new multilayered geography, coming to stand not just for Gràcia, but for Barcelona and Catalonia as a political whole within Spain. Even as this echoic memory of resistance was inscribed into an increasingly normative Catalan media apparatus (see Chapter 1), Picarol's participation in the later free radio of La Campana de Gràcia and in La Esquella (the little bell) theater group, as well as the movement of Ràdio PICA's on-air sound around the *vila* and his performance of his silenced voice on the steps of the Ajuntament, recuperated an echoic memory of resistance rooted in the acoustics of Gràcia's town square. That he has chosen for his last name Picarol, itself a reference to another kind of bell (a smaller *esquella* used for sheep or goats) that was also used as a pseudonym by Josep Costa (one of the collaborators of both *La Campana de Gràcia* and *L'Esquella de la Torratxa*), to represent his "mafia" shifts the mediatic inscription into a walking, living vocal chord in the style of Ocaña. It resounds in any public appearance he makes, as his voice continues to function like the bell to both provoke the authorities and form an alternate community, however momentary.

Today Gràcia is still somewhat removed from the tourist centers of the city, although increasingly it is being marketed as a hip alternative to the historic center, in the sort of appropriation of the counterculture Costa has described. It is perhaps not surprising, then, that the *ateneus* that were crucial for guiding the sentiment that bolstered the anarchists prior to the Civil War remain a site of collaboration and protest for *okupa* groups today, many of whom come from beyond Barcelona or move among similar spaces in the city in a kind of anticapitalist nomadism.[124] Their spaces are often defined by the free libraries and bookstores that, like the *Ones lliures* of La Campana de Gràcia and Ràdio PICA, promise to make circulation of alternative information possible without market interferences. As I will examine in Chapter 4, they recuperate both punk and Gràcia's public bell ringing when they perform direct-action poetry and sonic protest on the same streets around the Marieta bell tower that are now occupied by tourists, to insist that those who come to Barcelona in response to the globalized marketing of the city leave Gràcia's local spaces alone.

RESTLESSNESS AND DISPOSSESSION

By focusing consistently on minor, yet repeated, *meneos* that accompany normative soundscapes and sensibilities, the underground free radio and experimental sound scene in Barcelona have drawn our ear to the suffocating ways in which neoliberal ideologies of possession apply not just to physical property, but to market concepts that affect speech and access to communication, such as intellectual property or radio airwaves.

The radiophonic restlessness of and around Ràdio PICA does not return to some auratic concept of pure selfhood prior to alterity; rather, it recognizes the construct of alterity as a daily fractal through the larger (sonic) fabric of society.

After Franco's death, radio expert Lawrence Soley claimed that "clandestine" radio stations had disappeared from Spain by 1981 because "The introduction of democracy and the ability to openly criticize government, eliminates clandestine radio criticism."[125] But Ràdio PICA, subsequent free radio stations, and even Víctor Nubla's search for an acoustic universe through the Gràcia Territori Sonor organization and its annual LEM festival (the underground equivalent to the more popular Sónar festival) illustrate that the Cold War binary led to the incorrect supposition that Spain's turn to democracy would resolve any concerns about free speech or expression.[126] At stake was not just the voice of democracy but a wider understanding of aural space as overdetermined by capitalist and totalitarian ideas of possession and territory.

Listening in to the *meneo barcelonés* not just as a musical moment but as a lifelong practice of experimentation that, in its refusal to settle, challenges normativity as a structure of life, suggests an important corrective to tendencies among some scholars to hear the colonial ear in terms of an antagonism between European and non-European cultural experiences, even as it calls into play the old question about Spain's place in Europe. Tom McEnaney asserts that considering rebel radio in Cuba and Argentina allows one to "[complicate] the tendency in the European continental tradition that exclusively links the radio voice with fascism."[127] In a similar, though geographically distinct fashion, here we see that although free radio in Spain (and, more specifically, in Catalan Spain) explicitly associated state-controlled radio voices with fascism, that designation rested on an understanding of voice as being linked to economic and legal structures that, in their desire for modernity, produced a geography of the ear shaped ideologically by both dispossession and experimentation. As I describe in the next chapter, the concerns about democracy, voice, and the sonic occupation of space continue to play out in the political protests that circulate in the streets of Barcelona today.

4 Protest and the Acoustic Limits of Democracy

On September 10, 2012, then–Prime Minister Mariano Rajoy, of the right-wing Partido Popular, got himself in hot water when he referred to the Catalan independence demonstration that was to take place in Barcelona the next day—one of many that would form part of the *marxa cap a la independència* (march toward independence)—as an unnecessary "algarabía" (racket/confused shouting).[1] Although for much of the period after Franco's death, Catalan politics regarding central Spain largely revolved around having more economic control and policy-making autonomy, recently the tides had turned toward a drive for complete independence from Spain. The Diada protest in 2012 to which Rajoy was responding was organized around the slogan *Catalunya: nou estat de Europa* (Catalonia: Europe's new state) and drew between 600,000 and 1.5 million people, who waved the *senyera* and the *estelada* flags and chanted "in-inde-independència" (in-inde-independence).[2] By all accounts it was a peaceful event, but in the days that followed, local proindependence newspapers in Catalonia, like *El Punt Avui* and *Diari de Girona*, as well Spanish papers like *El País*, and even global platforms like the *Huffington Post*, seized on Rajoy's word not because he proclaimed the protest's noise and disorder even before the demonstration took place, but because of the negativity associated with the term. It has several meanings in Castilian, according to the Real Academia Española (RAE):

Del ár. hisp. *al'arabíyya*, y este del ár. clás. *'arabiyyah*

1 Gritería confusa de varias personas que hablan a un tiempo.
2 f. coloq. Lengua atropellada o ininteligible.
3 f. p. us. árabe (‖ lengua).
4 f. p. us. Enredo, maraña.

(From the Hispano-Arabic *al'arabíyya*, and this from classical Arabic *'arabiyyah*

1 Confused shouting from several people talking at the same time.
2 f. coloq. Stumbling or unintelligible language.
3 f. p. us. Arabic [‖ language].
4 f. p. us. Tangle, tangled mess.)

Derived from the Arabic term for its own language (*al'arabíyya*), the Castilian word hears the voices of those who speak their own, non-Castilian language as noisy, unintelligible, and, metaphorically, a tangled mess.[3] Wielded as a critique of the pro-Catalan independence demonstrators that were demanding democratic legitimacy for their people, Rajoy's use of *algarabía* inscribed the contemporary Catalanist movement in a long aural imaginary of the sound of "foreign" language that echoed the period of Muslim presence in the peninsula. For centuries this aural history had been defined by the ideology of the Reconquista, especially during the Franco regime; that is, the beliefs used to justify the Catholic expulsion of Muslims from Spain. Franco would later integrate those same beliefs into his propagandistic affirmation of the Spanish "race" as monolithic: white, traditional, heteronormative, and Catholic. At the same time, as José Luis Venegas has outlined, such echoic notions of race are complicated in practice: even Franco was able to justify his alliance with North African troops by appealing to their spirituality, which he opposed to the secular thought associated with the Second Spanish Republic, overlooking both racial and religious differences in the process.[4]

Through the etymological breadth of *algarabía*, as a term specifically associated with linguistic sound heard as noise, we can recognize how the contemporary soundscape is shot through with aural examples of what Wai Chee Dimock would call in literature *deep time*, and which I hear in the echoic memories that bring the past into the sound of the present: deep time "highlights . . . a set of longitudinal frames, at once projective and recessional, with input going both ways, and binding continents and millennia into many loops of relations, a densely interactive fabric."[5] The long temporal geography of the ear suggested by the term *algarabía*, along with its reference to a fabric that suggests a weave similar to (but more ordered than) the term's nonsonic equivalent in *maraña* (tangle), fuses transnational aurality with contemporary, local notions of belonging and exclusion into a tangle of sensations. As I wrote in the Introduction, going back as far as the *Requerimiento*, the legal mechanisms that hear vocal difference as mere noise, as opposed to reasoned discourse tied to one's own language, were often used by colonizers to justify violence because they assumed that the "unintelligible" sounds of others reflected a faulty ear that needed to be disciplined. The term *algarabía* reminds us that the start of colonial conquest in the Americas coincided with the official end of Arabic presence in Spain, elucidating the transatlantic construction and dissemination of the colonial-capitalist sleight of ear as being epistemologically grounded in the geography of the New World, but also rooted domestically in a Reconquista-driven hearing of "foreign" language as a tangled, loud noise that needs to be cleaned up, by physical force if necessary. In contemporary cases of protest, similarly, a dominant Western aural imaginary hears the letter of the law as a vehicle for social order, at odds with the communal noise of the streets. In this aurality, those who protest noisily do so because they are unable to understand the civilizing function of law and thus must be forcibly put in their place: Rajoy's premature use of *algarabía* to describe a demonstration that had not yet happened warned that the proindependence demonstrators were about to cross the line between meaningfully celebrating their heritage and showing they could not hear their own Spanishness effectively, which in turn would make their voices unintelligible to a reasoned ear.

For their part, pro-Catalan advocates heard Rajoy's use of the term as offensive and incorrect, and wielded their own form of rationality to denounce it. Besides being an arcane word reminiscent of the nineteenth century, wrote the Catalan philologist Eulàlia Lledó Cunill a few days after the event, the preferred Catalan equivalent would be *guirigall* (racket), since the Catalan word with the same etymology, *algaravia*, is not frequently used in Catalan.[6] Moreover, citing several of the definitions mentioned by the RAE (although eliding the reference to Arabic by subsuming it to the phrase "among other things"), Lledó suggests that Rajoy himself became entangled when he invoked a polyphonic term that was not as straightforward as *guirigall*, which only means "many people speaking at once." Màrius Carol wrote in *La Vanguardia*, Catalonia's leading newspaper, that the orderly, smiling procession of middle-class families participating in the largest Catalan protest since the Transition would better be characterized as a choir singing, because "ningú no va desafinar amb la veu ni mitjançant la seva conducta" (no one was out of tune in their voice or their conduct).[7] Moreover, he suggested, in his own rather arcane reference to Biblical history, that *algarabía* carries with it a misdirected cultural association: "En un ús saduceu de l'idioma va acabar per definir l'enrenou confús de persones que parlen alhora, un costum que ha acabat per ser tant o més espanyol que no pas musulmà" (In a Sadducee [a Biblical Jewish sect] use of the language, [the word] ended up defining the confusing commotion of people who speak at the same time, a custom that has ended up being at least, if not more, Spanish than Muslim). Once again invoking history and religion, he attributes the messy disorder of language associated with the noisy sound of Arabic to Spanish culture writ large. The gesture not only repudiates the application of the term to the Catalan demonstration, it turns the comparison back on Spain and Spanishness. In critiquing Rajoy's irrational use of the term, moreover, Carol subtly reinforces the value of *seny*, a specifically Catalan notion of common sense, as a deliberate form of thoughtful reason that separates Catalonia from the rest of Spain. Regardless of political position, what all parties seemed to agree on was that this term, with an etymological history of a local, non-Islamic ear hearing Arabic as noisy gibberish, was an insult.

Perhaps not surprisingly, they also all seemed to agree with the supposition of European superiority that underlies the term's etymology. This is relevant because it allows us to hear September 11, the Diada (the National Day of Catalonia), as an example of an event that is no longer solely inscribed in a geography of the ear enclosed by Spain's borders. The date commemorates the 1714 fall of Barcelona to the Spanish, rooting Catalan national pride in the loss of autonomy. But this memory is conflated in the contemporary period with the 2001 terrorist attacks that brought down the Twin Towers in New York City, ushering in an era defined by the so-called global war on terror. During the Transition, when globalization meant aspirations toward neoliberal development, Salvador Picarol and Víctor Nubla could proclaim that their uses of sound performed micro acts of poetic terrorism (see Chapter 3), but today similar rhetoric is heard by the state as evidence of criminal terrorism to be prosecuted. One need only consider the case of the rappers Valtònyc and Pablo Hasél, until recently in exile from Spain (and in Hasél's case, imprisoned for three years beginning in 2021) because they were under threat of arrest for writing song lyrics that harkened back to the days of punk, anarchy, and antisystem protest in the 1980s.[8] In a post-9/11 world, their lyrics resonate legally as an *algarabía* that is heard more for its relationship to *al'arabíyya* as a conservative code-word for terrorism than for their poetic attempt to signal the inequities of Spain's political and economic system.[9]

The media debate over the proper vocabulary to describe the protest was obviously political, illustrative of the tensions over Catalan identity and autonomy that have dominated not just Catalonia, but all of Spain, over the past two decades. This discussion was but one of many precursors to the Catalan referendum on independence that took place five years later on October 1, 2017. But the linguistic debate also revealed an aural imaginary of protest, grounded in colonial epistemology and shared by both sides, which associates certain public displays of community with negative aural constructs of disorder and unintelligibility that run counter to Western constructs of democratic voice.[10] By reframing the noise of protest as a choir singing, for example, those who supported the movement instead suggested that their use of sound would participate in a utopian public voice, which Brandon Labelle has called a "sonic agency" that employs "sonic sensibility" as a way of "nurtur[ing] modes of engaged attention" that will suggest compassion and empathy: "Sound and listening are highly adept as carriers of compassion and the forcefulness of one's singularity—the intensification of affective sharing (*this sound that goes right into my body*)."[11] Nevertheless, as I demonstrate below, such productions of compassion as embodied sensations are simultaneously inscribed in an aurality in which silence represents a positively inflected aural imaginary of language as rational. Moreover, there is a hierarchy of affects associated with sounds. It is helpful, perhaps, to consider a public noise campaign taking place in Barcelona in recent years, which explicitly ties the long-prized Catalan value of "civilization"—developed since the nineteenth century in ways that echo Barcelona's colonial enterprises in Cuba and Africa—to an association of hygiene with quiet. While walking the city in 2018, I saw multiple posters advocating for a quiet Barcelona, one that will not wake the children or otherwise disturb city dwellers. Drunk tourists, rowdy kids, or others who might disturb the peace, are instead encouraged to shush unwanted sounds in the same way they are expected to clean up for themselves in public squares (see figures 4.1–4.3, from 2018).

Although at times, in highly touristed areas, these posters appear in English, ironically, many are in Catalan, excluding most tourists from their linguistic address. Instead, they largely depend on drawings to communicate the idea of noise. The faceless, undefined shape of noise magnifying

4.1 Viu i deixa viure. This poster reads, "Live and let live. Don't let your freedom restrict that of others." The same image was used in another poster, which I discuss here, titled "Dret a dormir" (Right to sleep). Its tagline was "Tothom té dret a dormir. Vetllem perquè el soroll no sigui un malson" (Everyone has the right to sleep. Let's make sure noise is not a nightmare). Part of the *Compartim Barcelona* campaign in 2018. Illustration by Miguel Montaner, graphic design by Andrés Requena.

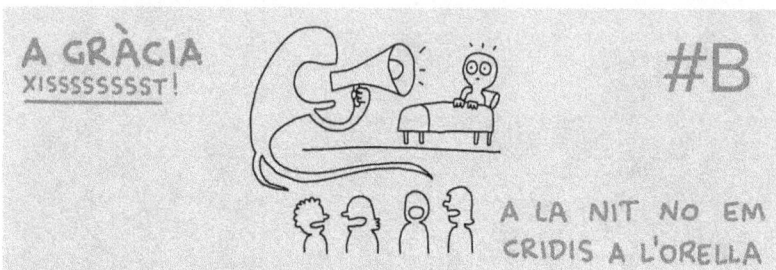

4.2 Evoluciona. The English version of these *Compartim Barcelona* posters was translated "*Homo wild—Homo rowdy—Homo plaza. In the plaza, evolve.*" The same images, changed to refer to the beach instead of the plaza, were still present around Barceloneta in 2024. Illustration by Javirroyo, graphic design by Andrés Requena.

4.3 A Gràcia, Xisssssssst! Some *Compartim Barcelona* posters specified that they were referring to Gràcia, as was the case with those I saw in the *vila* in 2018. The English version of this particular poster was translated "In Gràcia, sssssssssssh! Don't shout in my ear at night." Illustration by Javirroyo, graphic design by Andrés Requena.

voices through a megaphone in figure 4.3, for example, contrasts with the wide-eyed look of the person trying to sleep. Similarly, in figure 4.2, only the final image of "homo plaça" has facial features—a smile—while the under-developed litterer, who also is associated with smoking and music, makes a capital-letter sound of "EEE." In a sense, the poster echoes the overlap of animality and race in a colonial ear shown in the Massagran stories and comics I addressed in Chapter 1. The unfortunate insomniac in figure 4.1, meanwhile, is defined by his eyes, with a graphic depiction of sound as a razor-edged explosion outside the window suggesting a city under siege. All of the images reflect an aurality in which only the silent figure is defined as a person; the bothersome noisemakers or litterers are faceless, amorphous, or invisible. Even the visual representation of civilization seeks to strip vocal noise of the conditions of its material production.

The posters' representation of nighttime noise as pollution not only reifies the discourse around punk rebellion I interrogated in the last chapter, associating noise with garbage, bad manners, and rot, it contextualizes as retrograde or insulting any noisy *algarabía*, while respectful sonic agency can be heard as a source of positive fellow feeling (to indirectly cite Benedict Anderson's *Imagined Communities*). For the Diada to be accused of *algarabía*, when the city clearly values its ability to be silent as a sign of cleanliness, seems an affront to the cultural belief that began with the *noucentistes* and presumed that Catalonia could both become civil and instill civilization in others. In Chapter 1, I interrogated the vococentrism that is produced by colonial civilizing ideologies, an ideal of a bodiless and uniform voice that is grounded in a colonial violence that has been made invisible and inaudible. But as I illustrate here, that vococentrism also means a silencing of negative affect that the idea of *algarabía* helps us disentangle.

Thus, in this chapter I use the notion of *algarabía* to think through the *maraña* of protest sounds that obtained in a few recent intermedial soundscapes of Barcelona. Together, these various protests complicate a national geography of the ear, which can no longer be heard as merely self and other, civilized or barbaric, collective or individual, although, as the posters above illustrate, those discourses continue to circulate as a means of ordering sound and voice, noise and protest. My goal is not to attribute certain affects to particular political parties or ideologies, but rather to illustrate how affective auralities in the twenty-first century occupy the city space in ways that draw our ear to the relationships between media, democracy, and sound. In moments of protest, law and voice are both put to the test as a measure of a society's democracy. How protestors approach sound is revealing in what it says about what they think public occupations of space can and will do, and how those outside the movements will respond in turn.

The suggestion of acoustic barbarism evidenced in the exchanges around the Diada march—and the political insults it made possible— illuminates the *longue durée* of a dichotomy of sound and noise that is itself framed around an aurality of possession and dispossession, where possession is not just about property, but about the right to voice more broadly. As Athena Athanasiou writes in her discussion with Judith Butler, "the power of dispossession works by rendering certain subjects, communities, or populations unintelligible, by eviscerating for them the conditions of possibility for life and the 'human' itself."[12] This understanding of possession and dispossession presumes that the life and humanity of a community depend on the possibility not just of sounding a voice, but of that voice being heard as intelligible and, at least in the contemporary period, democratic: Those accused of *algarabía* are heard as making sound outside the parameters of that democratic intelligibility. Yet such intelligibility is not only measurable in sound, but in the sensations of affect sound harnesses. If the Diada protests reflected intense political debate between Spain and Catalonia starting in the '00s, smaller protests against gentrification by *okupes* and their allies have generated complaints of noisiness and public disturbance for a longer period. But they have largely been tolerated over the last three decades, especially by a socialist left who hears their squatting of buildings or public performances of illegal concerts spilling into the streets as indicative of "admirable ideals but whose politics, ultimately, is ineffectual and harmless."[13] In what follows, then, I use the construct of *algarabía* as a means of listening in to the *gritería* (shouting) of multiple people talking at once and the confusion to the ear caused by "unintelligible" voices, especially voices in protest, to interrogate how the spatialized sound of democracy is performed around political issues that are often at odds with one another.

Although echoic memories are a function of deep time, in this chapter I focus on a relatively short period of time—a little more than five years, revolving around 2017 as a point of inflection. I do so in order to illustrate how echoic memories are formed and reshaped in the different aural constellations with which bodies, discourses, and affects engage daily in the long wake of the Transition, the history of which, as Santos Juliá has argued, has been retaken as a political tool in recent years.[14] I have therefore opted to perform a close listening to three very different public occupations of space. First, I establish as a base the proindependence performances that have intentionally turned Barcelona into a public media spectacle over the last fifteen years; groups like Òmnium Cultural and the Assemblea Nacional Catalana (ANC, National Catalan Assembly) have used the presence of the sonically affective body in space as a means of affirming Catalonia's right to independence, doing so in a way that engages the media but also reaffirms, rather than questions, the affective association of silence and reason that underlies the contemporary understanding of democratic voice. I contrast that mediatized performance of national identity with a close listening of how small antigentrification protests by a group of writers who call themselves *poéticamente incorrectos* (poetically incorrect), and whom I observed and recorded in 2018, wrest control of acoustic space from tourists and capitalist developers, if only for a Saturday afternoon. Last, I broaden these national and local concerns out to the global by listening in to media coverage of the voice—or specifically, the whistle as non-voice—of one of the ISIS members responsible for a terrorist attack on Les Rambles shortly before the 2017 independence referendum, in order to explore the necropolitics of sound. I therefore blend mass media representations of echoic phenomena with my own observations of antigentrification activism in Gràcia to show how the daily complexities of echoic memory, especially sound filtered through the media, imply a *maraña*—an overlap of multiple discourses, performances, feelings, and ideologies that coexist simultaneously within a single space. Using my own ear as a filter, I listen in to how spatial occupations resound with an aurality that maintains democracy as a silent political ideal, perpetuating an exclusionary construct of community, even as, alongside it, the affective sounds of protest draw the ear to various dispossessions that are felt in the city. Taken together, these tensions of *algarabía* point more broadly to the biomediation of the ear, which, as I explore below, concerns how traditional and new media stage voice in ways that affect not just how collective sounds of protest are heard but how they mean within a vococentric aural imaginary of democracy, in which a written legal code legitimizes a certain kind of voice by contrasting language to the lived acoustic presence of the body.

STAGING DEMOCRATIC VOICE

In 2022, the Spanish government took a remarkable step: Supported by the PSOE, Unidas Podemos, Bildu, the PDeCAT, and PNV, the state approved a new Ley de Memoria Democrática (Law of Democratic Memory).[15] The law, which replaced the Ley de Memoria Histórica (Law of Historical Memory) of 2007, went further than the previous one, attempting to redress the many ills of Franco's regime, which would include disinterring hundreds of thousands of bodies buried in mass graves and establishing a DNA database of Franco's victims in order to identify the remains. The law also broadened the definition of "victim" associated with the dictatorship. No longer referring only to those who were murdered or imprisoned by Franco, "victim" also included LGBTQI+ people and those who were adopted at birth without the consent of their birth parents, as well as language and culture itself. In the new law, Catalan, Basque, and Galician, as well as Aragonese, Occitan, and Asturian were all recognized officially as victims of a "policy of persecution and repression": "Se consideran víctimas las comunidades, las lenguas y las culturas vasca, catalana y gallega en sus ámbitos territoriales lingüísticos, cuyos hablantes fueron perseguidos por hacer uso de estas" (The Basque, Catalan and Galician communities, languages and cultures in their linguistic territorial areas are considered victims, whose speakers were persecuted for using them).[16] In this new legalization of democratic memory, language, as a collective form of expression, was heard as a victim of political violence, making it a subject, not merely a representation of one.

The move may have been, in part, an attempt to pacify Catalan independentists who felt the Spanish government did not hear them as viable interlocutors, although by shifting victimhood from the people to the language, the move also shifted the politics of voice from an individual expression of identity or political representation to an aurality of identity as institutionalized sound. As Justin Crumbaugh has pointed out, since the late 1990s, contemporary peninsular politics has relied on media constructs of victimhood for legitimacy: "the discourse of victimhood is so ubiquitous that . . . nowadays, one can *only* enter the political sphere by speaking *through* the victim of one form of political violence or another."[17] For Crumbaugh that victimhood is linked to the symbolic gestures through which a lost life, often due to terrorism or other form of political violence, becomes what he considers, invoking Judith Butler, a "publicly grievable life," and how it is spun by the media into a political position.[18] But here the symbolism is no longer merely mediatic: it is enshrined in law, thus occupying a space in official discourse that gives it more weight than any talking head could. I wrote most of this chapter before the law was passed, so I cannot attend to its ramifications here, but I wish to use this development to lay out the stakes of language as a collective subject and object of democratic sound. In what follows, I attend to how the affective discourses and bodies that occupy Barcelona are intertwined in a biomediated space through which both conservative and liberal ideologies alike produce the sound of democracy through protest.[19]

At stake is the relationship of democracy to voice, a term that is used popularly to denote political representation, but which also refers to representation in a public sphere or, more in line with my purposes, what Michael Warner has referred to as publics and counterpublics. As he writes, "Dominant publics are by definition those that can take their discourse pragmatics and lifeworlds for granted, misrecognizing the indefinite scope of their expansive address as universality or normalcy. Counterpublics are spaces of circulation in which it is hoped that the poiesis of scene making will be transformative, not replicative merely."[20] Nevertheless, for Warner, writing before social media, the Arab Spring, and other global events that might have called into question his assertion, publics and counterpublics are essentially agentless: "Readers may scrutinize, ask, reject, opine, decide, judge, and so on. Publics can do exactly these things. And nothing else. Publics—unlike mobs or crowds—remain incapable of any activity that cannot be expressed through such verbs. . . . Counterpublics tend to be those in which this ideology of reading does not have the same privilege."[21] Warner ends his text with a description of agency as being either of mobs or crowds without sovereignty, or a public who can "acquire agency in relation to the state."[22] In other words, protest voice can only be a performative act until it is legitimated by law.

Yet, as I show below, not all protest can be conceived of as being legitimated only by law. That may be the case for the Catalan separatist movement, whose end goal was a replication of law, moved to a different territory, but it is not so for the *okupes* and antigentrification protestors I consider below, for whom most governed territories are defined by a geography of repression; nor is legal change the goal of the ISIS members who attacked Les Rambles only six weeks before the independence vote. Through the sounds of these contradictory movements, we can hear a validation of protest as an aural occupation of space with no clear endgame other than a radical affirmation of collective voice as sound.

As Mladen Dolar has explored, voice carries with it an acoustic aspect—what Barthes, again, called the grain of the voice—that is rarely interrogated. Because it is unquestioned, institutionalized social rituals of voice—court testimony, for example—give the impression of the peaceful coexistence of the letter and the law without accounting for voice as an acoustic performance: "it is the fiction of the universal accessibility of the letter and of its unchangeable nature which makes the law possible, as opposed to the fleeting nature of voices."[23] A similar unquestioning is at stake, I might suggest, in the notion of counterpublics without agency present in Warner's theory, which, grounded in reading, reduces their voices to silent signs rather than actionable expressions. Yet as Patrick Eisenlohr points out, the grain of the voice draws attention to the presence of the body "not controllable by human intentionality" that, in turn, produces an atmosphere "in which sonic energy—the sound waves' deviations from the ambient atmospheric pressure—is converted into different forms of energy in the body, creating new psychological and sensational phenomena in the process."[24] If we keep the role of sensation in mind, paying attention to the acoustics of voice calls into question the aural imaginary of silence that upholds democracy as a voting practice that, in turn, legitimizes the sovereignty of the state as the authorized display of public will.

The aurality of unity transmitted through the letter, Dolar argues, depends on a biopolitical notion of voice that Aristotle heard as being divided into *bios* and *zoe*. Quoting Giorgio Agamben's work on biopolitics, he argues that "*zoe* is naked life, bare life, life reduced to animality; *bios* is life in the community, in the *polis*, political life."[25] In other words, *zoe* is an animal sounding of the vocal cords that simply signifies a living being, while *bios*, as *logos*, connotes meaning and, later, law: "Voice is like bare life, something that is supposedly exterior to the political, while *logos* is the counterpart of *polis*, of social life ruled by laws and the common good."[26] This dual notion of voice has informed philosophical, legal, and cultural understandings of identity in the West, particularly as it relates to the construction of community limits, which are defined by *logos* as law. Yet, following Agamben, Dolar asserts that there is an overlap between the two, such that, in contemporary politics, voice is a void between *bios* and *zoe*, phone and *logos*, that is "precarious and elusive, an entity which cannot be met in the full sonority of an unambiguous presence, but *is not simply a lack either*."[27] For Dolar, "The moment this voice is taken as something positive and compelling on its own, we enter the realm where obnoxious consequences are quick to follow. In politics it quickly turns into His Master's Voice, supplanting law."[28] This means that modern democratic states subsume voice into silent election practices that merely support the sovereignty of the state:

> The electoral voice has to be a silent voice (a silenced voice?): it has to be given by writing (by crossing or encircling), and it has to be performed

in a small cabin . . . in complete silence. Furthermore, it has to be done one by one, so that the collective outburst of the acclamatory voice is broken down, . . . seemingly deprived of its . . . spectacular effects. It is the voice measured and counted, the voice submitted to arithmetic, the voice entrusted to a written sign, a mute voice deprived of any sonority.[29]

This is the silent voice to which the 2017 referendum for independence aspired and which was denied by Spain's Constitutional Court. But it is also the silencing voice that, depending on the geography, might convert the "collective outburst of the acclamatory voice" into a system (the Catalan language) that can be legally determined to be a victim, rather than a felt and embodied expression of voice that cannot be contained in a vote, legal or otherwise. The vote and the legal system it represents keep voice in its place.

Still, political demonstrations formed a fundamental part of the trajectory toward the independence vote on October 1, 2017. In staging the sound of voice as democracy, the separatist movement pushed at the acoustic limits of a democracy sustained by the fiction of a lettered voice, since public demonstrations linked the sound of vocal noise to a universal notion of democratic rights that, presumably, exceed the bounds of the ballot box. In these acoustic occupations of the city, the voice that is a void is not a lack. It is also the sounded excess of voice that is heard into vocal tone as affect, a reflection of the sentient body that overflows, and which is especially notable in moments of protest. At stake is what Francine Masiello calls the "senses of democracy":

When the discourse on democracy is altered—when public participation is engaged or foreclosed, when the concepts of the 'people' are redefined, when we catch sight of nations in distress or hear repeated calls for war, when we feel the weight of modernity pressing upon the walls of tradition—then . . . we reframe the sensorium and the uses of human perceptions. The sentient body is a placeholder for a large discussion of the effects of state practices on its people, and alternatively, it traces the ways in which the population resists or transforms social life. It also becomes a measure of aesthetic performance that captures the political drift.[30]

For Catalan separatists, the Spanish had unjustly captured a geographical sense of nation-statehood that had to be reclaimed by aligning national identity heard through the sound of language, with political sovereignty expressed by new state borders. The sentient body taking to the streets therefore was, as Masiello says, a "placeholder" for the legal realization of change that had not been allowed to take place. As Masiello points out, these senses of democracy are often at odds with the discourse of "civilized" reason that perpetuates democratic institutions. At the same time, as Mónica López Lerma argues, law is not pure reason, but "a form of perception and signification that allocates ways of sensing, feeling, being, moving, and thinking within a given society."[31] Within this frame, the discourses wielded by protestors across the political spectrum may be heard both as emotive expressions of silenced voice (captured in a vote) tied to (future iterations of) the law, and as *algarabía*, as we saw in the political debates over the status of the demonstrations with which I began the chapter.

Regardless of political position, then, a corporeal and sensed appropriation of *algarabía* pushes at the limits of democratic voice in the politics around secession. In a twisted embroiling of affect and sleight of ear, listeners on all sides who hear law as reason hear others' *algarabías* as *algarabías*, emphasis on *rabia* (rage), the lack of a sensible, intelligible ear. This is taken as a defense of their cause, rooted in a colonial aurality through which the other side must be heard as having a faulty ear, to be incapable, colloquially, of "listening to reason." What I call *sleight of ear*, through which a vocal affirmation of authority hears the ear of others as flawed, is thus at the center of this alternation between the sensation of democracy and the lettered recognition of democracy; the tension between the two plays out in the public performance of protest and the aurality of public space it produces in a mediatic public sphere.

It is here that the term *biomedia* may be useful for linking the affective, the mediatic, and the legal. As Eugene Thacker has argued, at the intersection of new media and biology, biomedia "are neither quite 'things' nor 'actions' but a process of mediation that enables 'life itself' to appear as such."[32] For Thacker, digital media is instrumental in producing this image of life itself, and focusing on biomedia means critiquing metaphoricity to focus on the biological as information. I would argue, however, invoking Lisa Gitelman's suggestion that "new media are less points of epistemic rupture than they are socially embedded sites for the ongoing negotiation of meaning as such," that the aural imaginaries of protest also produce "life itself" as a fundamentally democratic concern, where democratic life is equated to voice as a sense of the body that can be transduced into law as a linguistic text.[33] That is to say, biomedia is not just a question of informatics. As Patricia Clough writes, there is an affective aspect to the media as well that directly impacts the body. In particular, mediatic affect stages democracy by producing the values of life and death that are heard into acts associated with lawfulness and its extreme opposite, terrorism: "It is through this intersection of biomedia and new media, the biomolecular body and embodiment, that the flows of a political economy of affect are traversing, while they draw on and more profoundly enable a biopolitics of state racism and counter/terrorism that is reconstituting the norms of living and living as a human subject."[34] The mediatic production of the ear draws everyday sounds into an affective production of geographies that reaffirm or contest certain political positions or social categories such as race, gender, or ethnic and national identities, as they are associated with the sounds of voice.

Sara Ahmed, for her part, thus argues that affective economies produce ideologies of hate that differentiate some bodies from others, a process that never ends because the fearful subject awaits the arrival of other bodies, whether asylum seekers or terrorists, as part of the same threat to the nation: "The bodies of others are hence transformed into 'the hated' through a discourse of pain."[35] At the same time, she has argued the counterside to this production of immigration or alterity as a threat: that the "happy" narratives aligned with national identity often close off the possibility of feeling otherwise. "We become alienated—out of line with an affective community—when we do not experience pleasure from proximity to objects that are already attributed as being good."[36] When faced with an association of a moral end associated with goodness (which will make us happy), the feeling of disappointment that might emerge from not feeling happy "can lead to a rage directed toward those that promised us happiness through the elevation of this or that object as being good. We become strangers, or affect aliens, in those moments."[37] The media ear is a site of mediation for the production and recirculation of felt imaginaries of hate, fear, happiness, and solidarity associated with both the sounds of protest and, here, the sounds of nation (which, I have argued throughout this book, often circulate through the sound of language and accent as sensations rather than just semantic meaning).

In this context, we may consider the series of annual Diada protests in Barcelona that set the stage for the 2017 referendum. Despite the fact that for years the Spanish central government had refused to authorize such a referendum, Catalonia planned and executed a vote deemed unconstitutional by the Spanish Constitutional Court, in which more than

1 million people participated.[38] Prior to the event, organizations such as Òmnium Cultural, which began in 1961 as a group dedicated to preserving the Catalan language under Franco, and Assemblea Nacional Catalana (ANC), a group founded in 2012 after the *indignados* protests in Madrid made headlines around the world, regularly sounded bodies in the streets to create new acoustic imaginaries of democracy, seemingly as a way of pushing the boundaries of what is legitimized by Spain's Constitution. In so doing, they made their sense of loss heard not through legal recourse but through sonic occupations of space that were meant to mediatically signify a unified national identity other than the one delivered by the state. As I showed in the opening to this chapter, by striving to differentiate emotion—as Catalan "fellow feeling"—from noise, these demonstrations present national identity as a choral voice that, in an echo of Ernest Renan, expresses Catalonia's shared will to become a new European state by occupying its territory with an aural performance of national spirit. These demonstrations politically frame democratic voice as a corporeal expression and an occupation of public space in which the spatial expanse of voice reveals the limits of the silent/silencing modern democratic system that has become a Spanish geography out of sync with its Catalan territory. By way of comparison, at the same time, unionist protestors took to the streets of Madrid with Spanish flags and sang the *franquista* "Cara al sol" in opposition to independence, expressing a sense of rage at the potential loss of a unified Spain (and openly advocating for dictatorship in the process).[39] Both involve personalized senses of community that hear gibberish in the voices of others, that hear their voices and emotions as unintelligible. But the contrasts are not as simple as left versus right, Catalan versus Spanish. Support for, and rejection of, Catalan independence comes both from conservative and anarchist or left-wing groups. In just one example, the conservative secessionist ANC began demonstrating shortly after the left-wing *indignados* first made massive street occupations popular, adopting their strategies as they sought to occupy Barcelona annually on the Diada. As Raphael Minder has put it, "For left-wing militants, secessionism looked suspiciously like an attempt by the Catalan establishment . . . to reclaim its political leadership. For the secessionists, however, a left-wing movement that had originated in a Madrid square 'looked like the Trojan horse sent by Madrid to destroy all proindependence claims.'"[40] In other words, very few could hear the sounds of collective voice as anything other than what I termed in the Introduction *sleight of ear*: the use of the sound of voice as a political tool.

In the Diada movements that are my focus in this section, the production of protest as a happy experience, one that aligns with national identity as a shared good, is often evident. One of the earliest examples of this produced sound of affective nationalism was the video "lip dub for independence" performed in the town of Vic by over 5,700 people in 2010.[41] Instead of chanting or singing in their own voices, the group joyfully performed an exquisitely choreographed, ventriloquized choral performance that overtook the streets of the city center. In one long traveling shot through the city, participants dressed in all forms of traditional Catalan regalia, danced the *sardana*, built *castells* (human towers), and waved the *senyera* and *estelada* flags as they lip-synced the song "La flama" by Obrint Pas. The lyrics alone showed how the song was meant as a call to engage in the proindependence cause, but to do so cheerfully:

> Baixa al carrer i participa.
> No podran res davant d'un poble
> unit, alegre i combatiu
> . . . Viure mantenint viva
> la flama a través dels temps
> la flama de tot un poble
> en moviment

> (Go down to the street and participate.
> They can't do anything against a people
> united, happy and combative
> . . . To live keeping alive
> the flame throughout time
> The flame of a people
> in movement)

The song's idea of a positive occupation of the streets was reflected in the group dances of the *sardana* and the *ball de bastons* (Catalan stick dance), the visual representation of the *gralles* (Catalan wind instruments) playing, and devils—a common figure in Catalan celebrations—running through the crowd. The choreography ordered what could have been taken as an unintelligible *algarabía* of decentralized protest into a happy movement through Catalan space that, in its organization, echoed the protofascist *noucentista* aesthetic of a hundred years earlier, which valued eurythmics and other group performances of identity as a show of national superiority.

The single traveling shot suggested that, although the event was staged, the community was so well organized that no one was out of step, and performing a lip sync meant everyone "sang" in tune. Although the community's voices were ventriloquized, the thousands of participants' ability to "sing along" in their own language rejected the Spanish state's insistence on the continued use of Castilian in court systems and other legal venues, drowning out the language of central Spanish law with the now-systematized sound of Catalonia's collective language.[42] The crowded streets of Vic, filled with singers who were all in sync with each other, further suggested that Spain misheard as unintelligible a language that was a unifying and orderly voice of its people.

Larger manifestations centered in Barcelona and taking place on the Diada began in 2012 and replicated this shift from voice as representation to voice as occupation, conceived as part of a longer process of mobilization toward independence, known as *el procés* (the process). The Via Catalana in 2013 and the V in 2014 are two examples. In the first, an estimated 1.6 million people formed a human chain that stretched four hundred kilometers, symbolizing the path toward Catalan independence. The event began with the ringing of bells throughout various jurisdictions and included a coordinated singing of *Els segadors*, the Catalan national anthem, at an appointed hour. Although *El segadors* is, tonally, a discordant and somewhat serious, almost mournful song, its lyrics affirm both a future realization of victory, and defense of territory as a continued struggle that will make Catalonia's enemies tremble:

> Catalunya triomfant
> Tornarà a ser rica i plena.
> . . . Bon cop de falç,
> Defensors de la terra,
> . . . Que tremoli l'enemic
> En veient la nostra ensenya.

> (Triumphant Catalonia
> Will be rich and full again.
> . . . A good strike of the sickle,
> Defenders of the land,
> . . . Let the enemy tremble
> Seeing our banner.)

By singing together, the unity of voice in space created a collective shaping of community grounded in a four-hundred-kilometer-long territory, and returned the sonic occupation of space to several national memories in the process: the seventeenth-century Catalan Revolt to which the song refers; the nineteenth-century composition of the song, based on oral tradition, as part of Catalonia's political and cultural consolidation as a modern nation; the memory of Franco's dictatorship when the song was banned; and its reinstatement in 1993, once democracy and neoliberal development had been achieved. Alain Corbin has argued that in Europe, the ringing of bells historically signaled the inner and outer limits of a community: only those who fell within the acoustic range of the bells were considered part of a town's jurisdiction, forming a "notion of territoriality, one obsessed with mutual acquaintances."[43] In this case, the coordinated ringing and singing across communities signifies an expansion of collective voice across space, an expansion that presents itself as not relying on radio waves or technological artifacts to geographically extend resistance and challenge the state's messaging. Instead, the spectacle pushes the line between language and its acoustic, spatial performance into bodies in contact, in a way that preserves the idea of orderly voice even as it extends the silent voice of the ballot box into a spatial occupation not just of the city but of all of Catalonia. At the same time, rather than an underground occupation, such as those I describe below, this was a massive spectacle of national identity broadcast across all major networks, as the biomediated sound of shared subjectivity extended outside the territorial bounds of Catalonia. Similarly, the V the following year was a physical arrangement of bodies in the streets in the shape of a V meant to represent *via, voluntat, votar i victoria* (path, will, voting, and victory), embodying across the city the alliterative sound of a phrase geared toward solidarity and democracy that would one day end at the ballot box.

With the Via Catalana and the V, this sounding of song across space made voice into a linguistic and sung geographical occupation of territory, on a day that echoed Catalonia's defeat, but was now associated with a future triumph. Having been staged, though, the performance of emotion also carried out a process of what Patrick Eisenlohr calls *entextualization*, through which sound loses its spontaneous quality and "acquires textual qualities—that is, it becomes a relatively bounded, recognizable, and replicable chunk of discourse we call text."[44] Drawing on tropes of oppression as well as mutually understood rules about what lines governments can or should not cross, the singing of the national anthem "[exhibited] accumulated traces of the social relationships leading to its production while also indicating the anticipated moral and political futures that the ongoing process of textuality is directed to."[45] With access to mainstream media and a hierarchical structure behind the organization of a specific political agenda, moreover, organizers were able to contrast this massive occupation of space with the *ràbia* of the smaller occupations and the images of violent evictions of encampments by police that accompanied the *indignados* only a couple of years earlier. Both the *indignados* and pro-Catalan independence supporters felt dispossessed by Spanish democracy. But the intertwining of *bios* and media shows how Catalanist desires for independence, hot on the heels of the lived experiences of dispossession occasioned by the economic collapse of 2007, increasingly presented Catalonia as a victim of an abusive Spanish state by using acoustic occupation to counteract the letter of the Spanish law even as it upheld the same epistemological frame valuing reason and text as order over noisy sound.

The circulation of this territorial sounding to a mediatic public sphere, moreover, extended the protests outward in both time and space, through longitudinal extension and temporal dispersal. The Via Catalana and the V ended after a day, but organized videos in Catalan and English appeared on YouTube channels that supported the independence movement. An official video about the Via Catalana posted on the Assemblea YouTube platform, for example, presented high-resolution aerial shots and footage from Barcelona, Girona, and other places along the route, meant to demonstrate the force of what the video claimed, in subtitles, to be 2 million participants.[46] The video celebrates the effective organization of the event and includes images of people along the route holding hands, babies smiling, and children singing. It also features interviews in English, German, and French with non-Catalan participants, smiling, dressed in yellow (the color of the day), matter-of-factly explaining their peaceful request for Catalans to have the right to vote in a referendum on independence so as to become "a free and sovereign country." These stagings of democratic voice thus play toward the acoustic bias of sound as presence, even as they present the acoustics of that space as multilingual, inscribing Catalan into a polyglot Europe in which Spanish is not dominant. Still, they do not replace the silent voice on which the state depends for its authority. They instead become a sort of felt surplus of voice whose happy noise supplements—and reproduces—a legitimatized sense of democratic meaning as law.

The tension between the senses and reasonings of democracy clashed violently in 2017, however. The central government had prohibited a vote on independence, but the proindependence movement elected to hold the referendum anyway. In the months prior to October, Catalan officials secretly compiled a computerized voter registry, complete with addresses and Documentación Nacional de Identidad (DNI) numbers. Hackers—first in Spain and Europe, and then in Russia and Central Asia—cloned the voting websites onto foreign servers that would allow them to replicate the voting mechanisms used by the Spanish state without disruption from European authorities. With the support of the Generalitat, civilians hid paper ballots printed in France and smuggled them over the border into private homes so they could be used on the day of the referendum. On the morning of the vote, Spanish police arrived at certain polling stations and, *Requerimiento*-style, announced their intention to take the ballot boxes by force if they were not handed over peacefully. When they were not, in scenes that echoed the Revolta de les Quintes I described in the last chapter, the Setmana Tràgica, the *Cu-Cut!* incident, and other moments associated with Spanish military cruelty in Catalonia, the police reacted with violence. Images and videos of police beatings circulated across the internet. In an example of the geographical complexity of biomediation, *El País* reported that these were disseminated by 4,800 Russian bots that had also pushed out incendiary material both before and after the referendum.

In response, demonstrators abandoned the peaceful performance of voice and turned to banging pots and pans in protest, invoking the long-standing tradition of the *cassolada*, known in Castilian by different names, such as *cacerolada* or *cacerolazo*. Noisy protests such as these use sound as a prosthesis for voice, reaching into the physical depths of the ear through volume and timbre when legal arguments are ineffective. These noisy protests overflow the physical occupation of public spaces: protestors' presence acoustically turns corners, extending the reach of voice beyond what the eyes of those who are physically present can see.[47] By rejecting language entirely, these collective stagings of proindependence sentiment go a step further to uphold an imagined notion of democracy in which the sound of a collective body challenges the silent, legal mechanism of voice.[48] This acoustic, languageless transmission signals the community's unity otherwise than through the letter of the law or the written vote.

Yet, by staging the vote despite its being forbidden by the Spanish Constitution, the Catalan government also revealed their own echo of a colonial aurality, as they both heard Spanish law as unintelligible and asserted a need to replicate the silent structure of voting so that within a European frame (in which the silent election is privileged as law), the collective acoustics forged through massive demonstrations could be legitimated on paper. In other words, if dispossession resides in one's unintelligibility to the ear of the law, the Catalan independence cause heard itself as both dispossessed by Spain and capable of making itself heard because it had possession of alternative, yet still institutionalized, mechanisms of power—economic, networked, mediatic, and political—similar to those the Spanish state had.

The independence movement also directed its efforts to an outside ear, appealing to online publics that could not legally affect the outcome of the vote, and harnessing other affects that portrayed Catalans as what Ahmed called *affect aliens*, where affect and democracy are aligned. Following the violence with which Spain reacted to the referendum, the civic group Òmnium Cultural, headed by Jordi Cuixart, put out a video titled "Help Catalonia. Save Europe," which used the rhetoric of loss to appeal to the idea of a shared system of democratic values that had sustained the 2012 Diada years before. Produced in English, the video features a woman on the verge of tears discussing the October 1 events. She opens by saying, "Freedom, democracy and respect for human rights. These are Europe's founding values. Values that were achieved at great cost. Values that are under attack right now. Catalonia has always been at the forefront of Europe's values. Our country has always cared about social rights."[49] The gesture suggests that Catalonia is out of place in Spain, which is itself assaulting Europe when it assaults the Catalan people. Appropriating the foreignness associated with *algarabía*, then, the video asserts that the police forces that beat voters on the day of the referendum were akin to an invading army. The line "What crime had they committed? Going out to vote" successfully elides the embattled discussion over the constitutionality of the referendum and reinforces the idea that democracy depends on a geographically defined sound/spirit of a people rather than a legal construct to define its boundaries. But most importantly, the tearful wavering of the woman's voice suggests the sound of emotion is itself evidence of oppression. Unable to maintain the reasoned approach that had defined the happy, peaceful movements of years before, the overflow of emotion, bound up in the audible imaginary of government violence, becomes a threat to all of Europe.[50] Appealing to an outside audience, the video hears democratic self-governance as a human right, under threat of oppression, that justifies unconstitutional reenactments of democratic scenarios like the October 1 (1-O) vote.

The volumes and tones, the nuances of voice that make it neither a presence nor a lack, as Dolar claims, and that shade into everything from shouts to aphonia, tearful gulps to righteous indignation, moreover, can produce legal consequences for voices that use that physical sound as evidence of the state's lack of reason. Less than two weeks before the

October 2017 vote, for example, in an echo of the *Cu-Cut!* incident a century earlier, without a judge's order, the Guardia Civil (Spanish police) entered the Conselleria d'Economia of the Generalitat's Department of Economics and Finance, where they arrested fourteen people on charges of sedition for organizing the referendum, for, in essence, supplanting Spanish law with a different Master's Voice, a Catalan one. The actions provoked a spontaneous, day-long protest on the streets that grew to thirty-five thousand people. On September 21, the same groups who had produced proindependence spectacles, including the Via Catalana and the V, showed up to advocate for the Catalan officials to be released. Jordi Sànchez, the president of the Assemblea Nacional Catalana, and Jordi Cuixart, president of the Òmnium Cultural, led the protests. A report in *El País* described the voices of the two leaders:

> Jordi Sànchez, president de l'Assemblea Nacional Catalana (ANC), molt afònic, ha proclamat que "serem aquí . . . fins que tots surtin en llibertat. No són delinqüents." . . . Per la seva banda, el president d'Òmnium Cultural, Jordi Cuixart, també amb un fil de veu, ha afegit que "avui, ahir i els dies que faci falta el poble de Catalunya està parant un cop d'Estat. No en volem saber res més. No tenim por."

> (Jordi Sànchez, president of the Catalan National Assembly [ANC], very aphonic, has proclaimed that "we will be here . . . until everyone is released. They are not criminals." . . . For his part, the president of the Òmnium Cultural, Jordi Cuixart, also with a thread of a voice, added that "today, yesterday and however many days it takes the people of Catalonia are preventing a coup d'Etat. . . . We are not afraid.")[51]

Several months later, in March 2018, another article appeared in *El País* with the title "Jordi Cuixart: parlar fins a l'afonia" (Jordi Cuixart: talking to the point of aphonia).[52]

Known colloquially as "the two Jordis," a phrase that doubles the independence movement into a mirror image of power and sacrifice, both are described here as "aphonic" and "with only a thread of a voice." In this portrayal of the resistance of Catalan protestors, the articles suggest that the Jordis put their vocal cords to such strenuous use fighting for their democratic rights that they verged on silence. On one hand, saying the two Jordis are aphonic suggests that they were hoarse due to so much physical exhaustion of the larynx after shouting against the central government, especially the day before, as they tried to calm the crowds during twelve hours of tensions between the Policía Nacional and the Catalan populace. Yet, the speakers' near absence of voice also draws the ear of the reader to an aural imaginary in which voice and body are united in a wearing down of the oral sound of voice toward the silence of the law.[53] The aphonic voice is thus a sort of erosion, or wear and tear of the acoustic body, trailing temporarily into near silence, without even reaching it fully. Aphonia turns *zoe*, as an archival record of meaning and law as life, into a corporeal presence that is wearing itself out. The sound of a near-absent voice is not silence, then, but the vocal movement of the speaker toward a limit of dispossession that would be the death of the democratic body, produced by the state. As Ochoa Gautier writes, "The ephemeral nature of sound is supposed to be one of its defining qualities, but when sonic perceptions are troubling, or perceived as unwanted, then sound becomes endlessly unbearable, materialized on the body as a sign of the limits of listening as a dialogic practice."[54] Indeed, after the failed referendum, leaders of the independence movement began demanding dialogue with the central government, which Spain refused. Eventually the state would arrest both Jordis for sedition. Yet the image that was used by the courts to justify the sedition, paradoxically, was that of the Jordis standing on a police car with a megaphone, trying to persuade the crowds to leave, precisely so that violence did not occur. As the documentary *20-S*, directed by Jaume Roures, illustrated, the photographic image of "undemocratic voice" used to cement the legal case leaves out the longer, daylong negotiations and attempts by the accused to prevent a violent protest, reverting instead to an aural imaginary that hears incitement as angry sentience with a power to change the structures of life itself because it defies reason. When the government refused to speak more about independence, the Jordis and others who were exiled or jailed could also be heard as martyrs for the failed cause of independence and the undemocratic posture of the Spanish state, whose own sleight of ear had converted the image of the megaphone into an image of sedition.

At the same time, this ambiguity manifested in the aural erosion of the body opens the voice up to more interpretation by the listener than the dictatorial voice described by Dolar does, wherein the Master's Voice easily supplants law.[55] In the distorted reverberations of the Jordis' voice, or even in hearing protest as *algarabía*, the sound of voice is activated as a subject. The anti-independence listener, following the law as it is written, can hear into the Jordis' aphonia a threat to the lettered system of authority, while a separatist who hears voice as acoustic evidence of life itself can hear martyrdom wrought on the body. In either case, the letter of the sovereign law is challenged when anyone watching the footage, or present in person, hears the acoustic form of their voice as eroded by law. This sound of erosion may not have enough lettered support behind it to legitimately supplant law, but the spectacle of it circulated in the media draws attention to the sound of dispossession as a lived experience, even if that dispossession may itself be a performance.

In these various scenes of democratic protest, we become aware that the aural imaginary of democratic voice functions like Ernest Gellner's construct of nationalism, or, as quoted here, David Panagia's understanding of democratic politics: "We think that contrarieties cannot coexist unless they are made compatible; yet democratic politics perseveres in its insistence that any two or more people, groups, images, identities, subjectivities, and so forth, can persist simultaneously. One could go so far as to suggest that heterology is the ontological condition of democratic politics."[56] Although deep rifts exist between how different groups hear democracy's ideal manifestation—as a silent, orderly voting process, or as a collective voice of spatial occupation—this very rift perpetuates an echoic memory of reasoned voice as the foundation of democratic community. And unlike the Master's Voice that Dolar signals as an always-imposed voice from above, in a globalized context in which national identity must play not only to those within a nation, but to those outside it, here the epistemology of the democratic ear, and the sleight of ear it performs to maintain itself, is so widespread as to be occupiable by all sides. Precisely because protests strike the ear by interrupting the conspiracy of silence that surrounds the democratic registering and counting of voice, they challenge the silent letter of the law and draw it back into its own (often violent) authoritative function, in which nonconformity is heard as a sign of a deaf ear that must be arrested or otherwise physically reformed. This hearing is not necessarily one sided: the violence between protestors and police often plays out in protests as a result of this limit of sound as a dialogic practice, evincing how far the violence of the modern/colonial ear has spread beyond its original territorial confines. In politics, there is a mutual provocation that rests on each political position hearing the opposing side's voice as possessing a faulty ear. Simultaneously, the voice that is strained by its own erosion acquires power by drawing the listening ear into hearing the appeal to a silent vote as a political strategy, a sleight of ear carried out by the other side. There is a danger, however, to hearing these erosions of voice as "just politics." For there are still bodies present in these sentient sounds. More than that, to state the obvious, there are communities affected by these aural imaginaries of voice and the sleight of ear that produces them.

Given the mediatic and other resources available to the Catalan independence movement, it would be naïve to consider it a movement of minority resistance against an oppressive central government; supported by the Catalan government, it *is* the public sphere in Catalonia, even though independentism includes anarchist and marginal left-wing participants who would resist being included in that descriptor. Even as a construct of voice as a European democratic sound has accompanied the public demonstrations for and against independence, the daily aurality of a globalized city like Barcelona—the historical inequities linked to language, the role of accent, the staged and restaged scenarios that play out an echoic memory of colonial power—have complicated how publics and counterpublics hear their relationship to dispossession. This is a networked reality, what we might call intersectional and biomediated.

SOUNDS OF OKUPATION

Whether biomediation works metaphorically, affectively, or informatically to produce experiences of democracy felt as rational, there are economic realities that impact the living body, through the precariousness of dispossession, in ways that the mediated ear can only partially reflect. Barcelona, in point of fact, has a long history of protest in which sound is activated to draw attention to those dispossessions. When Franco established the Ministry of Housing in 1957, it operated under the principle that "we don't want a Spain of proletariats, we want a Spain of property owners."[57] The result was the construction of multiple neighborhoods like Besòs and Sant Idelfons, which drew immigrants to the area.[58] After Franco's death, with Barcelona going up for full EC (European Community) status, some of the places in which these communities thrived—the *barraques* of Montjuïc, for example—were razed to make way for Olympic venues and tourist attractions.

The people who occupied these marginal towns prior to the Olympics resonate in the voices and sonic occupations of activists who, since the 1980s, have consistently protested gentrification, which has become the purview not of homeowners, but of multinational development companies that partner with the government to rebuild the city for tourism and economic growth. As I mentioned in the last chapter, in the 1980s young people associated with the punk movement began squatting houses as a means of protesting government control and social norms, and in 1985 several of them formed the Colectivo Squat Barcelona. After the Olympics, the '90s inaugurated another round of *okupaciones* which would eventually network via the internet with antiglobalization movements around the world, such as the *zapatistas* in Chiapas, Mexico, and the World Bank protestors in Seattle, not to mention Barcelona.[59] In the '00s, as interest rates dropped and people found themselves in increasingly precarious positions with respect to housing, squatting became a more widespread means of intentionally making the scarcity of housing public by noisily occupying properties and staging loud protests in the neighborhoods where squatters took up residence. Early manifestations of the specific relationship between the worldwide housing bubble and the *okupa* approach to protest in Barcelona took the form of *acampades* (encampments) associated with the V de Vivienda movement, which began in 2006 and was led by young people frustrated that, as their slogan stated, "no vas a tener una casa en la puta vida" (you'll never own a house in your fucking life.)[60] The sotto voce rhythms of the free radio movement occupying the airwaves, and its public punk expressions I analyzed in the last chapter, with their explicit ties to anarchist ideology and resistance to any form of capitalist or hierarchal political system, thus fused with transnational networks of resistance to globalized capitalism whose members occupied public spaces with intentionally noisy, shouting voices or loud concerts held in violation of noise ordinances. In some cases, they staged marches down public avenues. In 2008, for instance, the V de Vivienda movement carried out street protests in dozens of cities across Spain. In other instances, as when I was visiting in 2019 during the #AbdelahNoSenVa campaign, announced on social media, squatters from around the city came to rally around a family on the verge of eviction, occupying the streets around the property. Many times, such movements end in violent and loud scuffles with police or riots where screams, the sounds of batons hitting protestors, and sirens punctuate the public space. As Stephen Vilaseca has pointed out, there is a distinction between *ocupes*, spelled with a <c>, who squat due to need, and the *okupes*, written with a <k>, who squat as an act of protest, usually against capitalism, globalization, or the government.[61] Many of the *okupacions* in the 1980s and 1990s were considered more performative, in line with the punk antisystem ethos. In recent years, however, squatting has also been associated with the literal dispossession of subjects from their homes in the wake of the financial disparities wrought by globalization and other structural problems; former mayor Ada

Colau, in fact, got her start in politics through her activism around housing. Despite being around for decades, though, the movement became perhaps best known globally in 2011, when antiausterity occupations broke out in Madrid and reached all the way to Wall Street.

The acoustic strategies employed by the Catalan independentists, in point of fact, echoed those utilized just a few years earlier in the major protests that took place during the housing crisis and the ¡DemocraciaRealYa! movements of 2011, even though politically their goals were quite different. (The 2006 V for Vivienda and 2014 V for *via, voluntat, votar i victoria* cry out for a comparative, close reading of their use of sound and space, one I cannot engage in here.) The *indignados* movement in Madrid was perhaps the best-known example of the antiausterity protests that took place in Spain, but Barcelona's history of *okupa* movements had already created a culture of acoustic occupation, first by being associated with and advertised by Ràdio PICA and its punk broadcasts, and later by using noise to signal disagreement with, and resistance to, gentrification around the city. Before Madrid made international headlines, more than five hundred people squatted the central Plaça de Catalunya beginning on May 15, 2011; after a violent eviction by police on May 27 failed to fully dislodge the camp, they finally left in early June.[62] Thousands returned on June 19 of the same year to show they were still in solidarity with the movement. In Madrid, it was not until May 17 that several participants who had previously squatted buildings as a means of protest decided to marshal social media to encourage people to spend the night in the Puerta del Sol; several hundred did. Through social media, the protests spread to the United States, culminating in the Occupy Wall Street occupation of Zuccotti Park.

In these movements, both in Spain and in Zuccotti Park, the sentient, acoustic body became especially important not only for creating community, but also for communication, as Zeynep Tufekci has explained in her book *Twitter and Tear Gas*. As she shows, the networked public sphere that is at stake in protest today is not merely about the power of online mechanisms, as some may think: "Rather, it's a recognition that the whole public sphere, as well as the whole way movements operate, has been reconfigured by digital technologies, and that this reconfiguration holds true whether one is analyzing an online, offline, or combined instantiation of the public sphere or social movement action."[63] In Zuccotti Park, protestors were denied permits to use bullhorns and loudspeakers due to sound ordinances, so protestors turned to other modes of occupying voice and space, such as the "human microphone," in which people repeated, in a chain, the messages from the back of the crowd to the front. In Spain, the act of occupying public spaces with bodies and voices drew attention to a housing crisis that was quite literally putting citizens on the streets.[64]

The *okupes*' cause is not just about housing, capitalism, or globalization, then; it is also about information and a rejection of the biomediation that takes place through what they view as systemic control of the media linked to global development strategies. Prior to the internet, squatters in Europe depended on bulletin boards and flyers to circulate information; thus arose newsletters such as *InfoUsurpa* and *ContraInfo*, which consolidated events and news through alternate sources to the mainstream media. At times, they could count on free radio stations like Ràdio PICA, Radio Bronka, or Contrabanda FM to share information about the cause.[65] Later, Indymedia and other global purveyors of alterative information hit the web; the online platform for *InfoUsurpa* continues to be a primary consolidator of weekly events, organized by country and city. Another widely used international site is squat!net. When I was there in Catalonia in 2018, the primary site for information about the Okupa movement was okuparBarcelona (okupesbcn .squat.net/medias.html), which held true to the old norms of counterinformation; in fact, it cited Ràdio PICA, Contrabanda, and Ràdio 4 as inspiration. By 2023 when I was editing this book, however, that website had been replaced by an orderly and rather bureaucratic page, oficinaokupacio.com, which offered juridical and legal advice as well as practical how-to guides on how to make decent cement and other tricks of the renovation trade, conforming in form if not in content with a rather neoliberal aesthetic that, somehow, seems to have lost the ludic tone early squats once had.

Through these globalized means of dissemination, the antisystem protest has expanded its reach, but ideologically the *okupes* perform a scenario of resistance that goes back over a century. As Miguel Martínez-López and Angel García Bernardos have suggested, what unifies the heterogenous groups of varying ages who participated in the antiausterity, *indignados* and Occupy movements in 2011 was the precariousness of life—*la precariedad de la vida*—which is associated principally with the difficulties stemming from a lack of stable work or sufficient pension, undocumented status that cuts short immigrants' ability to find sustainable work, or lack of accommodations for people with health limitations or who are not able bodied, among other concerns.[66] The precariousness of life thus has to do with the state's exclusion of certain subjects from the social mechanisms that, according to Foucault, are meant to sustain society's right to life but do so by treating some people as disposable. As he writes, even within socialist mechanisms that suppose the government's job is to preserve an entire population, there is always present a certain racism that allows some of a populace's subjects—and, we could add in the contemporary period, especially Indigenous peoples and undocumented migrants—to be more readily exposed to death so that others may live.[67]

While initially the gestures of the Barcelona punks were more ludic, intended to draw attention to the hypocrisies of the supposed freedoms granted by the turn to democracy after Franco, these later morphed into more direct political actions. The *okupes* thus formed communities of artists and activists who combine resistance with art. Squatting various buildings throughout the city, such as the Can Tunis factory or the Banc Expropiat in Gràcia, they turned abandoned spaces into communal living spaces or community centers that promote the ideals of the anti-institution, pro-community *okupa* counterculture. Such movements continue the work of the *ateneus*, such as the Ateneu Llibertari de Gràcia, which also offer a space for the circulation of counterinformation, often in the form of free book exchanges. Anarchist bookstores, such as El Lokal, in El Raval, or Aldarull, in Gràcia, perform the same function, while also creating an alternative *mattering map* of voices in the city.

The refrain that "Carcelona" is a city of death or incarceration reverberates in many of the ideas that circulate in the *okupa* community, as necropolitics and biopolitics become intertwined in a cultural and artistic dispossession. Punk is the inspiration for okupation, but today the hip-hop group Malamara performs at squatted buildings, street corners, and on metros; their song "Livin' in Carcelona" was part of the soundtrack for a 2013 documentary about police brutality against the *okupes* called *Ciutat morta*.[68] The group also produced a music video of the song, available online. Full of fast-paced rhythms and wordplay, the song turns the immigrant *algarabía* of El Raval—"una mezcla de idiomas, negro africano, franco marroquí, chileno, peruano, pakistaní, el que existe sin lugar" (a mix of languages, black African, French Moroccan, Chilean, Peruvian, Pakistani, the one who exists without a place)—into a series of sharpened barbs that strike the ear, as they rap that in El Raval, "nadie me ataca" (nobody attacks me), but the police constantly "nos atraca" (hold us up). Peppered with images of people who have lost eyes to tear gas and police beatings, in part of the video, they sing "ojo con tu ojo" (be careful with your eye) and denounce the "mucho abuso de poder" (much abuse of power) that protects businesses, banks, and tourists, rather than the local community. Another song, "Sin permiso" (Without permission), not only claims the music they perform on the metro is revolutionary, it refers to the notion of dispossession when they announce they are "interviniendo en el espacio público como derecho fundamental" (intervening in a public space as a fundamental right) because "el arte está en la calle, si no la calle no es" (art is on the streets, if not, the street does not exist). Consciously occupying the space of the bothersome beggar who must be cleaned up like so much trash so the tourists are not uncomfortable, in the video they make these claims as they walk through the metro holding out their hands for loose change. As they do so, they claim a right to the city, with walking and rapping as acoustic affirmations of their presence and humanity. In this performance of voice, the sentient body is democratic because it is aurally present in community, not because it has a place in the silent voting system.

Writing in the wake of the massive antiausterity protests that began in Greece in 2010, Judith Butler and Athena Athanasiou debate the relationship between dispossession as a material and legal neoliberal/colonial practice of occupation and dispossession as a possible form of resistance: "The logic of appropriation and dispossession, whether it be colonial or neocolonial, capitalist, and neoliberal, endures by reproducing a metaphysics of presence in the form of the violence inherent in improper, expropriated, and dispossessed subjectivities. . . . Taking cue from Derrida's notion of 'ontopology,' which links the ontological value of being to a certain determined *topos*, locality or territory, we might track the ways in which dispossession carries within it regulatory practices related to the conditions of situatedness, displacement, and emplacement, practices that produce and constrain human intelligibility."[69] As an *ontopology*, the sonic mattering maps that are created by protest are also sonic maps of voices overcome by dispossession, their sound converted into the "unintelligible" that is the other side to the silent letter of democratic voice. Athanasiou, in her writing in the book, goes on to state that this link between regulatory practices, identity, and the metaphysics of presence means the logic of dispossession is "interminably mapped onto our bodies . . . through situated practices of raciality, gender, sexuality, intimacy, able-bodiedness, economy, and citizenship," questions of identity that manifest in the sounds and practices of the body in space.[70] Butler and Athanasiou thus draw attention to the spatial practices that link not just capitalism, but all forms of oppression, to the body and its identity, and in our case, its sound.

From an acoustic perspective, this *human intelligibility* is informed not only by the dominant aural imaginary of language as place I explored in Chapter 2 but by an aurality that hears the performances of the *okupes* as threatening, as an *algarabía*, because they embody sound in ways that do not respect capitalist notions of private and public property. By squatting institutions associated with finance, like the Banc Expropiat in Gràcia, they reterritorialize spaces of private industry or (foreign) investment into community centers that counter the conspiracy of silence as an individual practice of voice. As Mike Goldsmith, Jonathan Sterne, and Karin Bjisterveld, among others, have shown, modern legal codes have almost always included noise ordinances, primarily in order to shield the upper class from underclass sounds they deemed unpleasant.[71] But when everyday, microlocal sound becomes a tool of protest, those who hear their own voices as dispossessed by the system are able to turn their body in space into a prosthesis for voice that reaches the ears of those in power. It may be heard as noise, but it is a constant noise that, like static on the line, interrupts the "happy" biomediation of the ear that circulates through neoliberal media and practices. As Stephen Vilaseca has argued, "Squatted houses and social centers are not spaces of belonging, but spaces of continually repeating encounters. They are spaces in which to share experiences and to participate in a 'doing with,' a 'commoning.' . . . In squatted social centers, new words, sounds, writings, images, and bodies drift and disconnect from the preestablished program of free market capitalism and consumption-based strategies of urban growth and design."[72] These groups set the possessive aspect of capitalism adrift by restlessly "commoning" both small and larger city spaces with sound.

Sonic presence thus becomes not about emplacement and reclaiming the lettered democratic voice, as it is in the Catalan independence movement, but about a globally inflected, transnational movement of squatters where antisystem protests in one place resonate with movements for justice in other places. In just one example, the Assemblea d'Okupes announced an event "in defense of the earth, water, and the climate," called Guardianes de l'Aigua, in June 2017. The event was held in support of the Standing Rock protests against the Dakota Access Pipeline in the United States. In addition to a direct-action performance called "Esperad lo inesperado" (Expect the Unexpected), it included musical performances by the South Sudanese artist collective Ana Taban, and Navajo hip-hop artist Nataanii Means, as well as a protest in front of the US consulate. Given the contention over water as just one of the economic and political points of tension between Catalonia and the rest of Spain, the event aligned with international support for a global environmental cause but also drew attention to the inadequacies of the Catalan government's focus on water distribution and taxation as an assertion of autonomy from Spain's central government.[73] Rejecting the idea that something so integral to human life—water—should be converted into a bargaining tool between local and central governments, controlled and distributed by the state or multinational industry, with music and multilingual voices, their occupation of space through performance echoed the protests for water rights in Bolivia prior to the election of Evo Morales, captured by the fictional 2010 Spanish film *También la lluvia*. In these moments, the people's rejection of the silent voice of the voting booth in a local space turns corners by calling attention to the transnational practices of dispossession that accompany the colonial ear.

BEING *POÉTICAMENTE INCORRECTOS*

There is an aurality to this precariousness of the living body, then, perhaps most recognizable in the conflictive sounds of *rabia* that take place in public moments of protest. Through concerts, marches, and poetry readings, the *okupes* move the territorial notion of possession into aural imaginaries of solidarity in which the communal overcomes the boundaries of language or nationality, as the sounds of their bodies move through the city.

In July 2018, less than a year after the 1-O vote, I had the opportunity to observe one of these occupations in person. I followed and recorded a protest by a group that has sought to display this struggle against the logic of possession through the staging of poetic happenings and direct-action protests around the city. Calling themselves the Bio-lentos, they regularly host open mic and poetry readings at the Ateneu Llibertari de Gràcia, titled "Sí a Poe," in which "recitar es escuchar" (reciting is listening). Aligned with the *okupes*, they have also performed several "Ratsodes contra la gentrificació" (Rhatsodes against Gentrification), at sites of major reconstruction in El Raval, Sant Andreu, and Gràcia. On July 28, they protested in Gràcia, referring to themselves as *poéticamente incorrectos* (poetically incorrect).

Announced via email, listservs, and on counterinformation websites, the *rhatsodes* (a play on *rhapsodes*) were intended to draw attention to the sites where speculators were buying out buildings, renting apartments to tourists, and displacing residents of Gràcia. The announcement declared, "La fórmula es sencilla, las calles y nuestros pasos, nuestras palabras, nuestros versos, desde ellas, unidas en una misma voz. . . . Se hace poesía al andar" (The formula is simple: the streets and our steps, our words, our verses, coming from them, united in one voice. . . . By walking, one makes poetry). The justification reformulated an old fairy tale about the Pied Piper of Hamelin:

> De consumidores hemos pasado a ser RATAS (así nos llaman cuando compran los edificios enteros, habitados). Ya no nos necesitan, ahora quieren echarnos de las grandes ciudades. Pero, si somos RATAS (desde luego, no somos Ratos, ni Ritas, ni tomamos la Ruta de Andorra o de Suiza), también sabemos organizarnos y morder.
>
> Ahora somos las ratas, convocadas por la flauta de los RAT-SODAS DE HAMELIN (CONTRA LA GENTRIFICACIÓN), las que salimos a la calle y la hacemos nuestra. No necesitemos gatos que nos representen.
>
> . . . Más que un acto, una fiesta donde la palabra pertenezca a quienes luchan, quienes sufren los conflictos, los desahucios, la gentrificación, etc.
>
> Recorreremos las zonas cero de la especulación y haremos nuestra la calle, deteniéndonos en esos espacios para usar la poesía como arma de denuncia.

> (From consumers we have become RATS [that's what they call us when they buy entire buildings, inhabited]. They don't need us anymore, now they want to throw us out of the big cities. But, if we are RATS [of course, we are not Ratos, nor Ritas, nor do we take the Route of Andorra or Switzerland], we also know how to organize and bite.
>
> Now we are the RATS, summoned by the flute of HAMELIN'S RHAT-SODES [AGAINST GENTRIFICATION], who take to the streets and make them our own. We don't need cats to represent us. . . .
>
> More than a ceremony, a party where the word belongs to those who fight, those who suffer conflicts, evictions, gentrification, etc.
>
> We will go through the ground zeros of speculation and make the street our own, stopping at those spaces to use poetry as a weapon of complaint.)[74]

In calling themselves *ratas* (rats), the poets tie themselves to the rhetoric of sanitation and cleanliness that, as Stephen Vilaseca has shown, is often linked to the city's tourist-friendly position, which privileges Western understandings of global economic possession—that is, what David Harvey calls the "'new' imperialism" of multinational, capitalist investment in urban property—over other models of community.[75] He argues that the slogan painted on the sides of garbage trucks and trashcans, *Barcelona neta* (Clean Barcelona), an updated version of the 1980s *Barcelona, posa't guapa* (Dress up, Barcelona) campaign that sought to woo foreign investors, is used to justify the eviction of *okupes* from public spaces during impromptu performances in public squares.[76] Like the punks before them, the *okupes* who resist Barcelona's inscription in the larger European apparatus thus seek to recuperate an echoic memory of noise as resistance. However, if initially resistance took the form of concerts and protests regarding military conscription or the European support of the United States in Nicaragua, today the capitalist transformation is felt directly in the gentrification and catering to tourists through elevated restaurant prices, Airbnb rentals, and other causes of economic dispossession that bring along with them the sounds of foreign voices, flamenco shows for tourists, and other acoustic incursions into local space.

In August 2017, before Les Rambles had suffered a terrorist attack that killed thirty-five people, many of whom were tourists, just one example of the antitourist sentiment that now thrives in antigentrification protests appeared on Ràdio PICA's Twitter wallpaper. The image showed dead tourists whose blood was spilled under a sign that said, with dark irony, *Sangría* (bloodletting/sangría wine). Although after the attack the poster seemed to be in bad taste, the homonymic joke it made nonetheless captured the local sentiment around tourism, despite the city's official support for it. A globalizing example of Jordi Costa's assertion that the counterculture has been usurped by the government (in this case through the selling of the city to tourists), the noise that allowed punks to occupy the city is usurped by foreigners, stripping noise of its rebelliousness and inscribing it within the movements of global capitalism; it is this possibility for making noise that the *poéticamente incorrectos* hoped to recoup.

The Bio-lentos' assertion that "se hace poesía al andar" (one makes poetry by walking) therefore suggests that there are different ways to occupy city space, with some being an abuse and others being a practice of microresistance. Echoing Certeau's notion that one can remake the city by walking it, their gesture uses the movement of unauthorized, embodied voice through public spaces to challenge the government control of legitimacy through their use of sound ordinances. On the other hand, by declaring themselves *poéticamente incorrectos*, the protestors affirm their rejection of the normative aural imaginary of poetry and voice as private encounters that support a conspiracy of silence. Theirs is a poetry in the tradition of Chilean poet Nicanor Parra, or, perhaps more aptly, Raúl Zurita, an *antipoesía* that begins with a recognition of its own dispossession and seems intent on making a physical mark on the very territory from which its voice is excluded.

The protest I witnessed began at the Ateneu Llibertari and ended at the *casa okupada* (squatted house) Ca La Trava. Over the course of several hours, the group of about thirty protestors made several stops, where they read poems centered on the dispossession caused by gentrification. In between they walked casually from point to point, carrying a banner and playing music that invoked Gràcia's long history of resistance to the state (described in more detail below).

Importantly, the group occupied public spaces without asking permission to do so; just before they began marching, the *mossos* drove by, causing several protestors to walk away so as not to draw attention to themselves. But once on the route, at each stop, the leader of the group would recite the Rhatsodes' rallying cry against gentrification printed on their poster, and then several poets would read their works. At the Plaça de la Vierreina, a man who identified as being from Germany, wearing a black suit jacket and tie over his shorts, read a poem titled "El inversor" (The investor):

> Todos lo saben, la ciudad está en venta: City for Sale
> Y este barrio será mío
> Ha llegado el momento de
> Estas cuatro calles apestosas
> No se trata de cortar cabezas, sino su modo de vida . . .
> Simplemente separar familias
> Hasta que se vayan a otro barrio . . .
> Todos lo saben
> City for sale
>
> (Everyone knows, the city is for sale: City for Sale
> And this neighborhood will be mine
> The moment has come
> For these four stinking streets
> It is not a question of cutting off heads
> But their way of life
> Just separate families
> Until they go to another neighborhood
> Everyone knows it
> City for sale)

As he pronounced the refrain "City for Sale!" he adopted the voice of a circus barker, his turn to English an aural cue that investment came from outside Spain, the levity in the cry a tonal contrast to the more subdued reality of the dispossessed also voiced in the poem. At the same time, by identifying as an outsider who spoke accented Castilian, he laid claim to a repossession of the city that hears presence not as a simple local/other form of alterity, but as an ideological gesture that crosses national boundaries in the name of social justice. The Plaça de la Virreina has a large playground in the center, so the sounds of children playing, chatting, and screaming punctuated the poetry reading.

Another duo performed a dramatic reading of a poem they called the "Balada de antidisturbios" (Ballad of the Riot Police). In the poem, they echoed the voice of "civilization" circulated on public posters to show what was at stake in the conflict over government control of housing and other city spaces. In their performance, this voice was punctuated not only by the readers (as they pantomimed actions suggesting the violence of the state), but by the sounds of the plaza. In a particularly pregnant moment, the voice of the law they articulated in their poem met a response from a baby:

una palabra ley
La letra con sangre entra
Yo pego
[baby cries]
Yo todo, tu nada
Soy el perro que muerde

(a law word
The letter enters with blood
I hit
[baby cries]
I everything, you nothing
I am the dog that bites)

This unforeseen moment of encounter between a performance of the violence of the letter of the law and a human being who is not yet inscribed into the aural imaginary that links voice to silence, and silence to law, could be conceived of as a moment of what Slavoj Žižek would call the Real poking through the symbolic, the living body puncturing the social fabric of voice with a sound that is not yet ideologically determined. A similar moment occurred when another speaker read a poem claiming, "Dice que el barrio es nuestro, pero el barrio es mío, compai / Es que siempre se lo lleva todo" (It says the neighborhood is ours, but the barrio is mine, pal / Everything is always taken away), as children screamed over the sound of his voice. The aural contrast between the narrative possession of the *barrio* through poetry and the sounds of the children's play draws attention to how the lived experience of the *barris* has been discarded by a narrative interested only in attracting tourists. Moreover, the first-person singular "I" used in both poems above to signify the sound of investment or law contrasts with the emotional "commoning" that transpires, if only temporarily, in the moment of the group's occupation of the park or the square. We may contrast that commoning with the "Follow the barrio" campaign led by the Ajuntament, which has completely rebuilt Poblenou, a working-class town outside central Barcelona, as a hip beach town geared toward foreign investors, students, and tourists. The English name of the government campaign, which evokes the phrase "Follow the money," makes clear the investment is not for locals, even as, ironically, it invokes a cinematic catchphrase popularized by a North American film about government corruption. This is an inverse example of the sound of language being stripped of its meaning, in this case in order to use the sound of English to signify modernization and the future, notwithstanding the content of the words.[77]

Carrying a microphone and speaker system in a shopping cart, the group also created their own public soundtrack, accompanying the walk from place to place with a playlist of anarchist songs. Most were by Chicho Sánchez Ferlosio, son of the founder of the Falange, Rafael Sánchez Mazas, who during his life became a voice of resistance to the regime. These included "La anarquía vencerá" and "El destierro." The first condemns the church and the government, and appeals to anarchist reason to defeat them:

Los burgueses nos acosan
con la biblia y el bastón
pero no han tenido en cuenta
el poder de la razón
compañero basta ya
compañero basta ya
la anarquía vencerá

(The bourgeoisie harasses us
with the bible and the staff

but they have not taken into account
the power of reason
comrade, enough already
comrade, enough already
anarchy will prevail)

The second song recalls the unjust treatment, but also strength, of a worker condemned to exile for his beliefs:

A la fuerza me han traído,
sin sentencia ni proceso,
pero el fuego no se apaga
aunque a mí me tengan preso.
El destierro no me importa
ni me asusto de la muerte:
la anarquía vencerá

(They brought me by force,
without sentencing or trial,
but the fire does not go out
even though I have been imprisoned.
I don't care about exile
Nor am I afraid of death:
anarchy will prevail)

Together these classic songs of resistance narrate the long history of the contemporary struggle against economic dispossession, and evoke an echoic memory of transnational solidarity.

At the same time, the group played more recent music linked to the legacy of the punk tradition, such as "Corazón indomable," by Los Muertos de Cristo, about Buenaventura Durruti, and "José República," by Skalariak. This last song traces the history of Republicanism in Spain, evoking an image of the "pueblo huérfano" (orphaned people), who "con dignidad nunca dejó de caminar, ni dejará" (never stopped walking with dignity, nor will they), even as the refrain speaks of the blood and murder of the people by the government: "Eh! un charco de sangre. . . . Eh! un hombre en el suelo. . . . Huele a asesinato ya hay duelo José República ha muerto" (Hey! A pool of blood. . . . Hey! A man on the ground. . . . It smells like assassination, there is a duel, José República has died). The song's critique of anti-Republican, antidemocratic violence echoed in a speech given by the father of Pedro Álvarez at the first stop on the march. A young man from Barcelona, Álvarez was allegedly killed by police in 1992, but no one has officially been charged with his death. His father briefly discussed his case and the need for justice; although the hip-hop group Malamara did not perform at this particular event, his arguments echoed with the group's denunciations of police violence in the city of Carcelona.

By broadcasting their music and performing their poems live in public, the protests reterritorialized the aural dispossession that divides private from public, and which not only informs capitalist investment strategies but also takes place almost daily when people walk the city wearing AirPods, listening to GPS directions on their phones, or otherwise using digital technologies that, at least partially, make them deaf to their surroundings. As Michael Bull has suggested, "The 'entitlement' to 'private space' is entrenched in Western thought, at least since the Enlightenment," and it is often associated with private property.[78] The result are cities that tend to lend themselves to "aural solipsism."[79] On the one hand, this means listening subjects can create what Bull calls their own "acoustic bubbles," privatizing their sound; on the other, it allows the biomediated ear to overcome the lived experience of the body, dispossessing community of its humanity: the act of walking the city while "plugged in" converts the bodies of those who are unhoused, displaced, or different into two-dimensional figures, mute to the person walking the city. By openly broadcasting music of resistance, making an *algarabía* that is also an *algarabia*, interrupting locals playing with their children in the playground or tourists drinking at local bars, the Bio-lentos occupy the ears of those around them with a sound of community in protest, if only for a few moments. At the same time, they make the notion of occupation, as an ideological resistance to property, an affirmation of the restless and nomadic sound of public voice as mobile and temporary, rather than held in place by state or capitalist structures, or by a silent vote.

Performing this scenario of movement and resistance against the tourist appropriation of the city also restages a scenario that is familiar to a transnational struggle for liberation that has defined the Global South/ Global North divide for decades, which ranges from antiglobalization to the internationalist socialism of the early twentieth century to Maoism. The readings in the march included voices of immigrants, such as a poet who spoke about his Colombian mother; he recognized the alignment of the Bio-lentos with the *sudacas*—a pejorative term for South American immigrants—who have become *els altres catalans* in recent years. The sounds of Spanish, English, and Catalan spoken during the march harnessed an echoic memory of global solidarity in dispossession that contrasted aurally with the monolithic global economics of capitalist possession. Returning to the shopping cart playlist that aurally created a forward- and backward-hearing echoic memory of democracy rooted in Republicanism, we may recall that Sánchez Ferlosio's alliance was not just with the Republican cause in Spain, but with Maoists and the Cuban Revolution. Another of his songs echoed by the Rhatsodes' choice of music is "Son para Turistas," based on a poem by the Afro-Cuban writer Nicolás Guillén. Framed by author notes that specify the dialogue is between tourists in a bar and Cantaliso with his guitar (a reference to Guillén's poem "Cantaliso en un bar"), the song turns the performance of culture for Yankee outsiders in Cuba into a critique of their presence in a local space:

> No me paguen porque cante
> lo que no les cantaré;
> ahora tendrán que escucharme
> todo lo que antes callé.
> ¿Quién los llamó?
> Gasten su plata,
> beban su alcohol,
> cómprense un güiro,
> pero a mí no. . . .

> (Don't pay me to sing
> what I will not sing to you;
> now you will have to listen to me
> everything I kept quiet about before.
> Who called you?
> Spend your money
> Drink your alcohol,
> buy a *güiro* [Cuban instrument],
> but not me. . . .)

Although they did not play this particular song, the Bio-lentos restaged the scenario Guillén described, when they performed their protest directly for tourists, illustrating the daily globalization of the ear in ways that the public media images of the Catalan independence movement do not. Informed by socialist thought of different ideological stripes and national origins, this aural imaginary of resistance promoted a transnational movement of liberation, at the same time as it performed this idea specifically within a particular neighborhood in Gràcia. It harked back to both the international support for the Spanish Republican cause and the transnational performance of protest in the *cantautor* tradition, all while addressing a public that was both local and international.

At another site during the Bio-lentos' walk, in the Plaça del Sol, a protestor wearing a black t-shirt that said "Acció Directa" (Direct Action) spray-painted a statue in the square with the phrase, in English, "Tourist go home." Along the route, he and others surreptitiously graffitied walls with terms like "Tourism kills the city" and "Ca La Trava resisteix" (Ca La Trava resists). With these acts, the walls speak, and voice becomes again an occupation of territory that also lays a claim to it. In a twenty-first-century city beset by tourism, the aural imaginary of dispossession that informs these small protests is in dialogue with echoic memories of a wider Global South that reaches across its own territorial boundaries and into differing political ideologies, present simultaneously, decades later, in a metropolis whose ruling class continuously lays claim to its Northernness. The Bio-lentos thus imagine a local space that is only a few city blocks wide as participating in a larger geography of dispossession where the boundaries between colonialism and globalization, democracy and dictatorship, are fluid and difficult to identify within a single mappable territory.

But if part of the liberation performed by sound can be heard in its spontaneity, that spontaneity also opens up the space to confrontation. An alternative to the contingent moment of interaction between the baby and the voice of the law took place toward the end of the protest, in a vocal confrontation between a tourist ear and the poetically incorrect voice. As the group arrived outside the occupied house, Ca La Trava, one woman began declaring a relationship between exhaustion and being dispossessed: "Estamos hartos de trabajar de sol a sol. Estamos hartas, de fabricar armas, del tráfico de esclavos, estamos hartas de estar hartas" (We're sick of working from sun to sun. We are tired, of making weapons, of human trafficking, we are tired of being tired). As she spoke, some tourists from a balcony nearby began mocking the poetry. "Oh no, tourism kills the city!" one called out in accented English. In that moment, the poet's voice of exhaustion, articulated as such, was reinscribed by a tourist's ear, as her complaint was converted into a call and response that performed the political tensions around gentrification in real time, without the mediation of the television, radio, or internet. Notwithstanding the size of the protest group, the tourists, safe on a balcony above, acoustically dominated the space of the small intersection, as their voices echoed off the building walls more loudly than did those of the poets and their small mic system below.

By physically occupying the public spaces of the city with a voice that is poetically incorrect, the Bio-lentos used the sound of the human body moving through a local geography to signal the dispossession of humanity that is the other side of the capitalist focus on property.[80] In these small moments of unpublicized protest that contrast with the media spectacle of Catalanist independence, we hear how globalization re-creates aural cartographies of dispossession that are not limned by state borders but rather by acoustic ontopologies in which migration bleeds into tourism, and possession bleeds into dispossession.

In his 2005 essay *Destrucción de Barcelona*, Juan José Lahuerta Alsina offers us a spatial way of hearing Barcelona echoically. Lahuerta invokes the *taconeo* (clicking of heels) of the renowned flamenco dancer Antonio Gades as he dances his way down La Rambla de Santa Mònica toward the port in Francisco Rovira Beleta's 1966 film *Los Tarantos*. The film focuses on the Roma population in Barcelona, highlighting the overt spectacle of flamenco that marked the beginning of Barcelona's process of selling itself to tourists (part of Franco's plan for opening Spain up to the world after nearly two decades of isolation). At the same time, it draws the ear to the class and linguistic differences between the southern migrants who had come to the city and whose cultural sounds threatened a purer, bourgeois Catalan ideal that longed for a lettered understanding—and public circulation of—the Catalan language. Lahuerta counters the noise of Gades's flamenco heel with the physical effect the shoes of prostitutes have had on doorways on the margins of Barcelona's famed Les Rambles. Standing in the same doorways for decades, catering largely to foreigners, their high heels have eroded almost perfectly symmetrical holes into the stone. The almost imperceptible sound produced by their feet brushing over the same spaces night after night—the underground equivalent of the flamenco heel—has left physical, tactile traces on the city, even if at first they go unnoticed: "La violencia a la piedra es la marca del cuerpo violentado" (The violence against the stone is the mark of an abused body).[81] While Gades's performance is a singular event that is also a spectacle for capitalist, foreign consumption, the near-silent clicking of the prostitutes' shoes on stone—the "invisible" and hushed sound of the same foreign consumption in the port by American sailors and others—resonates only over time as metaphorical echoes that leave physical markers on the spaces and bodies of the people who traverse the city daily. Contrasted with the touristic imaginary of the flamenco spectacle, the echoes of the shoes re-create the material space in ways that challenge the ideal Catalan image of the city. The aural space of El Raval invoked here allows us to recognize that the echoic memories of sound in Barcelona are not just national, cultural, or linguistic. They are also gendered, racial, and economic, and these factors reverberate with the built spaces themselves.[82] Although Lahuerta's example focuses on the sounds of sexualized bodies of women, we can extend the space to include the nightly sounds of cabarets and bars the region was known for, or their disappearance, which took place largely through the gentrification that began at the turn of the twenty-first century and has continued until today.[83]

This scene serves metaphorically to encapsulate the performance of protest I have interrogated here. As biomediation of the ear extends the spectacle of travel, speculation, and investment around the globe, we may hear the noisy occupation of the streets with poetry as equivalent to the imperceptible sounds of the prostitutes' heels engaging daily, but faintly, with a globalized neoliberal media apparatus.

THE TERRORIST'S WHISTLE

The global linkages that connect Barcelona outward through tourism, industry, and speculation also connect them politically to auralities that have changed throughout the West. The *maraña* of *algarabía*, after all, also means listening outward to a sense of sound that is not native, that is noisy, that is hard to hear. Perhaps one of the greatest changes to the biomediated ear in recent decades has been its framing in terms of global terrorism, rather than local or national groups. Robert A. Saunders has argued that terrorism is largely dependent on media for its power. If in premedia days "whispers in the bazaar, bardic ditties about assassinations, and hanging corpses" all functioned to spread the word of nonstate violence, today terrorist groups "'speak'" via news broadcasts, newspapers, and the internet.[84] But it is also true that the media perpetuates the sound of terrorism as something that can be predicted, heard into the sounds associated with certain voices. Just as the media can be wielded as a tool to promote Catalan language, it can contribute to how the listening subject hears the sounds of other cultures as either assimilable to Western values or not.

Shortly before the October 1 referendum, on August 17, 2017, a van plowed through Les Rambles, killing fourteen and wounding many others; two more civilians would die in the aftermath.[85] It was the latest in a series of "do-it-yourself" attacks with improvised weapons in European cities. Almost immediately, security forces and the press condemned the attackers as terrorists with an ideological affiliation with ISIS. The organization soon claimed responsibility for the attack.[86] As tends to happen, for a short period of time, coverage of the attack dominated the news media. Within a week, however, the Catalan independence movement was back on the front pages and political talk shows.

The day after the attack, an estimated 130,000 people assembled in the Plaça de Catalunya for a moment of silence in commemoration of the victims, who had been from at least twenty-four different countries.[87] The congregation in the plaza included King Felipe, then–President of Spain Mariano Rajoy, then–President of Catalunya Carles Puigdemont, and then-mayor of Barcelona Ada Colau. It was broadcast via radio and television around Spain and later appeared recut online in various forms. As is traditionally the case, the moment of silence connoted respect for the dead linked to the supposed suspension of political rivalry. But this performance of silence quickly became noisy with an aural imaginary of political struggle. What was most reported on, in fact, was the way in which the moment of silence morphed into approximately ten minutes of applause, followed by the chant "No tinc por!," which means "I am not afraid!"

El País, Spain's foremost Spanish-language newspaper, for example, ran the headline: "'No tinc por!,' grita Barcelona."[88]ABC *España* and RTVE, for their part, went with: "Barcelona grita: 'No tengo miedo.'"[89] In their moment of silence meant to mourn a loss, in other words, the community spoke, redirecting the frustrated *crida per la democracia* (shout for democracy) movement for independence into a call for shared humanity in the face of terrorism. Online, Twitter users responded to the "No tinc por" chant in real time, with the user @yeyodebote mocking the Catalan movement by writing "Me parece muy fuerte que el minuto de silencio se haya hecho en catalán" (It strikes me as appalling that the minute of silence was done in Catalan). Those who commented on the post poked fun at the writer for ascribing a language to silence. But the idea that a moment of silence could be framed as a sign of national identity affirms the idea that geographically inflected aural imaginaries and sleights of ear tinge even the most basic of sound perceptions. The aural imaginary of language conflict that has surrounded Catalonia's push for Catalan education and public use made the silence of loss noisy.[90]

Just three days after this moment of silence, however, another kind of noise came into play, one that performs a sort of scalar hearing in which a global color line is more threatening than an internal one. Newspapers reported that the driver of the van, a Moroccan-born man named Younes Abouyaaqoub, a resident of Ripoll, had been killed by police. After a four-day manhunt, he was identified as the terrorist in part because, as various news sources reported, a woman called police after she heard a man whistling outside a home. An interview with the popular morning show, "El programa de Ana Rosa," broadcast on the station *Telecinco*, was later quoted widely online and in print sources, including overseas: "Según un testigo, el hombre estaba silbando cerca de una zona de casas aisladas, como si esperase a alguien" (According to a witness, the man was whistling near an area of isolated houses, as if waiting for someone).[91] Other reports followed suit: "Y, según algunas fuentes, Abouyaaqoub estaba silbando junto a un chalé de forma sospechosa, como si se tratara de una señal convenida para contactar con algún cómplice" (And, according to some sources, Abouyaaqoub was whistling next to a cottage suspiciously, as if the whistle were a prearranged signal to contact an accomplice), wrote one online newspaper.[92] "En una revista de Telecinco recogida por *Europa Press*, Agustí ha explicado que quien vio al terrorista fue otro vecino que llamó a su puerta para decirle que había una persona con aspecto extraño, desconocida en la zona y que iba silbando, una 'cosa sospechosa, una cosa rara'" (In a *Telecinco* magazine picked up by Europa Press, Agustí [a witness identified only by his first name] explained that the one who saw the terrorist was another neighbor who knocked on his door to tell him that there was a strange-looking person, unknown in the area, and that he was whistling, a "suspicious thing, a strange thing").[93] *El Economista* actually chose the whistle for the headline: "El vecino que delató a Abouyaaqoub: 'El terrorista iba silbando, algo sospechoso'" (The terrorist was whistling, a suspicious thing).[94] A retrospective article in Spain's *El País* tracing the driver's trajectory tersely recounted the whistle as one of a series of fateful actions: "Younes llega a la parte trasera de una casa adosada en Subirats. El sol cae a plomo. Silba. Busca ayuda. . . . Pero es tarde. Le han visto" (Younes arrives at the back of a rowhouse in Subirats. The sun is setting. He whistles. He looks for help. . . . But it's too late. They've seen him).[95] In these accounts, no mention is made that the listening subjects who heard the sound knew about the terrorist attacks and thus had heightened awareness that the attacker may be nearby. Instead, stripping the sound of its immediate context, these articles perform a key function of echoic memory, naturalizing certain sounds as threats, or suspicious signs, because they are not voice. This aurality heard into the whistle was reaffirmed by witness descriptions of Abouyaaqoub that claimed he was visually out of place since he "looked strange," had a "Maghrebi appearance," and even stood out for wearing a long-sleeve shirt.[96] Such assertions normalize the idea that visual difference must be somehow related to criminality, an idea that echoes back to nineteenth-century pseudoscientific conceptions of the body while also enforcing the listener's acoustic assessment with a visual one.

As Hillel Schwartz has shown, often whistling is taken as "acoustic turbulence" in which "[t]hose who whistle under tense or tender situations are held to be whistling for help."[97] Nevertheless, in this case the terrorist's search for help was heard by another ear than that to which it was directed, and it was heard as odd and threatening. As Schwartz explains, "It would not matter a whit if a person were whistling idly or with malice aforethought . . . whistling is phonologically obstruent, physically turbulent."[98] He cites the case of Emmett Till, who was taught to whistle to overcome a stammer, but whose fateful whistling one day in 1955 was presented retrospectively as a catcall that resulted in his brutal beating and murder in the name of a white woman's honor. In the case of the Barcelona terrorist's whistle, the political circumstances are radically different from those described by Schwartz, including the obvious fact that Abouyaaqoub had committed a violent crime, while Till had not. Moreover, Emmett Till's accuser later admitted he had not actually whistled at her. Still, the aurality that hears whistling as turbulent, as verging on the nonhuman because it is not voice, was articulated as sufficient justification for Till to be killed. Abouyaaqoub's lack of voice, as well as his visual association with a "strange" body, was presented as the justifiable reason for suspicion of him as a killer. Again, I want to be clear that I am not positing that Till is the equivalent of Abouyaaqoub, merely noting that the imaginary of whistling in both cases was invoked to justify assertions of criminality.

If accent is heard as emplaced voice that is out of place, as I suggested in Chapter 2, here the subject's nonlinguistic form of communication is similarly heard as out of place because it is *not* voice: from at least the nineteenth century on, the human vocal apparatus "was consistently framed as comparatively more advanced than the vocal apparatus of non-human animals" and this distinction, moreover, "aligns human speech with power."[99] To my knowledge, the whistle is not necessarily a sound with a historic, racially charged connotation in Catalonia. In this case, its association with a body that occupies the other side of the sonic color line of voice from intelligibility emplaces it within a nonlegitimized form of communication. At best the whistle is an *algarabía* that is nonlocal; at worst it is nonhuman, tied time and again to a coloniality of the ear through which "barbarism" is anything outside the rule of the metaphoric sound of democracy as intelligible voice.[100] By not operating within the conspiracy of silence that legitimates certain kinds of voice as a means of upholding the law, the terrorist is already heard as deaf to understanding the law, and unable to speak because he is unable to understand what civilization looks and sounds like. It is, in a sense, the seemingly irrational sound of the whistle that caught the listener's attention.

In this way, the terrorist's whistle occupies a double space of alterity that calls attention to the fusion of body and discourse that is the site of production of a biopoliticized, and now increasingly biomediated, echoic memory of colonialism. When Foucault first argued in 1976 that biopower is the power of the sovereign state to regulate life, linking state discourse to the body of the subject, he argued that moments of racism "break out at a number of privileged moments" in which bellicosity is invoked in defense of one's biologically unified population.[101] He could not have known that a few short decades later most of the globe would be linked technologically to the internet, or that, in the wake of 9/11, multilingual, transnational discourses on threats against the sovereign state's regulation of life would circulate daily well beyond the geographical and political bounds of the individual state. Even so, the symbolic constructs of legitimacy linked to sovereignty he described continue to dominate how voices are heard, at times still under the auspices of the state, but even more so within an increasingly wider frame of authority associated with the European Union or global (largely Western) allied states.

Moreover, this seeming agency of the listener that the narrators of Abouyaaqoub's arrest describe carries with it a shift toward vigilance that turns biomediation into a perpetuation of the colonial ear on a global scale. As I explored in Chapters 1 and 2, there is a sonic color line in Spain that associates the accents of darker-complexioned people with the *moros*, a stereotyped construct. Over the past century, this "Reconquista" scenario of white Christians encountering the *moros* largely played out in the relationship between the Catalan bourgeoisie and *els altres catalans*. But this echoic memory of *algarabía* plays out as well with respect to the influx of immigrants from Morocco and other parts of Africa, as Najat El Hachmi has explored in her writing. Especially since 9/11 and, more particular to Spain, since the Atocha bombings of 2004, the sonic color line that associates accented voice with skin-tone has acquired an association with an extreme form of protest—terrorism. During the later years of Franco's regime and early days of the Transition, when terrorism was homegrown, linguistic rather than racial identity might have informed ideologies of fear, and the temporality of concern would have gone back only as far as the Civil War. But in the early 2000s, former president of Spain José María Aznar made claims from within my home university, publicized in Spanish media, that "el problema de España con Al Qaeda no empezó con la crisis de Irak, sino. . . . cuando España, recién invadida por los moros, rechazó convertirse en una pieza más del mundo islámico" (Spain's problem with Al-Qaeda did not begin with the Iraq crisis, but . . . when Spain, newly invaded by the Moors, refused to become another part of the Islamic world).[102] Although he was pilloried by the left for this comparison, the comment provides a perfect example of how centuries-old echoic memories of race and "unintelligible" sounds of voice manifest in the ideological portrayals of deep time that circulate in today's geopolitical climate.

For the last two decades, an Orientalist perspective on Islam has been accompanied by geopolitical notions of good and evil associated with terrorism. ISIS took credit for the attack on Les Rambles shortly after it occurred, so in this case one could argue that the suspicion was warranted. But a little more than a year before that, in July 2016, a Pew Research Center poll revealed that "in several European nations, unfavorable views of Muslims seem to have surged."[103] Although less prevalent than in some other European countries, in Spain the unfavorable view of Muslims rose 8 points to 50 percent that year. 54 percent of those polled in Spain feared the influx of migrants into Europe posed a threat, and 40 percent—the lowest number among Europeans polled—believed that Muslim refugees were directly linked to terrorism.[104] The fact that Spain was the most tolerant of the European countries aligns, generally, with how immigration of all types has been viewed in Spain since the Transition. Until recently, political scholars have overlooked the fact that many of the early institutions and even participants in Spain's democracy came from Franco's dictatorship, arguing that the extreme right had become largely nonexistent in Spain, with nativism being associated with a nostalgia for Franco and, thus, something to be avoided.[105] Their argument was that this lack of a right wing, plus Spain's long history of emigration, had led to more tolerance for immigrants than in the rest of Europe. Saying that Spain is relatively tolerant when 40 percent of the population believes Muslims are linked to terrorism, however, strikes me as an awfully optimistic reading. What is perhaps most telling in the poll are not the numbers (themselves, I would suggest, unreliable measures of social attitudes), but the seemingly unspoken agreement that, whatever side of the issue one is on, however informed or uninformed one may be, the "problem" of Islam is so obvious that it does not need to be articulated for readers. In other words, there is a silent sense, rather than an explicit articulation, of Orientalist discord that one can hear into the poll, a silence of rationale, or what Jacques Rancière calls "self-evident facts of sense perception," that speaks ideological volumes.[106]

In that silence, Islam becomes a mythology, in Barthes's sense of the word: "Myth deprives the object of which it speaks of all History. In it, history evaporates. It is a kind of ideal servant: it prepares all things, brings them, lays them out, the master arrives, it silently disappears: all that is left for one to do is to enjoy this beautiful [or, I add to Barthes here, the terrifying or threatening] object without wondering where it comes from."[107] And, as another caveat to Barthes, not wondering where the terrifying object comes from is the result of a wider postcolonial imaginary in which the other is both from anywhere else but here and, more terrifyingly, is silently here among us. Such is the largely conservative, but often popular understanding of global terrorism in the post-9/11 age, opposed to another great Western myth, that of democracy as freedom.

Given that Spain has a system whereby individual autonomous communities decide how to address immigration—meaning that each community decides how many resources to use and in what form they would like to dedicate them to the reception of new arrivals—there is a local aspect to how immigrants are treated in different parts of the country that is not captured by polls such as the one above. An estimated 600,000 Muslims reside in Catalonia, more than anywhere else in the country, and the website of Barcelona's Ajuntament, which prides itself on being open to immigrants, includes a page on "getting to know" the religions of Barcelona, including Islam. Antoni Verdaguer's film *Raval, Raval* from 2006 uses music to present the mixing of Pakistani, Senegalese, Moroccan, and other immigrants to Barcelona as a celebration of multiculturalism in spite of the fact that neighborhoods where they live are being torn down in favor of gentrified tourist spaces. But as Najat El Hachmi has perhaps more perceptively critiqued, beneath the idea of multiculturalism and tolerance are certain suppositions about language that reflect a persistent colonial epistemology tied to notions of civilization and barbarism.[108] If the Catalan community can imagine itself as a European community through its stagings of democratic voice, they can also be vigilant listeners who hear how those from communities outside Europe pose a threat to these ideals. In previous chapters I explored the role accent played in asserting that aural difference, but here vocal silence—as opposed to voting silence—becomes a marker of nonvoice, which is heard as alterity as well. The silent threat of terrorism is perceived as being pervasive, because it can be heard not just in the accent of certain voices, but in the voicelessness of immigrants on the other side of the sonic color line who do not attempt to speak the local language because they cannot, or who for other reasons remain silent (this is, indeed, how Vasallo figures the silent burka in the novel I analyzed in Chapter 2).

Voice, then, is the metaphoric foundation of political representation in the West, but the ability of silence and noise to signify everything from radical difference to universalized solidarity, through the sentience that is a void that is also not a lack, depends on ideological tensions that reflect the mediatized saturation of the ear and its perpetuation of the tension between reason and emotion. As Jürgen Habermas's public and private spheres bleed into what Michael Warner calls publics and counterpublics, the biomediated ear reflects an echoic memory of voice through which noise is anything that challenges the norm. Furthermore, in a globalized West that has been defined for two decades by a "global war on terror," the relationship between noise and democracy struggles to maintain its solidity, shaping a global sonic color line that obviates the daily reality of sound as a constant presence of conflicting volumes, timbres, and rhythms—what spatial theorist Henri Lefebvre has called the living body.[109] For Lefebvre, the living body holds multiple rhythms that are simultaneous and diverse, but which the state and the media seek to fix in place. Hearing a whistle as threatening is just one way in which the aurality of voice manifests in a biomediated fashion.

REDRAWING THE EAR (ECHOES)

A few days after the October 1 referendum and nearly two months after the ISIS attack, an advertising agency hung a white sheet on its building in Madrid with the question, in Catalan, "Parlem?" (We talk?). Another responded in Castilian: "Hablemos" (Let's talk). Social media produced other drawings and cartoons around the suggestion, which El País compiled into

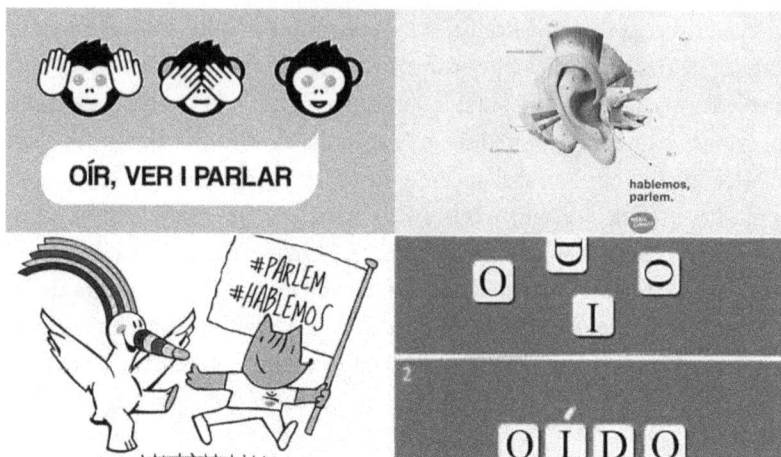

4.4 Oír, ver i parlar (three monkeys). Compilation of images posted to Twitter as part of the #Hablamos / #Hablemos / # Parlamos / #Parlem social media discussion from October 2017. Although hard to see in this reproduction, it is notable that the diagrammatic figure of the ear in the upper right is labeled "hablemos, parlem," and that the letters in the lower right that spell "ODIO" (hate) are rearranged to spell "OÍDO" (ear) with a handwritten accent mark. Reproduced by Patricia Gosalves in *El País* on October 6, 2017.

a montage (see figure 4.4).[110] One cartoon shows an anatomical drawing of an ear. A line proceeding from the inside of the ear ends in the bolded terms "hablemos, parlem," suggesting that speech originates in the ear. In another drawing, the three wise monkeys are rearranged so that hearing comes first, with "no" eliminated from the maxim to read: "Oír, Ver i Parlar" (Hear, See, and Speak). In yet another, titled "Escogemos las palabras" (We choose the words), white letter blocks with black capital letters that remind the viewer of the game Scrabble appear in two sections. In the first, they are somewhat scrambled, facing in different directions, to spell "ODIO" (HATE). In the second, those that were facing the wrong direction have righted themselves to spell "OÍDO" (EAR) with a hand-drawn white accent mark over the I. The suggestion that "hate" can be corrected to spell "ear" places the onus of dialogue on listening, at the same time as it suggests that the rearranging of language is not the simulacrum's flattening out of meaning, but rather a necessary move from language to body; the hand-drawn accent mark brings the body into the drawing in ways the typeface cannot.

In this campaign, if voice is the foundation for democratic processes, the physical reception of sound is the foundation for voice, suggesting not that legal concepts of voice have been eroded, but that the ear has not worked hard enough to listen in to and alter them with the body. Just as the Catalanist public spectacles in the 2010s celebrating the Diada advocated for a sonic construct of voice at a time when the Spanish government had forbidden Catalan citizens from participating in a vote on independence, just as Abouyaaqoub's whistle revealed a biomediated hearing of terrorism as nonvocal, just as the *okupes* rethink dispossession with sound as much as with their presence, the voice is rooted in the ear. And it is an ear that is as much an *algarabia* as it is an *algarabía*, an affective sensation that is also an echoic memory. By focusing on the words we can choose, this appeal to dialogue still hears voice as an aurality that avoids noise and disturbance. But in its visual appeal to emotion, it does so by drawing attention to the "presence that is not a lack" that is a part of the daily acoustic occupations of space.

The public space in which the media and the body coincide produces not just political performances, but life and death acts with real consequences, some of them violent. It is a biopolitical space in which voice most reveals its precariousness, depending as it does on a generous ear to affirm its legality and its legitimacy. It is a space in which a terrorist and a political activist may be heard by the law as the same, despite the broad social acceptance of voice as an expression of autonomy and difference. This space is one in which marginalized *okupes* who perform *cassolades* to draw attention to a housing crisis can be brushed off as noisy, while organized groups with resources can market their ideas globally as examples of freedom of speech and democratic intent.

Like all spaces of community, Barcelona carries in its soundscape echoes of anger and loss, of *algarabía* and *algarabia*, and the sounds of a shared frustration with voice conceived as a representation of the letter of the law. Upheld by an aurality of dispossession forged in a transatlantic setting five centuries ago, that itself spirals fractally into echoic memories that both predate and succeed the period, the senses and discourses of democracy and protest that circulate through the cities today confound geographies of nation. They signal the multiple and shifting directionalities of the sounds listening subjects encounter as they move through the world. Thinking through the relationship between the sentient listening body and the epistemologies of reason that undergird our naturalization of certain kinds of sound presents an opportunity for those of us whom Jonathan Sterne calls sound studies students to geographically expand our acoustic reach. But it also allows us to recognize our own cultural limitations when dealing with the sounds of loss and rage that overflow and exceed the ear's capacity for order and meaning, itself deeply rooted in language. Local geographies are always more than my humble ear can apprehend, though I have tried in this chapter to do just that. Accepting the *algarabía* as a daily reality of sound means recognizing the humble ear's constant need to move and adjust, reconsider its position when listening, and do everything possible not to become emplaced in a colonizing aural imaginary in which borderlines are firm and positions arc heard as permanent.

Coda

The Humble Ear and the Shape of Sound

There is a shape of sound in what Jon Mowitt calls the audit, the sonic equivalent of the gaze.[1] It changes with movement, through time and space, creating a geography of the ear that resonates personally and also intertwines with the echoic memories that pervade media, be they acoustic or textual, visual or sonic. I have argued throughout this book that the colonial ear persists in unexpected places as a result of these interactions, which cross national and linguistic boundaries as accents, persist in local neighborhoods as community radio, perform daily microresistances through art, perpetuate sonic color lines and the racial-historical deep time of protests' *algarabías*. I began this book by framing that ear within the troubling geography of North and South that persists in the construct of the Global South. But the recognition of the pervasiveness of the colonial ear even in the most local of places is particularly resonant within the Hispanic Studies field, which has recently taken a turn for some scholars toward what has been called the Global Hispanophone. This designation purports to carry out a decolonizing function by "steering away from Latin American and Iberian studies, as well as from Hispanism" and broadening the geographical scope beyond the Atlantic frame and into the Maghreb and Asia.[2] The approach "challenges the linguistic principle on which Hispanism was allegedly founded: the privileging of knowledge and cultural production in Spanish at the expense of all other cultural and linguistic traditions around it."[3] Nevertheless, I would suggest that within the frame of echoic memory that I have traced here, it is not so much a question of decolonizing Spanish, as it is of recognizing the daily practices through which sound engages the colonial ear, often—especially—through language. Exploring how similar processes of dispossession repeat even within seemingly minor languages that can also be institutional forces—in this case, Catalan—allows us to recognize that the geography of the ear cannot be framed solely according to a single Hispanophone, Francophone, Anglophone, or other monolingual understanding of geography. If linguistic decolonization is ever to even be on the horizon, I would suggest that it must be approached through a multilingual understanding of sound whereby knowledge is not only not monolingual, but, fundamentally, cognizant of its own faulty ear, not just that of others.

Trying to disentangle all of this, especially as someone who is not from Barcelona, who is a product of the very media I try to understand, I have come to realize that perhaps what is needed in sound studies is less an authoritative ear, and more a humble one, and it is with a humble ear that I wish to conclude this work. Traditionally an origin narrative would come at the beginning of a text; postcolonial theorists typically emphasize the locus of enunciation as fundamental to one's positionality. Within that scheme, one's intellectual, ideological, and historical makeup precedes her interaction with ideas coming from other places, be they geographical or epistemological. Elsewhere I have critiqued this concept for its privileging of place over temporality, as conceptually it might convert origin into destiny.[4] But I have also left my personal relationship to this book until the end, because it was only by working through the concepts of the aural imaginary and coloniality that I was able to make explicit the personal preoccupations that have informed this work.

AN ENDING, BY WAY OF A BEGINNING

One may suppose that it is possible to create a gap between how one hears intellectually and how one hears daily. Jonathan Sterne's suggestion that those of us who think about sound could be considered sound students addresses that question, framing positionality as relevant to the field(s) we engage.[5] As someone whose formative education included Gayatri Spivak's critique of Western intellectuals, however, I have always recognized my position as a North American scholar approaching Spain and Latin America from a place of my own inability to speak for or with anyone I study.[6] As a transatlanticist working within the Hispanist tradition, but with an inexplicable (or so I thought for many years) affinity for, and interest in, Catalan culture, I have long felt encumbered by the labels that describe my work and, by extension, my scholarly identity. With graduate students and undergraduates alike, I have discussed the impossibility of an apolitical position, of our need to recognize the ideological constructs that determine our thought. We have spoken at length of the impossibility of an Anglo-European comprehending and representing voices of the South, an argument epitomized by the ideas put forth by, among others, Bolivian decolonialist Silvia Rivera Cusicanqui.[7] Within that frame, I have at times felt that what I had to say about my objects of study could not be legitimate, since I cannot claim a right to Hispanic or Iberian identity of any sort. Still, my children were both fluent in Spanish before they could speak English, because I spoke to them only in my nonnative tongue. When strangers hear my daughter speaking to me in fluent Spanish and ask where we are from, I struggle to answer.

I was born in Canada to a Serbian father and an English mother. We immigrated to the United States when I was only a year old. In the current political climate, it seems relevant to say not only that we lived here without documentation for some time—my parents even bought a house while my father was between jobs and thus unsponsored—and not once, but twice when I was a teenager in the early 1990s, we drove over the border into Canada to visit my relatives, and we were stopped because I had an expired green card. Both times the border control officer told my parents they should really get that taken care of, otherwise they might not be able to let me through the next time. My father was olive-complected, bearded, and spoke with a thick accent. I did not have valid documentation. We were allowed into the United States anyway. The memory shocks when I consider how that scene could play out today. Maybe not, though. My father was darker and accented but carried a Canadian passport. The border was the United States and Canada. The global war on terror did not yet exist.

Still, it was only after completing this manuscript that I realized that although I was interested in sound as an object of study, the driving question of this book, that of the immigrant accent, is intensely personal. My father learned English in his twenties and never quite mastered either the language or the pronunciation: among many other details that have faded in my memory since his death, I remember most strongly that he routinely confused his articles. He also pronounced his *v*'s as *w*'s, and "th" became "t." In high school, my cousins, sister, and I compiled a list of all the funny expressions he and our other first-generation relatives used incorrectly, as they talked about putting their dirty laundry in the "clothes humper" or referred to our cousins as sisters, since in Serbian the word for cousin and sister was the same. My mother, on the other hand, had emigrated to Canada from London at age eight, and while she dropped her English accent in public, when she spoke to her parents it came back, albeit in a rather watered-down version; it seemed her childhood accent haunted her adult speech. When her parents died, so did her accent; I have never heard it since. With this background of accented experiences that, as a child, I recognized made my family inherently different from others, when I took up the study of Spanish, I was determined to perfect my pronunciation. I spent a summer at the age of twelve teaching myself how to roll my *r*'s. As a family, we spoke often of the fact that my sister, born in Virginia, was the only one in the family who was American. While this multiculturalism was a strong source of pride in my family, when I took up graduate studies in Spanish and Latin American literature, I suddenly became aware of the fact that I have no ancestral connection to the specific identities I study and teach about. At the same time, I used to pride myself when people in other countries with whom I spoke Spanish could not detect where I was from. Argentines thought I might be German, Colombians thought I was Spanish, Spaniards thought I was from the Caribbean, a Puerto Rican academic once thought I might be Catalan. For me, the most important thing was that no one thought I was from the United States.

And so, I realize now that I was surrounded by cultural assumptions about the sound of language within my family that became generalized when they were projected onto a societal level. My mother's insistence on correcting my father's grammar and pronunciation echoed her own need to sound English in front of parental figures who clung to their heritage and its ties to the British empire despite their choice to leave England. By the same token, the sound of my father's voice seemed to contribute to his sense of never quite fitting in. He used to say he was always at a disadvantage because, as he would put it, he "talked funny." He seemed to carry with him the sort of shame about not sounding quite right that reflected the notion of acculturation that was so in vogue when he immigrated to Canada in the 1960s; his stories of whispering Serbian to his sister in the supermarket so people didn't hear his foreignness informed how I understood Richard Rodríguez's (later, heavily critiqued) autobiography, *The Hunger of*

Memory, about learning to speak a public language (English) while leaving the private, family language (Spanish) at home.

Yet, in another turn that shows the daily relationship between politics and sound, my father's shame turned to outrage when his hometown of Belgrade was bombed by the United States in 1999, a case of what he might have described as American imperialism. Whatever the political and ethical reasons for the strikes, the personal reality was that my aging grandfather was alone after his wife died of a stroke the family attributed to the bombing, and travel restrictions and sanctions meant my father could not legally enter Serbia or send money to help (as immigrant communities always do, he and others eventually found ways, regardless). Instead, he took to calling a local AM news talk show to express his anger at the US government; Chris Core, the host at WMAL, took his number and, whenever the topic came up, would ask his producer to get "Mike from Vienna" on the line, if my father had not already called in first. My mother, at home, used to say she would hear that phrase over the air and cringe, wondering what he was going to say. To the radio producer, apparently the accented voice of my father came to represent the politics and culture of the Serbian diaspora in the Washington, DC, area, even though he had fairly little connection with the community. I never asked him, but I suspect my father felt he had acquired a sort of notoriety of which he was proud, perhaps because by participating on the radio, he was able to have a public voice in this country, one where his accent was authority, finally, rather than a sign of not belonging. But my father's brief encounter with public life on the radio contrasted with how his name betrayed the ways in which the sound of language created a dispossession within his linguistic identity. "Mike from Vienna" was his public, American persona, although my mother and close friends from his early years in Canada called him (with a Canadian accent) by his Serbian name, Miša, short for Miomir. Thus, he had the fairly common immigrant experience of sublating the foreign sound of his native language to local normative (now inversely accented) sound in a way that bifurcated the very linguistic essence of his identity.

I bring these personal stories to light because within them are the roots of my preoccupations with the geography of the ear and its relationship to the colonial not only as a material economic experience, but as a social and political sense of identity. Economically, my father lived the American dream of pulling himself up by his bootstraps, achieving upper-middle-class status after coming to Canada with forty-five dollars in his pocket and starting his own business. But his perceived sense of aural dispossession never left him. It is what I hear now in the messy globalized relationships through which the sound of identity plays out daily in very local circumstances, any time I meet someone with an accent, whether at home or abroad, and wonder which language to address them in.

Even so, it was not until I was at the end of writing this book that I realized, already in my forties, that when my father referred to Serbia's past, he was, ultimately, referring to a colonial setting. That thought had never crossed my mind before, especially given that my only familiarity with any history or politics of the Balkans really came through the media portrayal of Serbs, Croats, and Muslims that defined the 1990s for me. It was almost incomprehensible to anyone here that a Serb should defend Kosovo when 90 percent of the population was Albanian. I was ashamed that my father couldn't get with the program and realize that we were the bad guys. I was also ashamed that we *were* the bad guys. But the snippets of memory I have from his description of the Ottoman Empire's occupation of Serbia defined the sound of his voice for me. I was fascinated by his folkloric history of *ćela kula*, a tower of Serbian skulls built in 1809, after an uprising of Serbs against the empire led to their slaughter. The preservation of that monument, and others, as signs of foreign imposition and a celebration of resistance, were part of the lore my father told of his homeland. And now I realize that he too perceived his homeland as being threatened by empires, including the United States. It is just that those empires are inscribed in a European geography that seems unrelated to the kind of colonialism associated with the changing of the world in 1492. Within the geopolitics of the late twentieth century, the atrocities committed by Milošević and his kind secured in the public domain the impression that Serbs were barbaric, and that theirs was an old fight best left in the past. But read Misha Glenny's *The Balkans*, as I have done only recently, and it becomes clear that coloniality intertwined with the production of the modern nation-state in the Balkans repeats some fairly familiar territory for those of us who study the late Atlantic world.

Now, from the little I can glean from my humble ear (somewhat knowledgeable about Catalonia but wholly unprepared to speak authoritatively about the Balkans), fervent Catalanists share some tendencies with the nationalist ethnic groups that made up the briefly united Yugoslavia, reaching back to before 1492 to hear their origins, and thus hearing the last few hundred years as evidence of their oppression. And now I understand why, when I visited Belgrade in 2007 and had only a radio to keep me company, I stumbled across a Catalan language program. At the time I was fascinated and a bit perplexed by the broadcast. Of course! I now wish to say, ironically, echoing Jonathan Sterne's admonition that, à la Stuart Hall, "Of course" is the most ideological moment, because it is the moment when you are most unaware of your ideological inscription.[8] Of course Serbs and Catalans would feel they shared an experience of oppression within Europe. Their cultures have both been responsible for atrocities (Serbs during the war; Catalans through their role in the slave trade), but politically, when they reach into the deep time of history, they can hear themselves as colonized in their own right. The sound of their language carries all that weight of echoic memory with it. Or, at least, that colonial understanding of language can be harnessed for political reasons, as I have discussed in this book.

And so I am drawn back to why I chose coloniality and a struggle to think with decolonial theory to define this aurality. I chose coloniality to indicate a broader epistemological and sociocultural relationship to sound, one that has material effects but which does not necessarily take place in former colonies. This decision had its own disciplinary locus of enunciation to it, as I used the tools afforded to me by my formation in Hispanic Studies to think about the relationship of sound to place. But perhaps (I welcome all armchair psychoanalysts to think through this with me) coloniality also afforded me a way of hearing how the dynamics of dispossession and imperialism I heard in discussions of nationality in my own home repeated colonial patterns of relationality forged outside the geographical borders in which I was raised. Migration and travel have normalized movement, even as the echoic memory of emplacement as the sound of inclusion or exclusion remains. The multilingual tensions of the ear occur not just in Latin America, or Equatorial Guinea, but even in Eastern Europe and the United States, or more simply, any place where immigrants settle (and I use that term to invoke but also signal the limits of the idea of settler colonialism, and the role land and property play in producing the ear). *Sleight of ear, echoic memory*, the *geography of the ear* all circulate not just in the Americas or Barcelona but in these places too. As I showed in Chapter 1, the combined visuality and sonic production of race plays a crucial role in producing these geographies of the ear, a fact that forces us to recognize the impossibility of assuming that all colonial geographies of the ear play out equally in all spaces; binary conceptions of alterity are no longer sufficient for understanding these auralities.

Indeed, as Enrique Dussel has argued, we might consider this coloniality to reflect what he calls transmodernity: the webs of connections that reach far back into history and yet inform geographies that "se sitúan más-allá (y también 'anterior') de las estructuras valoradas por la cultura moderna europeo-norteamericana" (are situated beyond [and also "prior"] to the structures valued by modern European North American culture).[9] Dussel concludes that transmodernity implies the geographical construct of what he calls "la exclusión cultural de las víctimas de la Modernidad" (the cultural exclusion of the victims of Modernity).[10] We may choose to expand the geography of our frame of study into a Global Hispanophone or a Global South; we may choose to interrogate the relationship between the visual and the aural as it relates to race, but regardless, in the public sphere, this mode of thinking alterity in terms of victims persists, as I showed in Chapter 4, and it is, as Justin Crumbaugh has shown in a different context, a highly persuasive political tool.

Sound crosses national and transnational lines daily, in ways that this book has only scratched the surface of. More work must be done. Media platforms today are constantly expanding the fields in which we as scholars must work to understand sound as an inherently transnational, multilinguistic, interdisciplinary object of study. The Occupy Wall Street movement was inspired by the *indignados* movement in Madrid, but the *okupa* ethos that undergirded that movement has a long history that traces back to the punk squats of the 1980s and 1990s, all while merging with antiglobalization protests that were forged online in the 1990s, from Chiapas to Seattle to Buenos Aires.[11] The *caceroladas* (banging on pots and pans in protest) that, at least according to one testimony, drove one of Argentina's five presidents from office in 2001 during the economic crisis, resonate not only with the sounds of protest in Barcelona, but with the various *cacerolazo* apps available on mobile phones everywhere now, under names like "Cacerolazo Colombia," "Cacerolazo Chile," "Riot—Make Noise," or "Cassolada 2.0," as they are wielded in prodemocracy protests around the globe. These new technologies demonstrate how the aural imaginaries that question the local differences in law are circulated through a globalized circuit of media production, reception, and transduction that reveals just how extensive the colonial/globalized epistemologies that shape the sounds of community are, whether heard from a local or distant *locus auditivo*.[12]

Today it is virtually impossible to silence the sounds and voices that signal how experiences of dispossession occur. But hearing another's listening ear as faulty, suspecting that it has misheard, that it is in need of restructuring, remains in place as voice continues to be the predominant construct of democratic participation. The remixing of political sound into a variety of globally dis- and reconnected movements that share certain epistemological processes in their consideration of capitalism, colonialism, and aurality thus reveals a constant making and remaking of echoic memory, as an aural imaginary of modernity and capitalism is received corporeally and (re)produced cognitively and ideologically in ways that constantly move transnationally among globalized spaces. The Russian-made bots that intervened in the independence movement in Catalonia, and which I briefly touched on in Chapter 4, are nothing compared to what artificial intelligence programs can do now. It is yet to be seen how the trilingual sounds and accents of Barcelona so strikingly captured in Brigitte Vasallo's *PornoBurka*, which I analyzed in Chapter 2, will be affected as the mediatic relationships between voice and body become another creative tool in a politics of sound that circulates at different scales and in overlapping ways on a daily basis.

No one can hear globally. And as my own incomplete knowledge even of my personal audible past reveals, listening locally may be just as difficult because of the ideological, affective, and intellectual frames—the echoic memories—that come together in the ear. And so I end with a humility that hopes to have heard into the geography of the ear a culture and politics of sound that recognize limits as well as possibilities. Two decades ago Michelle Hilmes asked if there was a field of sound culture studies, and, if so, if it even mattered.[13] There is no question anymore that the field exists. What matters, in my humble opinion, is how to approach the geographies of the ear we bring to it.

Acknowledgments

I have many people to thank for helping me understand the Barcelona counterculture and underground. Coming from a literature background, it was new for me to rely so much on interviews and conversations, but without them this project would not have come together. My heartfelt appreciation first and foremost to Picarol, whose generosity made my visits to Barcelona not just illuminating, but fun. Thanks to him, I was fortunate to meet many people who graciously took time to share their perspectives: Víctor Nubla and Maria Vadell, of Gràcia Territori Sonor; DJ Shak Benavides, founder of TeslaFM; and Pepito Simó of the Llibrería Cap i Cua in Gràcia.

Thanks to the wonderful Sonia Betancort, I was also fortunate to meet Marc Caellas and Juan Pablo Roa, both of whom provided me with books on Barcelona and sound poetry that have been instrumental here. Without Sonia I also would not have gotten to know Francisco José Montes Fernández, who so generously shared his extensive knowledge of free radio in Europe with me. In a twist of globalized fate, moreover, through the Peruvian scholar Luis Millones I was able to enjoy debates and discussions on the past, present, and future of the Catalan independence movement with some of those closest to it.

Closer to home, I am extremely grateful to Javier Krauel, José Luis Venegas, Núria Vilanova, Leslie Harkema, Joanne Rappaport, and Gwen Kirkpatrick for reading portions of this text and offering, always, insightful feedback. Matthew Bush deserves a special shout-out for reading my work and being a sounding board for my ideas over the past decade. Thanks also to Cristina Sanz for her consistent support of this project, and to Alfonso Morales-Front, Víctor Fernández-Mallat, and Gabe Rodríguez for their phonological and sociolinguistic expertise. I am also indebted to Luis Cárcamo-Huechante for his suggestions regarding the shape this manuscript could take, and to Álvaro Kaempfer, who invited me to share some of my early work on protest with his students; their questions on democracy helped me see the material in Chapter 4 in a new light. I am also fortunate to work with wonderful colleagues throughout Georgetown, but especially in the Spanish and Portuguese department and the Comparative Literature program, all of whom make coming to campus a delight. Many of them helped me at various stages of this process, including Francisco LaRubia-Prado, Adam Lifshey, Vivaldo Santos, and Alejandro Yarza. Carole Sargent is truly a gift to faculty at the university. I am also grateful for the research grants I received in support of this book from the Georgetown Humanities Initiative, the Georgetown Americas Institute, and the Office of the Provost at Georgetown.

This project has taken a long time, and many graduate students have assisted me with this work. A special thanks to Diego Maggi for transcribing interviews (AI capabilities did not yet exist), Alexandra Mira-Alonso for editing the final manuscript, and Martina Thorne for collecting helpful documents on immigration in Catalonia. I was also fortunate to have the assistance of Xabier Fole-Varela, Kate Toll, Annie Robinson, Marina Young, Varun Biddanda, and Maggie Dunlap at various stages of this project.

I must also thank Courtney Berger for her extreme patience and generous feedback; Jonathan Sterne and Lisa Gitelman for their interest in my work; and the anonymous readers at Duke University Press for their support of this project and their very helpful suggestions.

Last, the warmest thanks go to my husband, Dave, and my children, who have accompanied me throughout this process. I never tire of the everyday sounds of their voices in our home, and my joy in observing my kids' seemingly innate travel ears only grows with each new adventure. *Los amo y los adoro, hoy y siempre.*

Notes

PREFACE

1 Maragall, *Antologia Joan Maragall*, n.p.; Bertrana, "La meva espurna," n.p.
2 Resina, *Barcelona's Vocation*, 94.
3 Resina, *Barcelona's Vocation*, 95.
4 *Crónicas* are a form of literary journalism unique to Latin America and
   Spain, combining social commentary with literary narrative style. They
   are often compared to New Journalism in the United States, although
   they predate it by almost a century. It is common for *crónicas* to be col-
   lected and published in book form, as in this case. See Gentic, *Everyday
   Atlantic*.
5 *Flâneuse* is the feminine form of *flâneur*, a nineteenth-century term used
   to describe cultural observers who walk the city but remain detached
   from it.
6 Said, *Orientalism*.
7 Sterne, *Audible Past*.

INTRODUCTION. ECHOIC MEMORIES OF DISPOSSESSION

1 R. Murray Schafer's soundscape, Don Ihde's phenomenology of sound,
   Emily Thompson's discussion of the acoustics of modernity, Dylon
   Robbins's *Audible Geographies*, and Steven Feld's acoustemology have all
   illustrated this point in different ways.
2 Kelly, *Sound*, 12.
3 In *Audible Geographies*, Dylon Robbins argues that place is both "a historical, specific location [and] a figurative, or discursive one, albeit with
   very concrete conditions and consequences" and that it is not just geographical or political, but sensorial (20). For me, thinking through Henri
   Lefebvre's concept of space, the geography of the ear is the epistemological frame that produces the place as a historical, sensorial, and political
   location; these cannot be separated out as such.

4   Even saying the city is "modern" is complicated. Brad Epps writes of four
     overlapping modernities at play in the city:

> One, marked by the triumph of bourgeois liberalism, that runs from
> the rise of industrialism, the demolition of the city walls, and Cerdà's
> planned expansion to the Universal Exposition of 1888 and beyond;
> another, marked by the growing contestation of bourgeois hegemony,
> that runs from the first bouts of Anarchist direct action or terrorism
> (depending on one's perspective) in the 1890s through the popular
> uprising against the mobilization of troops to Morocco in 1909 known
> as the "Setmana Tràgica" or "Setmana Gloriosa" (again, depending on
> one's perspective) and the revolutionary movements of the Civil War to
> the triumph of Franco; a third, under Franco, marked by a technocratic
> capitalism hostile to civil liberties and democratic process and largely
> oblivious or indifferent to historical and environmental preservation;
> and a fourth, generally called postmodern, in which neoliberal global
> capitalism grapples with environmentalism, historical memory, and the
> rights of citizens and neighbours. ("Barcelona and Modernity," 152)

5   In 1978, the new constitution declared that Castilian Spanish was the of-
     ficial language of the State and that all Spaniards had the responsibility
     to learn it and the right to use it. But it also established each autono-
     mous community as having the right to declare other official languages,
     and declared the plurilingual culture of Spain "patrimonio cultural."
6   See Chris Ealham's works *Anarchism and the City* (2010) and *Class, Culture,
     and Conflict* (2005) on class, politics, and neighborhood differences, espe-
     cially in Gràcia and El Raval, from the nineteenth century on.
7   This project of linguistic preservation was intimately tied to Catalans'
     own process of language formation and ideological positioning a
     century earlier, as I discuss in Chapter 1.
8   See Antoni Bassas in Casals i Martorell, *El català en antena*, 15, quoted in
     Chapter 1 of this book.
9   See current data from the Ajuntament at www.barcelona.cat.
10   Increasingly, polyglot societies navigate similar issues to those first con-
     fronted in the contact zones of the earliest European colonies around
     the globe, not because the polyglot sound is new but because, as a lived
     experience, it has continued to exist alongside attempts to produce mono-
     lingual national identities, which are increasingly fragile. See Gueneli,
     "Young, Diverse, and Polyglot"; Ruiz, *Slow Disturbance*; Robinson, *Hungry
     Listening*; and Dalov, *Sounds of Aurora Australis*. The Iberian Peninsula has
     itself been polyglot for over two millennia.
11   Stoever, "Splicing the Sonic Color Line," 64.
12   Ochoa Gautier, *Aurality*, 5.

13   There are many economic reasons for the independentist argument as well, but here I am interested in the cultural politics around language and national identity. See Crameri, *"Goodbye, Spain?"*

14   Catalanism and independentism are not, nor have they ever been, synonymous. They also play out differently in Barcelona and other regions of Catalonia. See Resina, *Barcelona's Vocation*; Minder, *Struggle for Catalonia*; and Crameri, *Catalonia*.

15   Minder, *Struggle for Catalonia*, n.p. In the same year, 170 prominent writers wrote a public letter demanding that Catalonia's bilingualism be revoked in favor of Catalan being the national language of the community.

16   Pratt, *Planetary Longings*.

17   Escobar, "Worlds and Knowledges Otherwise," 184.

18   Dussel, "Transmodernidad e interculturalidad," 14.

19   See Aníbal Quijano's "Colonality and Modernity/Rationality," on the construction of racial and geopolitical subjectivities based on the coloniality of reason, and María Lugones, "Decolonial Feminism," on gender.

20   Mignolo, "The Geopolitics of Knowledge," 228.

21   Venegas, "Uneven Souths," 532.

22   Venegas, "Uneven Souths," 536.

23   Calderwood, *Colonial Al-Andalus*, 167, 178.

24   As the edited volume by Dolors Poch Olivé demonstrates, accent in Barcelona is not just a question of regionalism, but of bilingualism, through which Castilian and Catalan linguistically interfere with one another. *El español en contacto*, 317.

25   Wolfe, "Settler Colonialism and the Elimination of the Native," 387.

26   Lafarga i Oriol, *Gràcia*, 12.

27   As I discuss in Chapter 1, *Catalunya ciutat* is a concept developed in the early twentieth century by the *noucentista* intellectual Eugeni d'Ors.

28   In addition to the wide variety of pronunciations that obtain throughout the Països Catalans, differentiating (among other places) the Balearic Islands from Valencia, El Pont de Suert from Benicarló, and Barcelona from all of them, there are also local linguistic tendencies, socially marked, including the *xava* and *bleda* pronunciations that mark Barcelona's linguistic soundscape. See Ballart Macabich.

29   Panagia, *Political Life of Sensation*, 3. This includes the "noise of utterance" (61).

30   Chávez, *Sounds of Crossing*, 8.

31   Halberstam argues that the asterisk in this term "refus[es] to situate transition in relation to a destination." *Trans**, 4–5.

32   Steingo and Sykes, *Remapping Sound Studies*, 5. Ironically, as the editors themselves acknowledge, approaches like this one often come from the North.

33   Martin-Márquez, *Disorientations*, 8–9.

34 Calderwood, *Colonial Al-Andalus*, 9.

35 See Rodrigo y Alharilla, "Cataluña y el colonialismo español (1868–1899)."

36 As Josep Maria de Sagarra would put it in *Vida privada* (1932), the Catalan language was embarrassing because it was the language of cooks, coachmen, and poets (n.p.).

37 According to F. Xavier Vila, Barcelona is socially trilingual (Spanish, Catalan, and English) but officially bilingual (Spanish and Catalan). Only around 19 percent of the population claims to speak 100 percent in Spanish, and only 7 percent say they speak no Castilian at all, only Catalan. "¿Quién habla hoy en día el castellano en Cataluña?," 147.

38 Lefebvre, *Production of Space*, 44.

39 Lefebvre, *Production of Space*, 47–48.

40 Lefebvre, *Production of Space*, 42, 44.

41 Feld writes against soundscapes because of their association with landscape as a "physical distance from agency and perception." "Acoustemology," 15. He stresses, instead, relationality as an ontological assumption that "life is shared with others-in-relation" (15). Nevertheless, I argue here that part of that relationality is necessarily linguistic, and it is in how we listen to that linguistic sounding within spaces that acoustemology as a sonic knowledge is produced and circulated to create spaces.

42 The show's title, *Poland*, is a reappropriation of the derogatory term used by some Spaniards to refer to Catalans as the "Polacks" of Spain.

43 On these debates, see Minder, *Struggle for Catalonia*, chap. 5. On these early debates in Aragon, see Bada Panillo, *El debate del catalán en Aragón*.

44 Despite actors paying close attention to the linguistic tics of the politicians they portray, Ugarte Ballester argues that accents or other differences from normative Catalan produce a *català deformat* (deformed Catalan) that can be used to represent all "foreign people," no matter their provenance. "El *Polònia* de TV3," 21.

45 Epps, "Barcelona and Modernity," 158.

46 According to Bob Snyder, echoic memory refers to the sensory memory of the brain through which sounds that hit the inner ear in a continuous stream are received as raw data and later coded and categorized to be retained as short- or long-term memories. Perceptions categorized as long-term memories provide an unconscious context for a listener's perceptions of a sound in the moment: "What we already know literally determines what we see and hear, which means that we see and hear what we look *for* more than what we look *at*." Snyder, *Music and Memory*, 11; emphasis in original.

47 Helmreich, "An Anthropologist Underwater," 622.

48 Feld, "Acoustemology," 13–14.

49 Pinchevski, *Echo*, 36.

50 Pinchevski, *Echo*, 36.

51 Ihde, *Listening and Voice*, 69.

52 Eidsheim, *Sensing Sound*, 7.

53 Feld, "Acoustemology," 14, sic.

54 See Rivera Cusicanqui, "*Ch'ixinakax utxiwa*."

55 Dolar, *Voice*, 541.

56 Robinson, *Hungry Listening*, 51.

57 "All linguistic practices feed into a single 'love of the language' which is addressed not to the textbook norm nor to particular usage, but to the 'mother tongue'—that is, to the ideal of a common origin projected back beyond learning processes and specialist forms of usage and which, by that very fact, becomes the metaphor for the love fellow nationals feel for one another." Balibar, "The Nation Form: History and Ideology," 98.

58 Erlmann, *Reason and Resonance*, 9–11.

59 Parry and Keith, *New Iberian World*, 290.

60 Faudree, "How to Say Things with Wars," 186.

61 Antonio Cornejo-Polar in *Escribir en el aire*, Ángel Rama in *La ciudad letrada*, and Joanne Rappaport and Tom Cummins in *Beyond the Lettered City* have illustrated that everything from grammars to maps to paintings and city planning documents produced *over time* the perpetuation of a lettered city that largely silenced oral cultures.

62 Ochoa Gautier, *Aurality*, 33.

63 Rabasa, "Thinking Europe," 51.

64 Ochoa Gautier notes, "The epistemological emergence of orality, as well as that of embodied musical others, arises at the same historical moment as the idea of autonomy in Western art music." *Aurality*, 102.

65 Ochoa Gautier, *Aurality*, 14.

66 As Roshanak Kheshti has argued, in listening, "we respond to the sounds with our feelings, and it is this affective investment that takes us out of our selves, into the aural imaginary where we engage in incorporeal material exchanges with the other," doubling aurality as the ear's capacity to mean into the ear's capacity to produce relationality among and within communities. "Touching Listening," 727.

67 Truax, "Acoustic Space," 254.

68 As Patricia Seed has pointed out, the reading of the text is a ritual derived from Islamic practices of submission in Spain; the word *requerimiento* is a translation of the Arabic term *da' ā*, meaning "to summon," "to implore," and to seek submission from another population all while "fighting according to proper legal principles." *Ceremonies of Possession*, 76, 72. What changes in the Americas is that in the Islamic use of the *da' ā*, those who took over lands in Spain hoped their new subjects would not convert quickly, because until they did they could be taxed; whereas here conversion is expected to be immediate and justifies violence in a way the Islamic text did not. Seed, *Ceremonies of Possession*, 79.

69   See Minder, *Struggle for Catalonia*, on that debate. Rodrigo y Alharilla, "Cataluña y el colonialismo"; Tsuchiya, "Monuments and Public Memory"; Piqueras, *Negreros*; and Fradera and Schmidt-Nowara, *Slavery and Anti-slavery* have clearly illustrated the large role Catalans played not only in the plantations of Cuba, Puerto Rico, and Equatorial Guinea, but in the illegal slave trade that supported them after abolition.

70   Diana Taylor, *Archive and the Repertoire*, 54.

71   Ngai, *Ugly Feelings*, 30.

72   In the process, the new rhetoric "sever[ed] the linguistic association between *requirement* and *da' ā* . . . [and] by relabeling the practices, potential linguistic reminders of its Andalusí Islamic origins were erased." Seed, *Ceremonies of Possession*, 95.

73   Mbembe, "Necropolitics," 172.

74   Ochoa Gautier calls these sounds "untamed vocality." *Aurality*, 167.

75   Segato, *La nación y sus otros*, 617.

76   See Eidsheim's *Race of Sound* on how the visual and sonic aspects of voice intertwine for the listener.

77   Cárcamo-Huechante, "Colonial Obliteration?," 246.

78   Ronaldo Radano and Tejumola Olaniyan have also written that one problem with invoking terms like "empire" is that, if overused, they descend into the realm of abstraction. *Audible Empire*, 2.

79   From the 1960s on, *Països Catalans* was meant to reflect a "sense of community identity, across administrative boundaries, based upon an historical and socio-linguistic reality." Costa Carreras and Yates, "Catalan Language," 6.

CHAPTER 1. TRAVEL, RACE, AND THE COLONIAL SLEIGHT OF EAR

1   Pratt, *Imperial Eyes*, 4.

2   Pratt, *Imperial Eyes*, 3.

3   I use *he* specifically in this case, since overwhelmingly the readers Pratt describes are men who identify as such.

4   Mowitt, *Sounds*, 5.

5   Giles, *Virtual Americas*, 6.

6   Giles, *Virtual Americas*, 2.

7   Feld, "Acoustemology," 185.

8   Balibar, "Nation Form," 103–4.

9   See Lippi-Green, *English with an Accent*, on the many ideological aspects of accent and dialect, written from a linguistics perspective.

10   Ochoa Gautier, *Aurality*, 15.

11   See Ochoa Gautier, *Aurality*, chap. 2.

12   Fuster, "Per a una cultura catalana majoritària," 65.

13 Casals i Martorell, *El català en antena*, 15. Literature played a huge part in the linguistic production of Catalanism as well. See King, *Escribir la catalanidad*; Crameri, *Catalonia*; and Resina, *Barcelona's Vocation*.

14 At the same time, as Marina Garcés explains in *Ciutat Princesa*, Catalonia's development of a new democratic bureaucracy in the 1980s and 1990s excluded some Catalan speakers from the propagation of their own soundscape by appropriating a hegemonic linguistic norm that heard accent and code-switching as indicative of a faulty ear. She had grown up speaking "un català ric i porós, farcit de castellanades . . . de frases dites en francès i de giragonses de l'Alt Empordà" (a rich and porous Catalan, full of Castilianisms . . . of phrases said in French and idiomatic varieties from Upper Empordà) and had a doctorate in philosophy, but could not apply to substitute teach because she lacked the paperwork to prove she had reached level C proficiency. Garcés, *Ciutat Princesa*, 49–50. Her perspective calls attention to how a new bureaucratic Catalan archive in construction counted and categorized speech, linking a hegemonic sound of Catalan to notions of modernity.

15 Balibar, "Nation Form," 104.

16 Tonkiss, "Aural Postcards," 305.

17 Bhaba, "DissemiNation," 297.

18 Eidsheim, "Marian Anderson."

19 Tofiño, *Guinea*, 36–37.

20 Goode, *Impurity of Blood*, 9.

21 Cassany, "Les aventures extraordinàries d'en Massagran," 49.

22 Several scholars have traced the colonial projects that have linked Spain to Africa. See especially Tofiño's capacious *Guinea, el delirio colonial*, for a detailed history of Spain's relationship with Equatorial Guinea; Calderwood's *Colonial al-Andalus*, on how Franco's recuperation of Al-Andalus in Spain, in part through music, produces Morocco's national identity; Eastman's *Missionary Nation*; Nerín i Abad's work on *hispanotropicalismo;* and Rodrigo y Alharilla's "Cataluña y el colonialismo."

23 Ramon Llull published the first book in Europe dedicated to children's education, the *Doctrina Pueril*, in Catalan in 1282. It is widely agreed that the first adult Catalan novel was not written until 1932, when Josep Maria de Sagarra published *Vida privada*. In other words, *Massagran* is only differentiated by the age of its presumptive audience.

24 Although the novels were published separately, I cite here from the double edition of the novels, put out by Editorial Bambú in 2018, under the simplified title, *Massagran*. The original *Aventures extraordinàries d'en Massagran*, edited by Josep Baguñà, was published in 1910 in the ninth volume of the magazine *En Patufet*. The sequel, *Noves aventures d'en Massagran*, appeared in 1913.

25 The four-part series starred Felip Peña, Nadala Batiste, Miquel Graneri, Carme Contreras, Ferran Poal, Oscar Molina, and Monserrat Puga.

26 See Castillón, "L'Any Massagran reivindicarà."

27 According to the website, the language, themes, and other content are guaranteed by the Corporació Catalana de Mitjans Audiovisuals (CCMA; Audiovisual Council of Catalunya) to be suitable for children (CCMA, "Coneix el sx3").

28 Folch i Camarasa, *Bon dia, pare!*, 21.

29 The title page for each story states, "Història completa i detallada de les trifulgues, peripècies i desoris d'un noi de casa bona" (Complete and detailed history of the tribulations, vicissitudes and disappointments of a boy from a well-off family).

30 Folch i Torres, *Massagran*, 192.

31 *Noucentisme* was an official movement of the Catalan bourgeoisie, defined by Eugeni d'Ors and directly tied to the political program of the Lliga Regionalista. The effort sought to produce an urban order that would be both aesthetically pleasing and a realization of a superior civil society. See Gentic, *Everyday Atlantic*; and Krauel, "Eugeni d'Ors's Politics," esp. 434–35.

32 Folch i Torres, *Massagran*, 311. (Hereafter cited in the text by page number.)

33 The series inspired the creation of the Castilian *Biblioteca Chiquitín*, published in Madrid. Although the Massagran stories were translated into Castilian, they do not seem to have had the same reception as they did in Catalonia.

34 Jardí Casany, *Els Folch i Torres i la Catalunya del seu temps*, 9.

35 Fuster, "Per a una cultura catalana majoritària," 59–60.

36 Fuster, "Per a una cultura catalana majoritària," 56.

37 For a history of Pompeu Fabra, see Costa Carreras and Yates, "Catalan Language."

38 See Vázquez Montalbán's *Barcelonas* for more on this history.

39 Folch i Camarasa, *Bon dia, pare!*, 114.

40 Cánovas del Castillo, *Apuntes*, 11.

41 Nerín i Abad, "Mito franquista," 17.

42 Quoted in Archilés, "When the Empire Is in the South," 244.

43 Nerín i Abad, "Mito franquista," 12. The seemingly civil aspect of the colonization project was tied to other European efforts in Africa: For example, the Asociación Española para la Exploración de África (Spanish Association for the Exploration of Africa) was a subsidiary of the Belgian Association Internationale Pour L'exploration et la Civilisation de l'Afrique Centrale (International Association for the Exploration and Civilization of Central Africa). Fernández-Fígares Romero de la Cruz, *La colonización del imaginario*, 75. See also Labanyi, "Raza," 27.

44 Bestard, "Prologue," 15.

45 Bassegoda, "L'excursionisme català," 31–32. Interestingly, the children's magazine *Cavall Fort* reasserts the importance of *excursionisme* for Catalan children in their 2014 book *Anem d'excursió per Catalunya*. See Quera, *Anem d'excursió per Catalunya*.

46 In Iberian and Latin American literary traditions, *costumisme/costumbrismo* refers to writings that depict the details of everyday life and its customs, often within a regionalist frame.

47 *Ateneus* were grassroots cultural centers, modeled after bourgeois clubs and institutions, but with a socialist or anarchist bent. They provided literature, leisure, and even co-op food stores to local working-class residents of the city. See Ealham, *Class*, chap. 2.

48 Known as La Moreneta because of the dark color of the statue's face, the Romanesque carving of the Virgin is displayed in the basilica. This, in turn, was built around a cave where, legend holds, in the ninth century, some local children reported seeing the Virgin. Since the late nineteenth century, the itinerary that leads tourists to the cave is marked by modernist artwork testifying to the intense relationship between this site and modern Catalan identity.

49 See Cicerchia, *Modernidad, nacionalismo y naturaleza*, 69–81.

50 Folch i Torres, *África Española*, 3, sic.

51 Folch i Torres, *África Española*, 96–97. (Hereafter cited in the text by page number.)

52 Folch i Torres, *Massagran*, 191. (Hereafter cited in the text by page number.)

53 The Valladolid debate, which took place in 1550–51 between Bartolomé de las Casas and Juan Ginés de Sepúlveda, concerned the moral responsibility of the Spanish toward the Indigenous.

54 Famously, Ginés de Sepúlveda argued that the Indigenous people of the Americas were not fully human, but rather were *homúnculos* (homunculi) and thus should be approached not as potential converts but as enemies against whom war would be justified.

55 Folch i Torres, *Massagran*, 178.

56 Folch i Torres, *África Española*, 24.

57 I use angle brackets (< >) to signal written expressions of a sound (graphemes) and forward slashes to signal phonemes (units of sound in a given language). Occasionally I use square brackets ([ ]) to signal allophones, which are specific pronunciations of the same phoneme. I am grateful to the linguists who assisted me in these representations. Any errors are mine.

58 This <k> is repeated in the Castilian translations.

59 Translating these scenes is quite difficult, as the sounds at play in the puns do not work in English in the same way as they do in Catalan or even in Castilian Spanish. Costa Carreras and Yates argue the morphological

and vocabulary differences between Eastern Central and Western Catalan can be compared to the difference between Peninsular and American Spanish, or British and American English, but such differentiations do not pick up on the political tendentiousness of the geographical question as it relates to Catalan speech. The issue is a thorny one and relates as much to how speakers from Barcelona hear rural speakers as accented, and vice versa, as it does to questions about how far the Països Catalans extend, whether or not Valencian and Catalan are related, and how much political interests about Catalonia vis-à-vis Spain influence how people hear these questions. "Catalan Language," 26. As I am not a trained linguist and, moreover, do not wish to pick a particular kind of pronunciation as representative of "barbarism," I have instead tried to replicate the idea of parsing, the faulty assembly of phonemes into morphemes, that takes place in the text.

60 Westergaard, "Microvariation."

61 Lightfoot further explains, "E-language is the amorphous mass of language out in the world, the things that people hear. There is no system to E-language; it reflects the output of the I-language systems of many speakers under many different conditions, modulated by the production mechanisms that yield actual expressions." *Born to Parse*, 37–38.

62 On the long-ranging trope of cannibalism in colonial narrative, see Jáuregui, *Canibalia*.

63 Mowitt, *Sounds*, 17.

64 I am grateful to my colleague, phonologist Alfonso Morales-Front, for his reading of these representations.

65 Harkness, *Songs of Seoul*, 17.

66 Harkness, *Songs of Seoul*, 18.

67 We could also argue that here accent is presented as having a qualic effect on the reader, which Harkness argues "refers to the actual instantiations of culturally conceptualized sensuous qualities that people orient to, interact in terms of, and form groups around." *Songs of Seoul*, 14. For Harkness, this qualia can be "tuned and manipulated phonically to align with a sonically experienced framework of value" on an individual level (15). Here, the fictional production of certain vocal sounds, as they relate to a proper or laughable diction or delivery of voice, effects a qualic tuning on the ear of the reader.

68 *Trencar l'olla* might best be translated as "break the piñata," but that translation signals the lack of an equivalent English term to describe this common game, and subsumes Catalan identity to a Hispanic one.

69 Connor, *Beyond Words*, 13.

70 Sterne, "Hearing," 67.

71 Ong, *Orality and Literacy*, 43–44.

72 LaBelle, *Lexicon of the Mouth*, 5.

73 Kahn, *Noise, Water, Meat.*

74 See Quijano, "Coloniality and Modernity/Rationality"; and Escobar, "Worlds and Knowledges Otherwise."

75 LaBelle, *Lexicon of the Mouth*, 5.

76 Folch i Torres, *Massagran*, 34.

77 The style in which they are drawn is, to me, so stereotyped it is hard to understand how defenders of the Massagran books could state that Junceda is celebrated for realism—which, according to Cassany, is evident in "l'ull sever de la turtle, o en la beatífica expresió de l'hipopòtam o en els trets blancs que es reconeixen rere el rostre pintat de negre d'un català" (the stern eye of the Tortuga, or in the beatific expression of the hippopotamus or in the white features that are recognizable behind the black painted face of a Catalan). "Les aventures extraordinàries d'en Massagran," 50.

78 Folch i Torres, *Massagran*, 333.

79 Eidsheim, "Marian Anderson," 663–64, sic.

80 The Massagran adventures appeared on the record with other stories, including traditional tales like "Cinderella" and "Snow White," inscribing Massagran firmly within the canon of children's fairy tales.

81 Davies, *La última escalada*, 9.

82 Except for a guard who yells out one line when he discovers Massagran trying to escape after he is captured by the *kukamuskes*, Penkamuska is the only person other than Massagran to speak in the recording.

83 We may contrast this representation of a barbaric accent to the way in which an English history professor's language is depicted in the comic book *Massagran i els pirates*. He is not accented, but he consistently uses the infinitive to express himself.

84 Vila, "¿Quién habla hoy en día el castellano en Cataluña?," 137.

85 Beginning in 1946, books could be published in Catalan if they demonstrated a folkloric or literary tradition, and public celebrations, such as the Festivities of the Enthronement of the Virgin of Montserrat in 1946 and 1947, were allowed to take place entirely in Catalan. Vilardell, *Books Against Tyranny*, 55. Music was considered harmless. In the 1960s, the melding of poetry and music into the *cantautor* tradition broke through the acoustic limitations of the Francoist public sphere. By 1970, occasional radio broadcasts were allowed in the language, and several publications, including three dedicated to children and teens, *Cavall Fort*, *L'Infantil*, and *En Patufet*, were published entirely in Catalan. Dowling, *Catalonia Since the Spanish Civil War*, 95.

86 Fuster, "Per a una cultura catalana majoritària," 65.

87 Domínguez Quintana, "Marés, Faneca y el torero."

88 No one I spoke with about the character in Barcelona—booksellers, librarians, or other academics—brought up the racism of the text.

As a point of comparison, we may consider the long-running British children's series *Noddy*, which featured Mr. Golly and other golliwogs, until the 1980s, when they were replaced by other characters. See Nikkhah, "Noddy Returns Without the Golliwogs."

89 Such representations are not limited to the Massagran tales. Madorell also illustrated a comic book series based on Joaquim Carbó's character Pere Vidal, which featured similar visual portrayals of Black characters. In 2010, a new musical based on Carbó's books featured the main character in blackface tap-dancing in the jungle as he held his hands aloft in an homage to the *sardana* while singing about the Catalan language.

90 Folch i Camarasa, *Les aventures*, 22. (Hereafter cited in the text by page number.)

91 See Medak-Saltzman, "Transnational Indigenous Exchange," and Miller, "Blushing at the Fair," on the role of Indigenous people in expositions.

92 Westerman, "Man Stuffed and Displayed." In 2019, DNA showed the man had in fact been from South Africa and therefore had been buried in the wrong place.

93 Folch i Camarasa, *"Bruixot blanc,"* 40. (Hereafter cited in the text by page number.)

94 On the history and politics of the 1992 Barcelona Olympics, including the role of Barcelona's mayor, Pascual Maragall, and Juan Samaranch, and the relationship between the games and the development of the city, see chapter 1 of Edgar Illas's book *Thinking Barcelona*.

95 See chapter 2 of Edgar Illas's book *Thinking Barcelona*.

96 Folch i Camarasa, *Olímpics*, 8. (Hereafter cited in the text by page number.) As with the above quote from the first novel, I have tried here to capture the spirit of the quote and its use of sound, even though I recognize it is not an ideal translation.

97 Until recently, there has been little discussion of the fact that much of Catalonia's wealth in the nineteenth century—the same wealth that bankrolled the Catalanist movement of which the Massagran stories are a part—came from the slave trade. Catalonia received special permission to do business directly with Cuba, Puerto Rico, and some regions of Africa and the Philippines in the eighteenth century. Zeuske, "Capitanes y comerciantes catalanes de esclavos," 67–68. When, under pressure from the English, Spain officially abolished the slave trade in 1820, Catalan slavers were able to use already-established connections, as well as Spain's antiliberal stance at the time, to navigate around antislavery laws and illegally fill a continuing demand for slave labor. Artur Mas, the former president of Catalonia, who pushed through the independence referendum in 2017, comes from a family of slavers who broke the law to continue trading. Given his family history, authors Rodrigo y Alharilla and Chaviano Pérez have commented on the irony that when he was

reinaugurated president in 2012, he used sailing metaphors to describe himself and his work leading Catalonia. *Negreros y esclavos*, 7–11.

98 Ihde, *Listening and Voice*, 66.

CHAPTER 2. OF IMMIGRANTS AND ACCENTS

1 Although today's *their* probably reflects the idea of personhood Ocaña expressed, I follow Nazario Luque's lead, alternating between *he* and *she*, since the constantly moving play between the two seems fundamental to Ocaña's portrayal and experience of gender. See Luque, *La vida cotidiana*.

2 Beyond this musical instrumentation and a single moment in which Ocaña says to a crowd "Viva Cataluña y el resto de España," there are few other aural indicators that this film is Catalan; as far as language goes, as José Luis Guarner pointed out, "lo único catalán de su lenguaje son los títulos de crédito" (the only Catalan of the film's language are the credits). Guarner, review of *Ocaña*.

3 For Fernàndez, this use of the *tenora* is "the campiest moment in the film," the equivalent of Ocaña lifting his dress to show his penis and, in the process, demonstrating the theatricality of the rest of the show; Fernàndez argues that by including this sound, Pons reappropriates Ocaña's performativity for a pro-Catalan cause. "Authentic Queen," 97.

4 Ocaña had already gained notoriety prior to the film for the "crime" of cross-dressing and stripping in public, a charge to which he responds in the film's interview.

5 Puig, "Yo también soy travesti," 15. Below I address the difference between *travesti* and *travestí*.

6 Fernàndez, "Authentic Queen," 70.

7 Yarza, *Francoist Kitsch Cinema*, 12.

8 Quoted in Benet, *El cine español*, 384–85.

9 Pujol, "Cinema of Ventura Pons," 193–94.

10 In the past two decades or so, political debates have been generated over whether or not this movement of people from outside Catalonia to the region constitutes immigration or not. As Joan Mena Arca aptly describes, political parties like Ciutadans—a party founded in 2005— claim that Andalusians and other workers from Spain who came to Catalonia over the course of the twentieth century are not immigrants, but rather Spaniards who simply moved for work. *No parlaràs mai un bon català*, 49. This retelling of the tale of migration affirms the unity of Spain as a nation, and thus the predominance of Castilian as a shared marker of identity. The gesture is often repeated in debates around linguistic immersion in Catalan schools.

11 Accent is, generally, the lexical, morphological, and phonological features of a certain way of speaking that distinguish one dialect of a language from another. Technically "Andalusian" is a recognized dialect of Spanish. However, I use the term *accent* more broadly to refer to sounds of difference that strike the ear as geographically locating the speaker in a place (or out of place), even when the specific dialect cannot be identified. Indeed, there are multiple pronunciation patterns even within "el andaluz" that are not uniform and thus cannot reduce "el andaluz" to any one way of speaking. See Lippi-Green, *English with an Accent*, on the relationship between accent and dialect.

12 Delgado, *Andalucía y el cine*, 82.

13 Josep-Anton Fernàndez argues that "Ocaña's performances are based on . . . a discourse of nostalgia" that responds to "the loss of Andalusia as a space uncontaminated and unaffected by modernity, prior to migration and to the misappropriation of its culture at the hands of an official discourse." "Authentic Queen," 77.

14 Eidsheim, "Marian Anderson," 645.

15 As Freeman writes in *Time Binds*, "Camp is a mode of archiving, in that it lovingly, sadistically, even masochistically brings back dominant culture's junk and displays the performer's fierce attachment to it" (68).

16 Freeman, *Time Binds*, 64; italics in original.

17 Monzó, "Inmigrantes para siempre."

18 Halberstam, *Trans**, 4–5.

19 Chion, *Voice in Cinema*, 6.

20 Vilarós seems to be referring to two of Francisco de Goya's best-known paintings, *La maja desnuda* and *La maja vestida*, to draw the comparison between Ocaña's exaggerated persona and his naked soul. Vilarós, *El mono del desencanto*, 186; sic.

21 Chion, *Voice in Cinema*, 126.

22 Derrida, *Of Grammatology*, 304.

23 Derrida, *Of Grammatology*, 307.

24 Derrida, *Of Grammatology*, 306.

25 Dolar, *Voice*, 544.

26 Dolar, *Voice*, 544.

27 See, as just one example, Bada Panillo, *El debate del catalán*.

28 On the literary debates in question, see Crameri, *Catalonia*.

29 Benet, *El cine español*, 398. A 2002 performance of Andalusian *coplas* by Isabel Pantoja in the Liceu Theater, the iconic site of Catalan identity, caused an uproar among Barcelona's elite because the performance was not in Catalan, and thus suggested association with the *xarnego* culture. Martínez, "Stick to the Copla!," 92.

30 With the backing of the Institut d'Estudis Catalans, in the 1930s Vandellòs published books such as *La inmigració a Catalunya* (1935). For more

on his and similar texts on *mestissatge* (miscegenation) in Catalonia from the 1930s, see García Cárcel and Simón i Tarrés, "Nación y familia."

31 Fernàndez, "Thou Shalt Not," 1.

32 Quoted in Crameri, *Catalonia*, 37.

33 Briefly, *seseo* refers to a feature of Peninsular Spanish pronunciation in which the sounds represented by the letters <c> (before <e> and <i>) and <z> are pronounced the same as <s>, as [s] rather than [θ]; in Spain, this feature is common in some parts of Andalusia but not, for example, in Madrid, making it a marker of a southern accent. Similarly, *yeísmo* refers to the absence of a distinction between the pronunciation of <ll> and <y>, a pattern that, while not exclusive to Andalusia, is more prevalent there than in other areas of Spain, such as Castilla y León and Asturias, where some older speakers still maintain the distinction.

34 Chion is citing Lacan when he writes "thingify difference," referring to Lacan's understanding of the *objet a* as a part object that is fetishized to represent the whole. *Voice in Cinema*, 1.

35 See Whittaker, *Spanish Quinqui Film*.

36 González Rodríguez, *Pensar*, 46.

37 The *locus of enunciation* is a postcolonial studies term, referring to the ideological and geocultural place from which one speaks; González Rodríguez's *locus auditivo* highlights the place from which one listens. For an early definition and discussion, see Mignolo, "Editor's Introduction."

38 In a previous work, I critiqued the locus of enunciation for its dependence on place and highlighted the need to consider temporality as a part of the formation of the subject. See Gentic, *Everyday Atlantic*.

39 Feld, "Places Sensed," 184.

40 Walden, *Sounding Authentic*, 6.

41 See Herrera Ángel, "El teatro andaluz costumbrista."

42 Delgado, Juan-Fabián, *Andalucía y el cine*, 16–17. *Andaluzada* refers to stereotyped notions of being Andalusian.

43 Vázquez Montalbán, *Crónica sentimental de España*, 162.

44 Woods-Peiró, *White Gypsies*, 246.

45 For more on the global politics of the Encuentros, see Benet, "Radical Politics."

46 Benet, *El cine español*, 335.

47 See Ayats and Salicrú-Maltas, "Singing Against the Dictatorship."

48 Chanan, *Politics of Documentary*, 220, sic.

49 Connor, *Dumbstruck*, 12.

50 Jarman-Ivens, *Queer Voices*, 19.

51 Jarman-Ivens, *Queer Voices*, 19.

52 Kahn, *Noise, Water, Meat*, 7.

53 Orringer, *Lorca in Tune with Falla*, 6.

54 Hirschkind, *Feeling of History*, 100–101.

55 Hirschkind, *Feeling of History*, 104.

56 Hirschkind, *Feeling of History*, 110.

57 *Zorongo* was a popular Andalusian dance in the eighteenth and nineteenth centuries. The song lyrics do not appear in every printed version of the play. For one version of the full text, see the Cátedra edition from 1999.

58 The *soleá* is a form of *cante jondo*, usually composed of three octosyllables and assonant rhyme schemes; according to Orringer, "The word in Andalusian dialect suggests solitude, aloneness" (Orringer, *Lorca in Tune*, xviii).

59 The line "moreno de verde luna" comes from the poem "Muerte de Antoñito el Camborio" from Lorca's *Romancero gitano*.

60 Orringer, *Lorca in Tune*, 68–69.

61 Robbins, *Audible Geographies*, Loc. 836.

62 Martínez, "Stick to the Copla!," 91.

63 Another vocal chord occurs in Ocaña's singing of Juanita Reina's "Yo soy esa" at the Cafè de l'Òpera, which echoes not only her song, but "Yo soy aquel" (I am that man), a powerful ballad performed in 1966 by Raphael for the Eurovision competition, which Teresa Vilarós has called "exquisitamente camp" (exquisitely camp). Vilarós, *El mono del desencanto*, 187. Thinking the performance through that international Eurovision hit, Ocaña also extends the geography of the ear related to gender outside Spain's borders.

64 Rodríguez, Anto, *¡Eres tan travesti!*, 163.

65 Chion, *Voice in Cinema*, 28.

66 I am referring to José Esteban Muñoz's concept of disidentification as linked to a "reconstructed narrative of identity formation that locates the enacting of self at precisely the point where the discourses of essentialism and constructivism short-circuit. Such identities use *and* are the fruits of a practice of disidentificatory reception and performance." *Disidentifications*, 6. Ocaña makes a similar statement when he is interviewed on the TVE-2 program *Popgrama* in 1977, in which he states, "Voy en camino de ser yo una persona. . . . Yo no quiero títulos, yo quiero intuitivamente . . . ser yo" (I'm on my way to being a person. . . . I don't want titles, I want intuitively . . . to be me). See Vila-San-Juan, *Barcelona era una fiesta*.

67 Rodríguez, Anto, *¡Eres tan travesti!*, 202–3.

68 Luque, *La vida cotidiana*, 74.

69 Luque, *La vida cotidiana*, 76.

70 A *saeta* is a religious song from Andalusia sung during Catholic processions. Often performed a capella, over time, they acquired *flamenco*

inflections and vocal flourishes. *Saetas* are associated with intense emotion, particularly sorrow. A *saetero* or *saetera* is the singer.

71 Robertson, *Guilty Pleasures*, 17.

72 Fernàndez, "Thou Shalt Not."

73 Roig, *L'òpera quotidiana*, 26. (Hereafter cited in the text by page number.)

74 The *parlar xava* (*xava* accent) in Barcelona denotes a pronunciation common in poorer neighborhoods of the city and associated with an interference of Castilian in Catalan pronunciation. Culturally, the term was perhaps first described in a 1910 novel, Juli Vallmitjana's *La Xava*, which took place in the working-class areas of the city long before the influx of Andalusian migrants. In the early 1980s, when Roig's novel was published, it clearly reflected the speech of the *altres catalans*.

75 The *ángel del hogar* (angel in the home) refers to the notion, first articulated in the nineteenth century and later emphasized by Franco and the regime, that, in Spain, a woman's ideal place is in the home.

76 Barbal, *Carrer Bolívia*, 13. (Hereafter cited in the text by page number.) Here, the character references his own fear of being heard as from elsewhere—as *xarnego*—and the fact that only cultivating a Catalan sound of voice will allow him to pass.

77 In his 2022 *No parlaràs mai un bon català*, Joan Mena Arca, of the Catalan political party En Comú Podem (In Common We Can), affirms that for children of immigrants such as himself, bilingualism of the sort Sierrita and Lina practice "és un esglaó més en la lluita per la justícia social i la igualtat d'oportunitats" (is one more link in the fight for social justice and equal opportunities) (15).

78 See, among others, Resina, "Double Coding"; King, *Escribir la catalanidad*; Crameri, *Catalonia*; and Woolard, *Double Talk*.

79 Marsé, *El amante bilingüe*, 48. (Hereafter cited in the text by page number.)

80 Thanks to Alex Mira-Alonso for her suggestions on how to translate certain parts of the text.

81 Vasallo, *PornoBurka*, 176. (Hereafter cited in the text by page number.)

82 The *ranchera* is a traditional Mexican genre of music. Here, the suggestion is a stereotyped reference to the idea that the Raval has been taken over by Latin American immigrants.

83 Once again, I have done my best to suggest a possible way of hearing the author's point, all the while recognizing that this is a truly insufficient translation.

84 Tonkiss, "Aural Postcards," 305.

85 Tonkiss, "Aural Postcards," 305.

86 Tonkiss, "Aural Postcards," 305.

1   Gonzalo, *La ciudad secreta*, 14. Gonzalo's reference to the *ciudad secreta* seems to echo Greil Marcus's work, which Nubla, Guillamon, and Riba also cite in their 1995 exploration of experimental music, *Alter músiques natives*.

2   Gonzalo, *La ciudad secreta*, 16.

3   Mercadé, *Odio obedecer*, 6. It is common for the Movida Madrileña social movement to stand in for all of Spain in discussions of punk and artistic expression after Franco. This is in part because, due to a change in the radio frequencies in Spain in the 1960s, FM opened up to younger broadcasters (older stations preferred to maintain their MW, or medium-wave frequency, usually used on the AM dial, programming instead), and in 1981 Radio Nacional de España (National Radio of Spain), which broadcast around the country, gave significant airtime to the Movida. For an introduction to the movement, see Guerra and Ripollés, "Post Dictatorships"; as well as Fouce and del Val, "La Movida."

4   Mercadé, *Odio obedecer*, 6.

5   Mercadé, *Odio obedecer*, 6. Gonzalo includes groups like Distrito V, Los Burros, Macromassa, Loquillo, Último Resorte, Shit S.A., and Decibelios, among others, in this list.

6   As Ealham explains, the libertarian ethos "prized the attributes of individual rebellion and heroism, generating a culture of resistance to the work ethic and the daily rituals of capitalist society." *Anarchism and the City*, 34.

7   Guillamon et al., "Experimental," 3. The text I cite here is an English translation, posted on the Gràcia Territori Sonor website, of *Alter músiques natives*. It was first published in 1995 with the support of the Generalitat de Catalunya.

8   Guillamon et al., "Experimental," 3. Josh Kun has coined the term *audiotopia* to describe musical maps that "reimagine the present social world . . . these maps point us to the possible, not the impossible; they lead us not to another world, but back to coping with this one." *Audiotopia*, 23. Here, the possible is the production of sound as resistance on a local, daily basis.

9   Guillamon et al., "Experimental," 3.

10   Brotons, "Libros."

11   Bourriaud, *Relational Aesthetics*, 17.

12   LaBelle, "Restless Acoustics," 276.

13   Quijano, "Coloniality and Modernity/Rationality," 176.

14   On the early radiophonic theories of the avant-garde in the Spanish-speaking world, see Birkenmaier, "Paul Deharme"; and Gallo, "Radiovanguardia."

15 Dimock, *Through Other Continents*, 77, sic.

16 Guillamon et al., "Experimental," 5.

17 Dimock, *Through Other Continents*, 77.

18 Nubla's GTS has continued this tradition by broadcasting a radio program called "Música y Geografía," on the licensed Radio Gràcia (107.7 FM), which promises "an acoustic trip" through space and sound, contrasting its sonic universe with the commercial sound of the surrounding airwaves.

19 Radiomai, "Manifiesto de Villaverde."

20 Guattari, "Popular Free Radio," 87.

21 On the history of El Raval, see Aisa i Mei Vidal, *El Raval;* and Ealham, "Imagined Geography."

22 See Ealham, *Class, Culture, and Conflict.*

23 Lafarga i Oriol, *Gràcia*, 44.

24 Mendizàbal, "Una posible geografía," 103.

25 Guillamòn, *La ciutat interrompuda*, 7–8.

26 As early as 1978, public accusations were made that radio licenses were being given out in an "irregular" fashion under the new socialist government, which beginning in that same year had opened up broadcasting in Spain to be in line with the Geneva agreement from 1975 that regulated radio throughout Europe.

27 See Prado, *Las radios libres.*

28 Díaz, *La radio*, 402.

29 Gonzalo, *La ciudad secreta*, 6.

30 Corporació Catalana de Mitjans Audiovisuals, "Llei de la CCMA."

31 RAC 105 emerged out of the resuscitated Ràdio Associació de Catalunya, founded in 1924 but effectively silenced under Franco.

32 Franco took control of Radio Barcelona and Ràdio Associació de Catalunya on January 16, 1939, renaming them Radio España number 1 and 2. Prohibited from broadcasting in Catalan, like all stations, they were required to broadcast information put out by Radio Nacional de España. Programs had to be transcribed onto paper and approved by censors before being broadcast; deviation from the script resulted in censorship. See Vilardell, *Books Against Tyranny.*

33 Costa, *Cómo acabar con la contracultura*, 26.

34 McNeill, "Barcelona," 328.

35 Caellas, *Carcelona*, 15.

36 Some young people of the period known as *pasotas* (apathetic people) preferred to "pass," embracing disillusionment as a form of ennui. Their voices reflected their uncaring attitude: "The words were mumbled and their vocabulary was pared to a bare minimum so that things, of whatever kind, were *chismes* (gossip) and situations or activities were dismissed as *rollos* (boring)." Hooper, *New Spaniards*, 3–4.

37   I am thinking of places like Somorrostro, Bogatell, or Pequín in port areas of the city, or those near Montjuïc, such as Can Tunis. See the documentaries *Barraques: la ciutat oblidada*, by Alonso Carnicer, Sara Grimal, and Eduard Sanjuan (2009), or *Can Tunis*, by Paco Toledo and José González Morandi (2007).

38   Harvey, "'New' Imperialism," 67–68.

39   Resina, "It Wasn't This," 242.

40   This followed the controversial 1968 Eurovision competition, in which the Franco regime forbade *nova cançó* singer Joan Manuel Serrat from performing in Catalan. Instead, he was replaced with Massiel. As Ayats and Salicrú-Maltas explain, the decision to sing in Spanish "almost jeopardiz[ed] the continuity of the [*nova cançó*] movement." "Singing Against the Dictatorship," 34. Serrat, nevertheless, made it big throughout Spain and Latin America because he signed with the industry and sang in Spanish. Many other Catalan artists, notably Lluís Llach and Raimon, refused similar offers.

41   Crameri, *Catalonia*, 50–51.

42   See Crameri, "Role of Translation."

43   See Novell, "Cantautoras catalanas," 139.

44   For the cultural and linguistic history of the Catalan rock movement, see Rull, "El model de llengua."

45   23-F refers to February 23, 1981, when Antonio Tejero and other members of the military stormed the Congress of Deputies during the swearing-in of the new, democratically elected president. The coup was put down and the leaders sentenced to thirty years in prison. The free radio stations Radio Nova in Paris, Radio Panik in Brussels, Onda Verde in Madrid, and Radio Klara in Valencia all started in the same year.

46   Joni D., *Que pagui Pujol!*, 36.

47   Eventually the record store Gay y Company—inspired by Salvador Picarol's earlier flea market music stall—became an outlet for purchasing foreign albums.

48   Zeleste, started by Víctor Jou, was the founding site of *rock laietana* or *Ona laeitana*, so named for its center in the Via Laietana in Barcelona. The style fused Mediterranean folk sound with jazz and rock, producing such singers as Jaume Sisa and Pau Riba. See Gómez-Font, *Zeleste i la música laietana*.

49   Llansamà, *Harto de todo*, 19. For more on protopunk or prepunk in the late 1970s in Spain, a good beginning is the website Shit-Fi.com, which has an article titled "*Drogas, sexo, y un dictador muerto*: 1978 on Vinyl in Spain."

50   Loquillo, *En las calles de Madrid*, n.p.

51   Loquillo also made waves on social media in 2024 when he was shown carrying the Spanish flag and calling for a return to Franco, leading some to speculate that he had been a *facha* (fascist) all along.

52　See García, "La piratería acaba."

53　Gonzalo, *La ciudad secreta*, 197.

54　Joni D., *Que pagui Pujol!*, 76.

55　Villagrasa, *Una història de Ràdio* PICA, 23.

56　Quoted in Joni D., *Que pagui Pujol!*, 71.

57　On the new plan, see Gorostiaga Alonso-Villalobos, "Nuevo Régimen."

58　As Naik Miret and Pau Serra del Pozo have pointed out, beginning in the 1960s, infrastructure projects to provide housing for immigrant working-class communities promoted travel by creating highways and metro lines, but also created mental and physical barriers between the communities through which they ran. "El papel de la inmigración extranjera," n.p.

59　Kelly, *Sound*, 12.

60　Picarol confirmed in a 2018 interview with me that he had plans to pass down the station to someone else. In 2021, in celebration of its fortieth year, much more online content about the site became available. Playback required hearing an ad, which led to the following disclaimer popping up on the website: "Frecuentamente al conectar Radio PICA on-line, durante unos segundos, suena publicidad intrusiva asquerosa, que no tiene ninguna relación con la emisora" (Frequently when Ràdio PICA is connecting online, for a few seconds, disgusting intrusive advertising sounds, which has nothing to do with the station). Noncommerical broadcasting seems to have fully reached its limit.

61　For a complete list of programs broadcast on Ràdio PICA, see Roca, *Això es Ràdio* PICA; and Hernando, "40 años de Radio PICA."

62　See Soley, *Free Radio*.

63　See Rodríguez, *Fissures in the Mediascape*, and Prado, *Las radios libres*, for more on this station.

64　On the history of free radio stations in Barcelona, see Aguilera and Sedeño, *Comunicación y música I*; Salvattore, *Història-anàlisi*; Villagrasa, *Una història de Ràdio* PICA; Reguero Jiménez and Camps Durban, "Medio siglo"; and Pascual Lizarraga, *Movimiento de resistencia*, on free radio in the Basque Country.

65　See Guattari, "Popular Free Radio," on European free radio. The burden for free radio, as opposed to pirate radio, was its resistance to advertisements of any kind, and the daily struggle this lack of funding or official support implied for its continuation.

66　The manifesto also specifies that free radio stations identify as nonprofessional, self-managed, politically autonomous but open to all opinions, and not for profit.

67　See Montes Fernández, "Radiodifusión pirata commercial," for the history of pirate radio stations in Europe.

68　On pirate or clandestine radio, see Remedi, "Production of Local Public Spheres"; and Soley, *Free Radio*.

69 Villagrasa, *Una història de Ràdio PICA*, 67.

70 For photographs of Picarol, Trashmike, and other members of the station on the boat they pretended was theirs, see the photochronicles of Xavier Mercadé, *Odio obedecer*.

71 Ràdio 4, also broadcasting in Catalan out of Barcelona, was started in 1976 during the Transition, but since it was part of Radio Nacional de España, it is not considered autonomous like RAC 105 or Catalunya Ràdio.

72 Joni D., *Que pagui Pujol!*, 76.

73 Salvans, *Gestió del caos*, 55.

74 For more on the comic in Spain, see Dopico, *El comic underground español*, and Luque, *La vida cotidiana*.

75 Picarol, "Correspondencia," *Ajoblanco* no. 6, 1975, 37.

76 Picarol, "Correspondencia," *Ajoblanco* no. 4, 1975, 34. Writing as the "Grupo artístico Picarol," in 1975, Picarol also published an article in which he accused the Ayuntamiento of discrimination and corruption for not allowing the establishment of a stall on Les Rambles to sell artwork, when book publishers were allowed to do so. Picarol, "¿Watergate en el Ayuntamiento?," 21. Picarol eventually pulled away from the magazine, even as in later years Víctor Nubla wrote about the radio station for the publication, and ads for Ràdio PICA continued to appear there. In question was a larger debate that began between *Ajoblanco* and *Star*, which reflected the question of how best to preserve a counterculture ethos of being antisystem.

77 In previous years the counterculture had been informed and inspired by Americans who brought the musical sounds of the hippies to the US military base in Roda. Those sounds of free love, so easily inscribed into the *gusto socialdemócrata*'s embrace of the counterculture past, differed from the strains of transnational anarchist thought and international punk music that came out of Barcelona, even if at times it merged with the Andalusian rock to create new musical styles. Picarol, for one, told me his interest in punk and hardcore emerged from his being tired of the hippie sound.

78 "Intro," 1.

79 Vilaseca, *Barcelonan Okupas*, 100.

80 Guillamon, *La ciutat interrompuda*, n.p.

81 In his text "The Unified Field Theory," Nubla cites Vian in order to illustrate the difficulty of consolidating experimental music such as his because Macromassa, Picarol, and others began from the personal, rather than from anything that could be conceived through a label, a genre, a school, or a trend: "Who can find a word strange when hardly anyone knows the names of the flowers that can be found in an ordinary garden?" Guillamon et al., *Alter músiques natives*, 136.

82 Schmurz, "58 de julio," 41.

83 There is some debate among aficionados of punk about whether or not Trapero del Rio belongs in the punk classification; as Quique from Skatalá explains, several of the bands now taken for punk were, as far as he was concerned, "pre-punk o rock macarra." (pre-punk or rock *macarra*). *Macarra* is a complicated term, but here the suggestion is that bands like Trapero were not as hard as later punk would be, still entwined with the 1970s sound one could associate more with the New York Dolls than the Sex Pistols. Llansamà, *Harto de todo*, 352. On the idea and expression of *macarra*, see Manrique, *¿De qué va el rock macarra?*; and Domínguez, *Macarras ibéricos*.

84 The song was performed as part of the Nicaragua Rock concert in May 1986 and reported on in *El País*. A cassette of the concert includes songs by Electroputos, La Polla Records, Cicatriz, L'Odi Social, Últimos de Cuba, and Kortatu. The cover includes a note that says, "No paguis per aquesta cinta mes de 30 pessetes" (Don't pay more than 30 pesetas for this tape).

85 Joan Manuel Serrat's request to sing in Catalan at the 1968 Eurovision competition, and his subsequent replacement by Massiel, highlighted the tensions around music and identity prior to Franco's death; that they continued during the Transition and after speaks to how dispossession as a colonial framework of access and power, as well as an identitarian reflection of voice, affects national politics.

86 Ayats and Salicrú-Maltas, "Singing Against the Dictatorship," 39.

87 Bellmunt, *Canet Rock*.

88 Allimant and Castellano, "Transitional Music," 109–10.

89 Rafael Duyos built his own synthesizers since Moog and other imported brands were too expensive. See Gonzalo, *La ciudad secreta*, as well as Blánquez Gómez and León, *Loops 1*, for more on these experimentations.

90 Labanyi, "History and Hauntology," 74.

91 The pact, along with the 1977 Ley de Amnistía (Amnesty Law), attempted to turn the country's attention to its future democracy by pardoning any crimes committed by government officials during the Franco regime. In the 2000s, however, human rights groups began to push back against the law, and cultural debates about the value of remembering rather than forgetting resulted in the 2007 Ley de Memoria Histórica (Law of Historical Memory) and subsequent 2020 Ley de Memoria Democrática (Law of Democratic Memory).

92 See Guillamon et al., *Alter músiques natives*. Víctor Nubla played at Zeleste on more than one occasion, including through the formation of the band Naïf, in 1983, which was a joint project between him, Enric Cervera, and Albert Giménez. See Gonzalo, *La ciudad secreta*, chap. 23.

93 Riba, *Los '70 a destajo*.

94 By the time the 1975 Canet Rock festival featured Maria Mar del Bonet, Jaume Sisa, who once claimed in Sala Zeleste that he was from the "profunditats subterrànies" (subterranean depths) of Poble Sec, was forbidden by the Franco government from playing. *Vibraciones* magazine claimed that the sound of his absence was louder than the sound of the music, since the acoustic arrangements of the site were poorly organized and thus insufficient for the space.

95 Another important and popular broadcast was *La bola de cristal*, hosted by the former Kaka de Luxe guitarist, Alaska, from 1985 to 1988. See Viñuela, "Popular Music in *Televisión Española*."

96 Hernando, *Los silencios de la radio*, n.p.

97 Brecht, "Communication Apparatus." Brecht wrote in 1967 that radio "is a pure instrument of distribution; it merely hands things out." "Radio as a Means," 25. He argued that it should become a mode of two-way communication, rather than simple top-down distribution.

98 Joni D., *Que pagui Pujol!*, n.p.

99 Vázquez Montalbán, *Crónica sentimental de la Transición*, 104.

100 The penal code violation dated to the 1940s but was updated in 1970 as the Ley de Peligrosidad y Rehabilitación Social (Law of Danger and Social Rehabilitation). The 1970 law also forebade homosexuality, although that aspect was overturned in 1983. The entire law was abolished in 1995. On the Vulpess case, see de la Cuadra, "Archivado el sumario."

101 Llansamà, *Harto de todo*, 472.

102 *El Sol*, "Ramón, reportero," 4.

103 Ramón invented written *greguerías* first, one-line witty puns that are similar to aphorisms, but which he defined as "humor + metaphor." When he began broadcasting them, he called them *greguerías onduladas* because they went out over the airwaves (*ondas*).

104 Barth, "Literature of Exhaustion," 71.

105 Llansamà, *Harto de todo*, 472.

106 Gonzalo, *La ciudad secreta*, 197.

107 Mercadé, *Odio obedecer*, 63.

108 See Villagrasa, *Ràdio PICA*, 68.

109 Applications had to go through the Corporació Catalana de Ràdio i Televisió (Catalan Corporation of Radio and Television), which was funded by the Catalan government, and in line with the overall project of modernization that held sway throughout the 1980s and 1990s in Barcelona.

110 When I visited GTS in 2018, Nubla pointed out the prevalence of red and black around the space as a sign of the group's adherence to anarchist thought.

111　This commentary used to be front and center on the Ràdio PICA home page, radiopica.cat, although it has since been removed (accessed February 23, 2022).

112　Rodríguez, "Jóvenes latinos y geografías nocturnas," 214.

113　Sánchez Fuarros, "Music and Migration in Multicultural Spain," 147.

114　Nubla parodies this festival in his novel, *El regal de Gliese*.

115　Babiano Sánchez and Contel Ruiz, *El campanar de Gràcia*, 42.

116　Corbin, "Auditory Markers," 118.

117　Corbin, "Auditory Markers," 122.

118　Babiano Sánchez and Contel Ruiz, *El campanar de Gràcia*, 66.

119　The exceptions were during the later years of the Franco dictatorship when the *festa* (festival) disappeared for a time. See Sánchez and Contel Ruiz, *El campanar de Gràcia*, 42–48.

120　From Gràcia's page on the website of the Ajuntament de Barcelona, https://ajuntament.barcelona.cat/gracia.

121　Some of the first satiric publications were *almanacas* (almanacs), since the 1855 Llei d'Imprenta required most publications to pass through bureaucratic mechanisms to be approved for publication. Almanacs were free to be published as long as they included astronomical information provided by the Observatorio Nacional (National Observatory). Catalan publications like *Singlots poètichs* and *Lo Xanguet* took advantage of the loophole. Capdevila, *La Campana de Gràcia*, 18.

122　Capdevila, *La Campana de Gràcia*, 53.

123　Torrent, "La campana de Gràcia," 30. The article published in 1970 seems to make its own use of this tongue-in-cheek reference to sound as resistance when it republishes this supposed thought by the editor. Including Catalan sentences of resistance in the piece produces a clash against the ear of the censor, as the *campanades* (bell chimes) and *esquellotes* (cowbell rings) had against the Spanish troops.

124　I do not mean to suggest that it is the fault of those who come from abroad that these protests exist. As Marc Caellas writes, those who critique the *okupes* as infiltrating the city from outside are playing the same xenophobic blame games as always, no matter whom they are critiquing: "Ya sean los madrileños centralistas, los extremeños gandules o los moros ladrones, el recurso es el mismo: acusar al foráneo" (Whether people from Madrid are centralists, people from Extremadura are slackers, or North Africans are thieves, the strategy is the same: accuse the foreigner). *Carcelona*, 73.

125　Soley, *Free Radio*, 248.

126　In yet another indication of the playful nature of signification at stake in these artists' projects, LEM, run by Gràcia Territori Sonor, is an acronym whose letters stand for different things each year.

127　McEnaney, *Acoustic Properties*, 214.

CHAPTER 4. PROTEST AND THE ACOUSTIC LIMITS OF DEMOCRACY

1   See Trueba, "Algarabía."
2   The *senyera* is the flag of Catalonia; the *estelada*, which incorporates the red and yellow bars of the *senyera*, represents independence and dates from the period when Spain lost its last colonies in 1898. The official video by the Assemblea Nacional de Catalunya (ANC), titled "Catalunya, nou estat d'Europa—Catalonia, new state of Europe," captures many of these sounds and images.
3   On metaphor in sound, see Moreno, *Musical Representations*.
4   Venegas, *Sublime South*, 17–18.
5   Dimock, *Through Other Continents*, 3–4.
6   Lledó Cunill, "Paraules clau," n.p. When referring to the term as "nineteenth-century," she seems to be implying both that it is arcane and that it carries with it a philological understanding of the relationship between culture and language. Her defense of *guirigall* suggests that anyone listening to Catalan culture must do so according to the rules and norms of the Catalan language, which, in a polyglot society, not everyone will do.
7   Carol, "'L'algarabía' es una planta," n.p.
8   For more on these specific cases, see Surribas Balduque, "La ficción," and Maldonado, "Estas son las canciones." Valtònyc has been somewhat embraced by the Catalan right for his antimonarchist songs, which have been heard by the Spanish government as threats against the king and the state.
9   At stake here is the scale of legal structures, and the shifts between local and global that affect them. For more on the scale of law, see Riles, "View."
10 I am drawing on decolonial theory's assertion that coloniality implies a Cartesian notion of rationality to understand binary divisions of subjects and objects, which developed alongside a transatlantic valorization of reason over emotion. See Quijano, "Coloniality," and Escobar, "Worlds and Knowledges."
11 Labelle, *Sonic Agency*, 7.
12 Butler and Athanasiou, *Dispossession*, 20.
13 Vilaseca, *Barcelonan Okupas*, xxv.
14 Juliá, *Transición*. See Introduction.
15 These are the following political parties, respectively: the Partit Socialista Unificat de Catalunya (PSOE, Unified Socialist Party of Catalunya), Unidas Podemos (United We Can), Euskal Herria Bildu (Basque Country Unite), the Partit Demòcrata Europeu Català (PDeCAT, Catalan European Democratic Party), and the Partido Nacionalista Vasco (PNV, Basque Nationalist Party).

16 Boletín Oficial del Estado, "Ley 20/2022."

17 Crumbaugh. "(Still) Miguel Ángel Blanco?," 366.

18 Crumbaugh, "(Still) Miguel Ángel Blanco?," 366. The context in which Crumbaugh makes this claim—the "We Are All Miguel Ángel Blanco" demonstrations—is quite different from the one I am studying here, but the larger point is applicable in its understanding of emotion as spectacle.

19 See Crameri's *"Goodbye, Spain?"* on the development of political positions and their manifestation in the cultural sphere over the last decade or so.

20 Warner, "Publics and Counterpublics," 88.

21 Warner, "Publics and Counterpublics," 89.

22 Warner, "Publics and Counterpublics," 89.

23 Dolar, *Voice*, 112–13.

24 Eisenlohr, *Sounding Islam*, 80, 84.

25 Dolar, *Voice*, 106.

26 Dolar, *Voice*, 106.

27 Dolar, *Voice*, 121; italics mine.

28 Dolar, *Voice*, 121.

29 Dolar, *Voice*, 112.

30 Masiello, *Senses of Democracy*, 3.

31 López Lerma, *Sensing Justice*, 6.

32 Thacker, "Biomedia," 126.

33 Gitelman, *Always Already New*, 6.

34 Clough, "Affective Turn," 11.

35 Ahmed, *Cultural Politics of Emotion*, 118.

36 Ahmed, "Happy Objects," 37.

37 Ahmed, "Happy Objects," 37.

38 On November 9, 2014, under the leadership of Artur Mas (of the Convergència i Unió [Convergence and Union] party), an earlier referendum on the political future of Catalonia also took place, in which voters were asked if they would like Catalonia to be a state, and if so, if they would like it to be independent. In that case, as well, the Constitutional Court of Spain prohibited the referendum, but it took place regardless.

39 See, for example, *Público*, "El saludo fascista."

40 Minder, *Struggle for Catalonia*, n.p.

41 The video, titled "Lip dub per la independència—world record (oficial)," was promoted on YouCat Canal, a website dedicated to Catalan independence. It received over two and a half million views. For more details, see Crameri, *"Goodbye, Spain?,"* 29–30.

42 In *A Voice and Nothing More*, Dolar argues the state uses vocally sounded voice, such as court testimony, to support the letter of the law. The Spanish state requires court proceedings be conducted in Castilian, which pro-Catalan proponents argue is an imposition of a foreign language on local jurisdictions.

43 Corbin, "Auditory Markers," 117.

44 Eisenlohr, *Sounding Islam*, 64.

45 Eisenlohr, *Sounding Islam*, 64.

46 This number was contested by the Spanish state.

47 A general strike on October 3, 750,000 strong, invoked a catchy rhyming phrase that turned the idea of criminality against the Spanish state, as protestors chanted about Rajoy, himself guilty of corruption, "Este delincuente no es mi presidente" (This delinquent is not my president). *La Vanguardia*, "Protesta masiva."

48 During the pandemic, however, as Almudena Grandes chillingly portrayed in her posthumous novel, *Todo va a mejorar*, such shows of solidarity through public noise-making can also become signs of inscription in the ideological state apparatus.

49 Òmnium Cultural, "Help Catalonia."

50 Òmnium Cultural, "Help Catalonia." The video was strikingly similar to one produced in Ukraine in 2014, called "I am a Ukrainian," in protest of Russia's invasion of the country. Also featuring a tearful woman decrying, in English, the Russian abuse, the video included much of the same language as the Catalan one. A Whisper to a Roar, "I Am a Ukrainian." Once the Catalan video was known, the woman, an actress from Barcelona, quickly participated in a parody on the TV3 show "Està passant," in a video titled "Help Rajoy," which took the opposing political side to that of the first, ending, "Help the Spanish government take control of Catalonia. Save Rajoy." The polyphony of political leanings wrapped up in the reverberations of the same actress's voice point toward how the Western democratic ear blurs the sounds of difference into an empty vococentrism, rooted in entextualization, wherein the physical sound of voice is hollowed out by repetition so that a silent, lettered construct of rational democracy can prevail.

51 Congostrina, "Concentració."

52 Zanón, "Jordi Cuixart."

53 In an interview with Younes Abouyaqqoub's parents after he was killed by police, *El País* invoked a similar aural gesture by describing their voices as "un hilo" (a thread) and "casi rota" (almost broken). Carretero, "Los padres de Younes."

54 Ochoa Gautier, *Aurality*, 33.

55 Dolar, *Voice*, 121.

56 Panagia, *Political Life of Sensation*, 3.

57 França, *Habitar la trinxera*, 19.

58 See Miret and Serra del Pozo, "El papel de la inmigración extranjera."

59 Erik Harley's *Promishuevismo* is a ludic, recent example of a similar movement.

60 See França, *Habitar la trinxera*, for more on the *habitatge digne* (adequate housing) movement in Barcelona.

61 Vilaseca, *Barcelonan Okupas*, 1.

62 See Rovisco, "'Europe from Below'?," as well as Martínez López and García Bernardos, "Occupation of Squares," on the relationship between squatting and the M15 (May 15, 2011) antiausterity protests of *Democracia Real Ya* and Acampada Sol.

63 Tufekci, *Twitter and Tear Gas*, 6.

64 In Spain, unethical mortgage practices often targeted immigrants and people of lesser economic means, resulting in unemployment rates of close to 22 percent in the country.

65 In an echo of the free radio movement, around the same time that the Occupy protests were taking place globally, the Okupem les Ones (We Occupy the Airwaves) drive in Barcelona sought to fill television stations with alternative information on the social issues underlying the protests.

66 Martínez-López and García Bernardos, "Occupation of Squares," 161.

67 Foucault, "Society Must Be Defended," 77.

68 On the representation of law in the film, see Vialette, "On-Screen Trial."

69 Butler and Athanasiou, *Dispossession*, 18.

70 Butler and Athanasiou, *Dispossession*, 18.

71 See Goldsmith, *Discord*, especially the chapter "The First Champions of Silence"; Bjisterveld, "Listening"; and Sterne, "Hearing," 70.

72 Vilaseca, *Barcelonan Okupas*, xiv–xv.

73 Because, as an autonomous community, Catalonia controls its own energy distribution, but at the same time water scarcity is a feature of the landscape, there have been disputes throughout the twentieth and twenty-first centuries about who has access to the water running through the region (and, at times, when or if to share the resource with surrounding communities outside Catalonia). For just one discussion of this problem, see Otero et al., "Water Scarcity."

74 Their reference to Ritas and Ratos is an allusion to Rodrigo Rato and Rita Barberá, both Spanish politicians accused of fraud and money laundering.

75 Harvey, "'New' Imperialism," 67–68.

76 Vilaseca, *Barcelonan Okupas*, 39.

77 The phrase "Follow the money" became popularized by the film *All the President's Men*.

78 Bull, "To Each Their Own Bubble,'" 276.

79 Bull, "To Each Their Own Bubble,'" 277.

80 In 2020, in another fusion of space, voice, and protest informed by COVID-19, they performed "Líneas de Metro." Voicing their resistance to the problems of "desahucios, trabajo esclavo, sanidad, educación, pensiones, movilidad, turismo, videovigilancia, alimentación" (evictions, slave labor, health care, education, pensions, transportation, tourism, video surveillance, food security), they explained in an email that they

planned to stop at various stations to resist a government they see as oppressive regardless of political party: "La alternancia de la izquierda o la derecha institucional en el poder es sólo otra variante del juego del poli malo / poli bueno" (The alternation of an institutional left or right in power is just another version of the good cop/bad cop game).

81   Lahuerta, *Destrucción de Barcelona*, 22.

82   As such, the colonial ear is not limited to spaces or times defined historically by explicit colonial infrastructures, but rather by what Rafico Ruiz, drawing on Kember and Zylinska's concept of *living mediation*, calls an infrastructural mediation. *Slow Disturbance*, 4.

83   Lahuerta makes this connection by discussing José Luis Guerín's documentary *En construcción* (2001), which captures the gentrification of the city by focusing as much on the noise of the excavations that unearthed the city's Roman past as on the contemporary dispossession of musical nightlife occasioned by the demolition of decades-old nightclubs in order to build new high rises.

84   Saunders, "Media and Terrorism," 429.

85   A similar, related attack took place in Cambrils a few hours later, killing one woman.

86   The attackers had links to Morocco and Melilla, both of which have a long history of tension with the Spanish state and whose presence in Catalonia reflect the issues of immigration I discussed in Chapter 1.

87   International Institute for Strategic Studies, "Barcelona Terrorist Attack."

88   Montañés, "'No tinc por!'"

89   *ABC España*, "Barcelona grita."

90   Microphones also picked up the almost manic sound of recording in progress, acoustically betraying their supposed invisibility through the sound of shutters clicking, and audibly highlighting the role the media played in the production of the event.

91   Calahorrano, "'Los Mossos.'"

92   Anguera de Sojo, "La Policía sospecha."

93   *ABC Cataluña*, "Un vecino."

94   *El Economista*, "El vecino."

95   García Bueno, "Seis días."

96   See Anguera de Sojo, "La Policía sospecha"; Strange, "Barcelona Attack Suspect"; and *ABC Cataluña*, "Un vecino."

97   Schwartz, "Whistling," 343.

98   Schwartz, "Whistling," 343.

99   Blake, "'Vocal Apparatus' Colonial Contexts," 40.

100   Europa Press, "Atemptat," n.p.

101   Foucault, "Society Must Be Defended," 76.

102   Del Pino, "Aznar."

103   Taylor, "Anti-Muslim Views."

104   The supposition that Muslims are automatically other to "Europeans"
       in itself overlooks the large numbers of people who live in Europe and
       practice the religion.

105   See Encarnación, "Politics of Immigration."

106   "I call the distribution of the sensible the system of self-evident facts of
       sense perception that simultaneously discloses the existence of some-
       thing in common and the delimitations that define the respective parts
       and positions within it." Rancière, *Politics of Aesthetics*, 12.

107   Dolar, *Voice*, 112–13.

108   In her memoir *Jo també soc catalana*, she presents Catalan as a sister
       language to her native Amazigh, because both have been oppressed
       politically by other languages in their home nations (Castilian Spanish
       and Arabic, respectively) and looked down upon for being "llenguatge
       oral, només, bàrbars, ens diuen" (nothing but an oral language, barbar-
       ians, they call us). El Hachmi, *Jo també soc catalana*, 27. She also critiques
       native Catalan speakers who address her and her son in Castilian
       because they assume foreigners cannot speak Catalan: when she cor-
       rects a shopkeeper by speaking Catalan to her, the woman nevertheless
       responds "bueno, bueno." This refusal to hear her as a Catalan speaker
       prompts Hachmi to invoke the same civilization/barbarism dichotomy
       used to justify the normalization program of which she herself is an
       example of success: "Per un moment em passa pel cap dir-li que 'bueno'
       és un barbarisme, però ja hi ha prou tensió entre totes dues" (For a mo-
       ment it crosses my mind to tell her that 'bueno' is a barbarism, but there
       is already enough tension between the two of us). El Hachmi, *Jo també soc
       catalana*, 51.

109   Lefebvre, *Rhythmanalysis*, n.p.

110   Gosálvez, "#Hablamos?"

CODA. THE HUMBLE EAR AND THE SHAPE OF SOUND

1   Mowitt, *Sounds*.

2   Campoy-Cubillo and Sampedro Vizcaya, "Global Hispanophone," 8.

3   Campoy-Cubillo and Sampedro Vizcaya, "Global Hispanophone," 7.

4   Gentic, *Everyday Atlantic*.

5   Sterne, "Sonic Imaginations," 3.

6   Spivak, "Can the Subaltern Speak?"

7   See Rivera Cusicanqui, "*Ch'ixinakax utxiwa*."

8   Sterne, *Audible Past*, 20.

9   Dussel, "Transmodernidad e interculturalidad," 28.

10   Dussel, "Transmodernidad e interculturalidad," 28.

11   See Wolfson, *Digital Rebellion*.

12   See González Rodríguez, *Pensar la música*.

13   Hilmes, "Sound Culture Studies?"

Bibliography

ABC *Cataluña*. "Un vecino que avisó de la presencia de Younes Abouyaaqoub: 'He hecho lo que debía.'" August 23, 2017.

ABC *España*. "Barcelona grita 'No tengo miedo.'" August 27, 2017.

Adell Argilés, Ramón, and Miguel Martínez López, eds. *¿Dónde están las llaves? El movimiento okupa: prácticas y contextos sociales*. Madrid: Catarata, 2004.

Aguilera, Miguel de, and Ana Sedeño. *Comunicación y música I: lenguaje y medios*. Barcelona: Editorial UOC, 2008.

Ahmed, Sara. *The Cultural Politics of Emotion*. 2nd ed. Edinburgh: Edinburgh University Press, 2014.

Ahmed, Sara. "Happy Objects." In *The Affect Theory Reader*, edited by Melissa Gregg and Gregory J. Seigworth, 29–51. Durham, NC: Duke University Press, 2010.

Aisa i Mei Vidal, Ferran. *El Raval: Un espai al marge*. Barcelona: Editorial Base, 2006.

Allimant, Israel Holas, and Ramón López Castellano. "Transitional Music: Popular Music as Agent of Social Change in Chile and Spain." *Journal of Iberian and Latin American Research* 26, no. 1 (2010): 100–114. https://doi.org/10.1080/13260219.2020.1778768.

Anderson, Benedict. *The Age of Globalization: Anarchists and the Anticolonial Imagination*. New York: Verso, 2013.

Anguera de Sojo, Iva. "La Policía sospecha que existe un piso franco o más integrantes de la célula terrorista." *El Independiente*, August 21, 2017.

Archilés, Ferran. "When the Empire Is in the South: Gendered Spanish Imperialism in Morocco at the End of the 19th Century." In *European Modernity and the Passionate South: Gender and Nation in Spain and Italy in the Long Nineteenth Century*, edited by Xavier Andreu and Mónica Bolufer, 241–61. Leiden: Brill, 2022.

Assemblea Nacional de Catalunya. "Catalunya, nou estat d'Europa—Catalonia, New State of Europe." YouTube video. 2012. https://www.youtube.com/watch?v=_xFDj27y_xE.

A Whisper to a Roar. "I Am a Ukrainian." YouTube video. 2014. https://www.youtube.com/watch?v=Hvds2AIiWLA.

Ayats, Jaume, and María Salicrú-Maltas. "Singing Against the Dictatorship (1959–1975): The Nova Cançó." In *Made in Spain: Studies in Popular Music*, edited by Sílvia Martínez and Héctor Fouce, 28–41. New York: Routledge, 2013.

Babiano Sánchez, Eloi, and Josep María Contel Ruiz. *El campanar de Gràcia: Símbol i testimoni d'una vila*. Barcelona: Ajuntament de Barcelona, 2015.

Bada Panillo, José R. *El debate del catalán en Aragón (1983–1987)*. Calaceite: Associació Cultural del Matarranya, 1991.

Balibar, Etienne. "The Nation Form: History and Ideology." In *Race, Nation, Class: Ambiguous Identities*. Translated by Chris Turner. New York: Verso, 1988.

Ballart Macabich, Jordi. "Variació fònica al català de Barcelona: L'accent xava." *Treballs de Sociolingüística Catalana* 23 (2013): 133–51.

Barbal, María. *Carrer Bolívia*. Barcelona: Edicions 62, 2005.

Barth, John. "The Literature of Exhaustion." In *The Friday Book: Essays and Other Nonfiction*, 62–76. New York: G. P. Putnam's Sons, 1984.

Bassegoda i Hugas, Bonaventura. "Les primeres publicacions de l'excursionisme català." *Muntanya* 879 (2008): 30–35.

Bellmunt, Francesc, dir. *Canet Rock*. Profilmes, 1975.

Benet, Vicente J. *El cine español: una historia cultural*. Barcelona: Paidós, 2012.

Benet, Vicente J. "Radical Politics and Experimental Film in Franco's Spain: 'Los Encuentros de Pamplona,' 1972." *Hispanic Review* 82, no. 2 (2014): 157–74.

Bertrana, Aurora. "La meva espurna a la flama de la sardana," n.d. Fons Bertrana, Memòria Digital de Catalunya. Universitat de Girona. https://mdc.csuc.cat /digital/collection/abertrana/id/13848/rec/1.

Bestard, Joan. "Prologue." In *Modernidad, nacionalismo y naturaleza. Anar a la terra: el excursionismo catalán (1876–1923)*, edited by Ricardo Cicerchia, 13–15. Rosario, Argentina: Prohistoria Ediciones, 2011.

Bhabha, Homi. "DissemiNation: Time, Narrative and the Margins of the Modern Nation." In *Nation and Narration*, edited by Homi Bhabha, 291–320. London: Routledge, 1990.

Birkenmaier, Anke. "Paul Deharme: Proposition for a Radiophonic Art." *Modernism/Modernity* 16, no. 2 (2009): 403–13. https://doi.org/10.1353/mod.0 .0090.

Bjisterveld, Karin. "Listening to Machines: Industrial Noise, Hearing Loss, and the Cultural Meaning of Sound." In *The Sound Studies Reader*, edited by Jonathan Sterne, 152–67. New York: Routledge, 2012.

Blake, Iris Sandjette. "The Vocal Apparatus' Colonial Contexts: France's *Mission Civilisatrice* and (Settler) Colonialism in Algeria and North America." In *Sonic Histories of Occupation: Experiencing Sound and Empire in a Global Context*, edited by Russel P. Skelchy and Jeremy E. Taylor, 25–49. London: Bloomsbury Academic, 2022.

Blánquez Gómez, Javier, and Omar León, eds. *Loops 1: Una historia de la música electrónica en el siglo XX*. New York: Reservoir Books, 2018.

Boletín Oficial del Estado (BOE). Ley 20/2022, de 19 de octubre, de Memoria Democrática. BOE-A-2022-17099, BOE 252. https://www.boe.es/eli/es/l/2022/10/19/20/con.

Bourriaud, Nicolas. *Relational Aesthetics*. Translated by Simon Pleasance and Fronza Woods. Dijon: Les Presses Du Réel, 2002.

Brecht, Bertolt. "Radio as a Means of Communication: A Talk on the Function of Radio." *Screen* 20, no. 3–4 (1979): 24–28.

Brecht, Bertolt. "The Radio as a Communication Apparatus." In *Radiotext(e)*, edited by Neil Strauss and David Mandl, 15–17. Brooklyn, NY: Autonomedia, 1993.

Brotons, Jesús. "Libros: *Cómo caza un dromedario*. Una entrevista naturalista a Víctor Nubla." Interview by Jesús Brotons. *Vice*, May 25, 2012. www.vice.com/es/article/av9m54/libros-como-caza-un-dromedario.

Bull, Michael. "'To Each Their Own Bubble': Mobile Spaces of Sound in the City." In *Mediaspace: Place, Space and Culture in a Media Age*, edited by Nick Couldry and Anna McCarthy, 275–93. New York: Routledge, 2003.

Butler, Judith, and Athena Athanasiou. *Dispossession: The Performative in the Political*. Cambridge: Polity, 2013.

Caellas, Marc. *Carcelona*. Barcelona: Melusina, 2011.

Calahorrano, Katty. "'Los Mossos' abaten a Younes Abouyaaqoub, autor del atentado de La Rambla." *Radio Pichincha*. August 21, 2017.

Calderwood, Eric. *Colonial Al-Andalus: Spain and the Making of Modern Moroccan Culture*. Cambridge, MA: Harvard University Press, 2018.

Calderwood, Eric. "Spanish in a Global Key." *Journal of Spanish Cultural Studies* 20, no. 1–2 (2019): 53–65. https://doi.org/10.1080/14636204.2019.1609234.

Campoy-Cubillo, Adolfo, and Benita Sampedro Vizcaya. "Entering the Global Hispanophone: An Introduction." *Journal of Spanish Cultural Studies* 20, no. 1–2 (2019): 1–16. https://doi.org/10.1080/14636204.2019.1609212.

Candel, Francesc. *Els altres catalans: Edició no censurada*. Edited by Jordi Amat. Barcelona: Edicions 62, 2008.

Candel, Francesc. *Los otros catalanes*. Barcelona: Ediciones Península, 1965.

Candel Ruiz, Josep Maria. *Gràcia desapareguda*. Baix Llobregat: Efadós, 2022.

Cánovas del Castillo, Antonio. *Apuntes para la historia de Marruecos*. Madrid: Imprenta de la América, 1860. https://www.bibliotecadigitaldeandalucia.es/catalogo/es/catalogo_imagenes/grupo.do?path=89200.

Capdevila, Jaume. *La Campana de Gràcia: La primera publicació catalana de gran abast (1870–1934)*. Barcelona: Pagès editors, 2014.

Cárcamo-Huechante, Luis. "A Trope of Colonial Obliteration? A Critical Note on 'Colonial Latin America' and Related Conversations." *Colonial Latin American Review* 32, no. 2 (2023): 243–48. https://doi.org/10.1080/10609164.2023.2205246.

Carol, Màrius. "'L'algarabía' es una planta." *La Vanguardia*, September 15, 2012. https://www.lavanguardia.com/encatala/20120916/54350453566/l-algarabia-es-una-planta-marius-carol.html.

Carretero, Nacho. "Los padres de Younes: 'Cuando vimos su foto en la televisión no nos lo creíamos.'" *El País*, August 24, 2017.

Casals i Martorell, Daniel. "El català a la ràdio durant el tardofranquisme: Aproximación al programa 'En totes direccions' (1971) de Ràdio Barcelona." *Llengua y Literatura* 17 (2006): 103–34. raco.cat/index.php/LlenguaLiteratura/article /view/304945.

Casals i Martorell, Daniel. *El català en antena: 20 anys construint el model lingüístic de Catalunya Ràdio*. Benicarló: Onada Edicions, 2003.

Cassany, Enric. "Les aventures extraordinàries d'en Massagran." In *El patrimoni de la imaginació: Llibres d'ahir per a lectors d'avui*, edited by Mònica Baró, Teresa Colomer, and Teresa Mañà, 47–51. Balearic Islands: Institut d'Estudis Baleàrics, 2007.

Castillón, Xavier. "L'Any Massagran reivindicarà la dimensió humorística de Josep Maria Folch i Torres." *El Punt Avui*, December 15, 2009.

Cavarero, Adriana. *For More Than One Voice: Toward a Philosophy of Vocal Expression*. Palo Alto, CA: Stanford University Press, 2005.

Chanan, Michael. *The Politics of Documentary*. London: British Film Institute, 2019.

Chang, Julia H. *Blood Novels: Gender, Caste, and Race in Spanish Realism*. Toronto: University of Toronto Press, 2022.

Chávez, Alex. *Sounds of Crossing: Music, Migration, and the Aural Poetics of Huapango Arribeño*. Durham, NC: Duke University Press, 2017.

Chion, Michel. *The Voice in Cinema*. Translated by Claudia Gorbman. New York: Columbia University Press, 1999.

Cicerchia, Ricardo. *Modernidad, nacionalismo y naturaleza. Anar a la terra: el excursionismo catalán, 1876–1923*. Rosario, Argentina: Prohistoria Ediciones, 2011.

Clough, Patricia. "The Affective Turn: Political Economy, Biomedia, and Bodies." In *The Affect Theory Reader*, edited by Melissa Gregg and Gregory J. Seigworth, 206–28. Durham, NC: Duke University Press, 2010.

Congostrina, Alfonso L. "Concentració davant del TSJC per reclamar la llibertat dels detinguts." *El País*, September 21, 2017. cat.elpais.com/cat/2017/09/21 /catalunya/1505991093_838513.html.

Connor, Steven. *Beyond Words: Sobs, Hums, Stutters and other Vocalizations*. London: Reaktion Books, 2014.

Connor, Steven. *Dumbstruck—a Cultural History of Ventriloquism*. New York: Oxford University Press, 2000. https://doi.org/10.1093/acprof:oso/9780198184331 .001.0001.

Corbin, Alain. "Auditory Markers of the Village." In *The Auditory Culture Reader*, edited by Michael Bull and Les Back, 117–25. New York: Berg, 2003.

Cornejo-Polar, Antonio. *Escribir en el aire: Ensayo sobre la heterogeneidad socio-cultural en las literaturas andinas*. 2nd ed. Lima: Latinoamericana Editores, 2003.

Corporació Catalana de Mitjans Audiovisuals (CCMA). "Coneix el SX3." July 8, 2024. https://www.ccma.cat/tv3/sx3/coneix-sx3/#criteris-programacio.

Corporació Catalana de Mitjans Audiovisuals (CCMA). "Llei de la CCMA." June 9, 2024. https://www.3cat.cat/corporatiu/ca/com-funciona/regulacio/.

Costa, Jordi. *Cómo acabar con la contracultura: Una historia subterránea de España.* Barcelona: Taurus, 2018.

Costa Carreras, Joan, ed., and Alan Yates, trans. "The Catalan Language." In *The Architect of Modern Catalan: Pompeu Fabra (1868–1948)*, 3–28. Amsterdam: Institut d'Estudis Catalans, Universitat Pompeu Fabra, John Benjamins Publishing Company, 2009.

Crameri, Kathryn. *Catalonia: National Identity and Cultural Policy, 1980–2003.* Cardiff: University of Wales Press, 2008.

Crameri, Kathryn. *"Goodbye, Spain?" The Question of Independence for Catalonia.* Liverpool: Liverpool University Press, 2014.

Crameri, Kathryn. "The Role of Translation in Contemporary Catalan Culture." *Hispanic Research Journal* 1, no. 2 (2000): 171–83.

Crumbaugh, Justin. "Are We All (Still) Miguel Ángel Blanco? Victimhood, the Media Afterlife, and the Challenge for Historical Memory." *Hispanic Review* 75, no. 4 (2007): 365–84.

Dalov, Beatrice. *The Sounds of Aurora Australis: A History of Australia's Musical Identity.* Liverpool: Liverpool University Press, 2021.

Davies, J.M. *La última escalada.* Alicante: Bilbioteca Virtual Miguel de Cervantes, 2015.

de la Cuadra, Bonifacio. "Archivado el sumario sobre una canción de las Vulpes." *El País*, October 10, 1986.

Delgado, Juan-Fabián. *Andalucía y el cine, del 75 al 92.* Sevilla: El Carro de La Nieve, 1991.

Delgado, Luisa Elena. *La nación singular: fantasías de la normalidad democrática española (1996–2011).* Madrid: Siglo XXI, 2014.

del Pino, Javier. "Aznar asegura que 'el problema de España con Al Qaeda' empezó con 'la invasión de los moros' y la Reconquista." *Ser*, September 22, 2004.

Derrida, Jacques. *Of Grammatology.* Baltimore: Johns Hopkins University Press, 1976.

Díaz, Lorenzo. *La radio en España, 1923–1993.* Madrid: Alianza Editorial, 1992.

Dimock, Wai Chee. *Through Other Continents: American Literature Across Deep Time.* Princeton, NJ: Princeton University Press, 2009.

Dolar, Mladen. *A Voice and Nothing More.* Cambridge, MA: MIT Press, 2006. https://doi.org/10.7551/mitpress/7137.001.0001.

Domínguez, Iñaki. *Macarras ibéricos: Una historia de España a través de sus leyendas callejeras.* Madrid: Akal, 2022.

Domínguez García, Helena. "Activismo mediático en los albores de internet: El caso del movimiento okupa de Barcelona." *Commons: Revista de Comunicación y Ciudadanía Digital* 4, no. 2 (2015): 93–123.

Domínguez Quintana, Rubén. "Marés, Faneca y el torero enmascarado o cómo asaltar el mito nacional y monolingüe en *El amante bilingüe* de Juan Marsé."

Bulletin of Hispanic Studies 89, no. 3 (2012): 293–301. https://doi.org/10.3828 /bhs.2012.21.

Dopico, Pablo. *El comic underground español, 1970–1980.* Madrid: Cátedra, 2005.

Dowling, Andrew. *Catalonia Since the Spanish Civil War: Reconstructing the Nation.* Liverpool: Liverpool University Press, 2022.

Dussel, Enrique. "Transmodernidad e interculturalidad: Interpretación desde la Filosofía de la Liberación." *Erasmus: Revista para el Diálogo Intercultural* 5, no. 1/2 (2003): 65–102.

Ealham, Chris. *Anarchism and the City: Revolution and Counter-revolution in Barcelona, 1898–1937.* Oakland, CA: AK Press, 2010.

Ealham, Chris. *Class, Culture, and Conflict in Barcelona, 1898–1937.* New York: Routledge, 2005.

Ealham, Chris. "An Imagined Geography: Ideology, Urban Space, and Protest in the Creation of Barcelona's 'Chinatown,' c. 1835–1936." *International Review of Social History* 50, no. 3 (2005): 373–97. https://doi.org/10.1017/ S0020859005002154.

Eastman, Scott. *A Missionary Nation: Race, Religion, and Spain's Age of Liberal Imperialism, 1841–1881.* Lincoln: University of Nebraska Press, 2019.

Eidsheim, Nina Sun. "Marian Anderson and 'Sonic Blackness' in American Opera." *American Quarterly* 63, no. 3 (2011): 641–71. https://doi.org/10.1353/aq.2011 .0045.

Eidsheim, Nina Sun. *The Race of Sound: Listening, Timbre, and Vocality in African American Music.* Durham, NC: Duke University Press, 2019.

Eidsheim, Nina Sun. *Sensing Sound: Singing and Listening as Vibrational Practice.* Durham, NC: Duke University Press, 2015.

Eisenlohr, Patrick. *Sounding Islam: Voice, Media, and Sonic Atmospheres in an Indian Ocean World.* Oakland: University of California Press, 2018.

El Economista. "El vecino que delató a Abouyaaqoub: 'El terrorista iba silbando, algo sospechoso.'" August 23, 2017.

El Hachmi, Najat. *Jo també soc catalana.* Barcelona: La Butxaca, 2004.

El País. "Òmnium copia un vídeo de las sangrientas protestas de Ucrania en 2014 para pedir ayuda europea." October 17, 2017. https://elpais.com/elpais/2017 /10/17/videos/1508244649_625336.html.

El Sol. "Ramón, reportero." November 22, 1929.

Encarnación, Omar Guillermo. "The Politics of Immigration: Why Spain Is Different." *Mediterranean Quarterly* 15, no. 4 (2004): 167–85. https://doi.org/10.1215 /10474552-15-4-167.

Epps, Brad. "Barcelona and Modernity." In *The Barcelona Reader: Cultural Readings of a City*, edited by Enric Bou and Jaume Subirana, 145–62. Liverpool: Liverpool University Press, 2017.

Epps, Brad. "Between Europe and Africa: Modernity, Race, and Nationality in the Correspondence of Miguel de Unamuno and Joan Maragall." *Anales de La Literatura Española Contemporánea* 30, no. 1–2 (2005): 97–132.

Erlmann, Veit. *Reason and Resonance: A History of Modern Aurality*. Brooklyn, NY: Zone Books, 2010.

Escobar, Arturo. "Worlds and Knowledges Otherwise." *Cultural Studies* 21, no. 2 (2007): 179–210.

Europa Press. "Atemptat. La síndica de Barcelona condemna l'atac: 'No podem claudicar davant aquesta barbàrie.'" VilaWeb, August 18, 2017. https://www .vilaweb.cat/noticies/atemptat-la-sindica-de-barcelona-condemna-latac -no-podem-claudicar-davant-aquesta-barbarie/.

Faudree, Paja. "How to Say Things with Wars: Performativity and Discursive Rupture in the *Requerimiento* of the Spanish Conquest." *Journal of Linguistic Anthropology* 22, no. 3 (2013): 182–200.

Feld, Steven. "Acoustemology." In *Keywords in Sound*, edited by David Novak and Matt Sakakeeny, 12–22. Durham, NC: Duke University Press, 2015.

Feld, Steven. "Places Sensed, Senses Placed: Towards a Sensuous Epistemology of Environments." In *Empire of Sense: The Sensual Culture Reader*, edited by David Howes, 179–91. London: Routledge, 2004.

Fernàndez, Josep-Anton. "The Authentic Queen and the Invisible Man: Catalan Camp and Its Conditions of Possibility in Ventura Pons's *Ocaña, retrat intermitent*." *Journal of Spanish Cultural Studies* 5, no. 1 (2004): 83–99.

Fernàndez, Josep-Anton. "Thou Shalt Not Covet Thy Roots: Immigration and the Body in Novels by Roig, Barbal, and Jaén." *Romance Quarterly* 53, no. 3 (2006): 223–35.

Fernández-Fígares Romero de la Cruz, María Dolores F. *La colonización del imaginario: Imágenes de África*. Granada: Editorial University of Granada, 2004.

Folch i Camarasa, Ramon. *Aventures encara més extraordinàries d'en Massagran*. Barcelona: Editorial Casals, 2003.

Folch i Camarasa, Ramon. *Bon dia, pare!* Barcelona: Editorial Laia, 1980.

Folch i Camarasa, Ramon. *Els Jocs Olímpics d'en Massagran*. Barcelona: Editorial Casals, 1991.

Folch i Camarasa, Ramon. *En Massagran i el bruixot blanc*. Barcelona: Editorial Casals, 1987.

Folch i Camarasa, Ramon. *En Massagran i els negrers*. 3rd ed. Barcelona: Editorial Casals, 2010.

Folch i Camarasa, Ramon. *En Massagran i els pirates*. 2nd ed. Barcelona: Editorial Casals, 2009.

Folch i Camarasa, Ramon. *Les aventures extraordinàries d'en Massagran*. Barcelona: Editorial Casals, 1981.

Folch i Torres, Josep Maria. *África Española*. Barcelona: Establecimiento Editorial de Antonio J. Bastinos, 1911.

Folch i Torres, Josep Maria. "Les aventures d'en Massagran. Espectacle en quatre actes." *La Escena Catalana* 74 (1921).

Folch i Torres, Josep Maria. *Massagran*. Barcelona: Editorial Bambú, 2018.

Foucault, Michel. "Society Must Be Defended." In *Biopolitics: A Reader*, edited by Timothy C. Campbell, 61–81. Durham, NC: Duke University Press, 2013.

Fouce, Héctor, and Fernán del Val. "La Movida: Popular Music as the Discourse of Modernity in Democratic Spain." In *Made in Spain: Studies in Popular Music*, edited by Sílvia Martínez and Héctor Fouce, 125–34. New York: Routledge, 2013.

Fradera, Josep M., and Christopher Schmidt-Nowara, eds. *Slavery and Antislavery in Spain's Atlantic Empire*. New York: Berghahn Books, 2013.

França, João. *Habitar la trinxera: Històries del moviment pel dret a l'habitatge a Barcelona*. Barcelona: Fundació Periodisme Plural, 2018.

Freeman, Elizabeth. *Time Binds: Queer Temporalities, Queer Histories*. Durham, NC: Duke University Press, 2010.

Fuster, Joan. "Per a una cultura catalana majoritària." *Nadala* 14 (1980): 52–80.

Gallo, Rubén. "Radiovanguardia: Poesía estridentista y radiofonía." In *Ficciones en los medios en la periferia. Técnicas de comunicación en la literatura hispanoamericana moderna*, edited by Wolfram Nitsch-Matei and Chichaia-Alejandra Torres, 273–89. Cologne: Universitäts- und Stadtbibliothek Köln, 2008.

Garcés, Marina. *Ciutat Princesa*. Barcelona: Galaxia Gutenberg, 2018.

García, Ángeles. "La piratería acaba con la industria discográfica." *El País*, December 8, 1984. elpais.com/diario/1984/12/09/cultura/471394805_850215.html

García Bueno, Jesús. "Seis días de horror y fuga." *El País*, August 27, 2017.

García Cárcel, Ricardo, and Antoni Simón i Tarrés. "Nación y familia en el pensamiento político y social catalán." In *Familia y mentalidades. Historia de la familia. Una nueva perspectiva sobre la sociedad europea*, edited by Ángel Rodríguez Sánchez and Antonio Peñafiel Ramón, 41–52. Murcia: EDITUM, 1997.

García Lorca, Federico. *La zapatera prodigiosa*. Madrid: Cátedra, 1999.

Gellner, Ernest. *Nations and Nationalism*. 2nd edition. Ithaca, NY: Cornell University Press, 2006.

Gentic, Tania. *The Everyday Atlantic: Time, Knowledge, and Subjectivity in the Twentieth-Century Iberian and Latin American Newspaper Chronicle*. Albany: State University of New York Press, 2013.

Giles, Paul. *Virtual Americas: Transnational Fictions and the Transatlantic Imaginary*. Durham, NC: Duke University Press, 2002.

Gitelman, Lisa. *Always Already New: Media, History, and the Data of Culture*. Cambridge, MA: MIT Press, 2006.

Goldsmith, Mike. *Discord: The Story of Noise*. Oxford: Oxford University Press, 2012.

Gómez-Font, Àlex. *Zeleste i la música laietana: Un passeig per la Barcelona musical dels anys setanta*. Barcelona: Pagès Editors, 2009.

González Rodríguez, Juan Pablo. *Pensar la música desde América Latina: problemas e interrogantes*. Santiago: Ediciones Universidad Alberto Hurtado, 2013. https://doi.org/10.4067/S0716-27902013000200010.

Gonzalo, Jaime. *La ciudad secreta*. Madrid: Munster Books, 2013.

Goode, Joshua. *Impurity of Blood: Defining Race in Spain, 1870–1930*. Baton Rouge: Louisiana State University Press, 2009.

Gorostiaga Alonso-Villalobos, Eduardo. "Nuevo régimen administrativo de la radio en España (1978)." *Revista de Administración Pública* 87 (1978). https://www.cepc.gob.es/publicaciones/revistas/revista-de-administracion-publica/numero-87-septiembrediciembre-1978-2.

Gosálvez, Patricia. "#Hablamos? #Parlem?, el mensaje más ilustrado." *El País*, October 6, 2017.

Guarner, José Luis. Review of *Ocaña, retrat intermitent*, dir. Ventura Pons, Producciones Zeta. Website of Ventura Pons. Archived September 21, 2014, at https://web.archive.org/web/20140921144410/http://www.venturapons.com/catala.html.

Guattari, Félix. "Popular Free Radio." In *Radiotext(e)*, edited by Neill Strauss and Dave Mandl, 85–89. Los Angeles: Semiotext(e), 1993.

Gueneli, Berna. "Young, Diverse, and Polyglot: Ilker Çatak and Amelia Umuhire Track the New Urban Sound of Europe." In *Minority Discourses in Germany since 1990*, edited by Ela Gezen, Priscilla Layne, and Jonathan Skolnik, 172–95. New York: Berghahn Books, 2022.

Guerra, Paula, and Fernán del Val Ripollés. "Post Dictatorships, Cosmopolitanism, Punk, and Post-punk in Portugal and Spain from 1974 to 1984." *Popular Music and Society* 44, no. 5 (2021): 610–27. https://doi.org/10.1080/03007766.2021.1948755.

Guillamon, Julià. *La ciutat interrompuda*. Barcelona: La Malgrana, 2001.

Guillamon, Julià, Víctor Nubla, and Pau Riba. *Alter músiques natives*. Barcelona: Generalitat de Catalunya, 1995. http://www.gracia-territori.com/petxina/pdf/alter_native_music.pdf.

Guillamon, Julià, Víctor Nubla, and Pau Riba. "Experimental Music in Catalonia from 50's to 90's. Alter Native Music." Website of Gràcia Territori Sonor. http://www.gracia-territori.com/petxina/pdf/alter_native_music.pdf.

Halberstam, Jack. *Trans*: A Quick and Quirky Account of Gender Variability*. Oakland: University of California Press, 2018.

Harkness, Nicholas. *Songs of Seoul: An Ethnography of Voice and Voicing in Christian South Korea*. Oakland: University of California Press, 2014.

Harley, Erik. *Pormishuevismo: un movimiento artístico*. Barcelona: Blackie Books, 2023.

Harvey, David. "The 'New' Imperialism: Accumulation by Dispossession." *Socialist Register* 40 (2004): 63–87.

Helmreich, Stefan. "An Anthropologist Underwater: Immersive Soundscapes, Submarine Cyborgs, and Transductive Ethnography." *American Ethnologist* 34, no. 4 (2007): 621–41.

Hernando, Javier. *Los silencios de la radio*. March 21, 2021. www.javierhernando.net/los_silencios_de_la_radio.htm.

Hernando, Javier. "40 años de Radio PICA." *Ojos de Músico Extraviado*, March 9, 2021. http://ojosdemusicoextraviado.blogspot.com/2021/03/40-anos-de-radio-pica.html.

Herrera Ángel, Rafael. "El teatro andaluz costumbrista: Los hermanos Álvarez Quintero." *Gibralfaro: Revista de Creación Literaria y Humanidades* 79 (2012): 10.

Hilmes, Michelle. "Is There a Field Called Sound Culture Studies? And Does It Matter?" *American Quarterly* 57, no.1 (2005): 249–59.

Hirschkind, Charles. *The Feeling of History: Islam, Romanticism, and Andalusia.* Chicago: University of Chicago Press, 2020.

Hooper, John. *The New Spaniards.* New York: Penguin, 2006.

Ihde, Don. *Listening and Voice: Phenomenologies of Sound.* Albany: State University of New York Press, 2007.

Illas, Edgar. *Thinking Barcelona: Ideologies of a Global City.* Liverpool: Liverpool University Press, 2012.

International Institute for Strategic Studies. "The Barcelona Terrorist Attack." *Strategic Comments* 23, no. 8 (2017): iii–iv.

"Intro." NDF 1 (1983): 1.

Jardí Casany, Enric. *Els Folch i Torres i la Catalunya del seu temps.* Barcelona: Publicacions de l'Abadía de Montserrat, 1995.

Jarman-Ivens, Freya. *Queer Voices: Technologies, Vocalities, and the Musical Flaw.* New York: Palgrave MacMillan, 2011.

Jáuregui, Carlos A. *Canibalia: Canibalismo, calibanismo, antropofagia cultural y consumo en América Latina.* Madrid: Iberoamericana Vervuert, 2008.

Joni D. *Que pagui Pujol! Una crònica punk de la Barcelona dels 80.* Barcelona: Editorial La Ciutat Invisible, 2011.

Juliá, Santos. *Transición: Historia de una política española (1937–2017).* Barcelona: Galaxia Gutenberg, 2017.

Kahn, Douglas. *Noise, Water, Meat: A History of Sound in the Arts.* Cambridge, MA: MIT Press, 1999.

Kelly, Caleb. *Sound.* Cambridge, MA: MIT Press, 2011.

Kheshti, Roshanak. "Touching Listening: The Aural Imaginary in the World Music Culture Industry." *American Quarterly* 63, no. 3 (2011): 711–31.

King, Stewart. *Escribir la catalanidad: Lengua e identidades culturales en la narrativa contemporánea de Cataluña.* Martlesham: Tamesis, 2005.

Krauel, Javier. "Eugeni d'Ors's Politics of Aesthetics: A Revision of His Authoritarianism." *Journal of Spanish Cultural Studies* 23, no. 1 (2022): 429–47.

Kun, Josh. *Audiotopia: Music, Race, and America.* Oakland: University of California Press, 2005.

Labanyi, Jo. "History and Hauntology; or, What Does One Do with the Ghosts of the Past? Reflections on Spanish Film and Fiction of the Post-Franco Period." In *Disremembering the Dictatorship: The Politics of Memory in the Spanish Transition to Democracy*, edited by Joan Ramon Resina, 65–82. Amsterdam: Rodopi, 2000.

Labanyi, Jo. "Raza, género y denegación en el cine español del primer Franquismo: El cine de misioneros y las películas folclóricas." *Archivos de la Filmoteca: Revista de Estudios Históricos Sobre la Imagen*, no. 32 (1999): 22–42.

LaBelle, Brandon. *Lexicon of the Mouth: Poetics and Politics of Voice and the Oral Imaginary*. New York: Bloomsbury, 2014.

LaBelle, Brandon. "Restless Acoustics, Emergent Publics." In *The Routledge Companion to Sounding Art*, edited by Marcel Cobussen, Vincent Meelberg, and Barry Truax, 275–85. London: Routledge, 2017.

LaBelle, Brandon. *Sonic Agency: Sound and Emergent Forms of Resistance*. London: Goldsmiths Press, 2018.

Lafarga i Oriol, Joan. *Gràcia: De rural a urbana; Historia d'un territori*. Barcelona: Taller d'Història de Gràcia, 2001.

Lahuerta, Juan José. *Destrucción de Barcelona*. Barcelona: Mudito and Co., 2005.

La Vanguardia. "Protesta masiva frente a la sede del PP en Barcelona." October 3, 2017.

Lefebvre, Henri. *Rhythmanalysis: Space, Time, and Everyday Life*. Translated by Stuart Elden and Gerald Moore. New York: Continuum, 2004.

Lefebvre, Henri. *The Production of Space*. Translated by Donald Nicholson-Smith. Cambridge: Blackwell, 1991.

Lightfoot, David W. *Born to Parse: How Children Select Their Languages*. Cambridge, MA: MIT Press, 2020.

Lippi-Green, Rosina. *English with an Accent: Language, Ideology and Discrimination in the United States*. New York: Routledge, 2012.

Llansamà, Jordi. *Harto de todo: Historia oral del punk (1979–1987)*. Barcelona: BCore, 2011.

Lledó Cunill, Eulàlia. "Paraules clau." *Huffington Post*, September 19, 2012.

López Lerma, Mónica. *Sensing Justice through Contemporary Spanish Cinema: Aesthetics, Politics, Law*. Edinburgh: Edinburgh University Press, 2021.

Loquillo. *En las calles de Madrid*. Barcelona: Ediciones B, 2018. Kindle.

Lugones, María. "Decolonial Feminism." *Hypatia* 25, no. 4 (2010): 742–59.

Luque, Nazario. *La vida cotidiana del dibujante underground*. Barcelona: Anagrama, 2016.

Maldonado, Lorena. "Estas son las canciones por las que el rapero Valtònyc irá a la carcel." *El Español*, February 20, 2018.

Manrique, Diego A. *¿De qué va el rock macarra?* Madrid: La Piqueta, 1977.

Maragall, Joan. *Antología Joan Maragall*. Barcelona: Calambur, 2018. Digitalia. https://www-digitaliapublishing-com.proxy.library.georgetown.edu/a/57914.

Marcus, Greil. *Lipstick Traces: A Secret History of the 20th Century*. 2nd ed. Cambridge, MA: Belknap Press, 2009.

Marsé, Juan. *El amante bilingüe*. Barcelona: Editorial Planeta, 1990.

Martin-Márquez, Susan. *Disorientations: Spanish Colonialism in Africa and the Performance of Identity*. New Haven, CT: Yale University Press, 2008.

Martínez, Sílvia. "Stick to the Copla! Recovering Old Spanish Popular Songs." In *Made in Spain: Studies in Popular Music*, edited by Sílvia Martínez and Héctor Fouce, 90–100. New York: Routledge, 2013.

Martínez-López, Miguel, and Ángel García Bernardos. "The Occupation of Squares and the Squatting of Buildings: Lessons from the Convergence of Two Social Movements." *ACME: An International Journal for Critical Geographies* 14, no. 1 (2015): 157–84.

Masiello, Francine. *The Senses of Democracy: Perception, Politics, and Culture in Latin America*. Austin: University of Texas Press, 2018. https://doi.org/10.7560/35033.

Mbembe, Achille. "Necropolitics." In *Biopolitics: A Reader*, edited by Timothy C. Campbell, 161–92. Durham, NC: Duke University Press, 2013.

McEnaney, Tom. *Acoustic Properties: Radio, Narrative, and the New Neighborhood of the Americas*. Evanston, IL: Northwestern University Press, 2017.

McNeill, Donald. "Barcelona: Urban Identity 1992–2002." In *The Barcelona Reader: Cultural Readings of a City*, edited by Enric Bou and Jaume Subirana, 323–46. Liverpool: Liverpool University Press, 2017.

Medak-Saltzman, Danika. "Transnational Indigenous Exchange: Rethinking Global Interactions of Indigenous Peoples at the 1904 St. Louis Exposition." *American Quarterly* 62, no. 3 (2010): 591–615.

Mena Arca, Joan. *No parlaràs mai un bon català: Les mentides sobre la immersió lingüística*. Vic: Eumo Editorial, 2022.

Mendizàbal, Enric. "Una posible geografía de las identidades de Barcelona: El caso del barrio de la Vila de Gràcia." *Finisterra* 45, no. 90 (2010): 91–109.

Mercadé, Xavier. *Odio obedecer: La escena alternativa en los 80; Punk, rock y hardcore*. Barcelona: Quarentena Edicions, 2011.

Mignolo, Walter D. "Editor's Introduction." *Poetics Today* 15, no. 4 (1994): 505–21.

Mignolo, Walter D. "The Geopolitics of Knowledge and the Colonial Difference." In *Coloniality at Large. Latin America and the Postcolonial Debate*, edited by Mabel Moraña, Enrique Dussel, and Carlos A. Jáuregui, 225–58. Durham, NC: Duke University Press, 2008.

Miller, Douglas K. "Blushing at the Fair: Indigenous Workers, Academics, and Competing Modernities in the 1893 World's Columbian Exposition." *Reviews in American History* 48, no. 2 (2020): 283–89.

Minder, Raphael. *The Struggle for Catalonia: Rebel Politics in Spain*. London: Hurst and Company, 2017.

Miret, Naik, and Pau Serra del Pozo. "El papel de la inmigración extranjera en el cambio social y urbano de el Besòs i el Maresme, un barrio periférico de Barcelona. Interrogaciones a partir de un estudio exploratorio." *Estudios Geográficos* 74, no. 274 (2013): 193–229. https://doi.org/10.3989/egeogr.

Montañés, José Ángel. "'No tinc por!,' grita Barcelona." *El País*, August 18, 2017.

Montes Fernández, Francisco José. "Radiodifusión pirata comercial." *Anuario Jurídico y Económico Escurialense* 41 (2008): 715–78.

Monzó, Quim. "Inmigrantes para siempre." *La Vanguardia*. January 26, 2011.

Moreno, Jairo. *Musical Representations, Subjects, and Objects. The Construction of Musical Thought in Zarlino, Descartes, Rameau, and Weber*. Bloomington: Indiana University Press, 2004.

Mowitt, John. *Sounds: The Ambient Humanities*. Oakland: University of California Press, 2015. https://doi.org/10.1525/california/9780520284623.001.0001.

Muñoz, José Esteban. *Disidentifications: Queers of Color and the Performance of Politics*. Minneapolis: University of Minnesota Press, 1999.

Nerín i Abad, Gustau. "Mito franquista y realidad de la colonización de la Guinea española." *Estudios de Asia y África* 32, no. 1 (1997): 9–30.

Ngai, Sianne. *Ugly Feelings*. Cambridge, MA: Harvard University Press, 2005.

Nikkhah, Roya. "Noddy Returns Without the Golliwogs." *The Telegraph*, October 18, 2009. https://www.telegraph.co.uk/news/6359248/Noddy-returns-without -the-golliwogs.html.

Novell, Pepa. "Cantautoras catalanas: De la *nova cançó* a la *nova cançó d'ara*; El paso y el peso del pasado." *Journal of Spanish Cultural Studies* 10, no. 2 (2009): 135–47. https://doi.org/10.1080/14636200902990679.

Nubla, Víctor. "Aquí no es allá el día de hoy." Website of Víctor Nubla. www.hronir .org/puzzleaqui.htm. Originally published 1993.

Nubla, Víctor. *El regal de Gliese*. Barcelona: Editorial Males Herbes, 2012.

Nubla, Víctor. "The Unified Field Theory." In *Alter músiques natives*, by Julià Guillamon, Víctor Nubla, and Pau Riba, 136–38. Barcelona: Generalitat de Catalunya, 1995.

Ochoa Gautier, Ana María. *Aurality: Listening and Knowledge in Nineteenth-Century Colombia*. Durham, NC: Duke University Press, 2014.

Òmnium Cultural. "Help Catalonia. Save Europe." YouTube video. 2017. https:// www.youtube.com/watch?v=OAEYtrIAfEI.

Ong, Walter. *Orality and Literacy: The Technologizing of the Word*. New York: Routledge, 1982.

Orringer, Nelson. *Lorca in Tune with Falla*. Toronto: University of Toronto Press, 2014.

Otero, Iago, Giorgos Kallis, Raül Aguilar, and Vicenç Ruiz. "Water Scarcity, Social Power and the Production of an Elite Suburb: The Political Ecology of Water in Matadepera, Catalonia." *Ecological Economics* 70, no. 7 (2011): 1297–308.

Panagia, David. *The Political Life of Sensation*. Durham, NC: Duke University Press, 2009.

Parry, John H., and Robert G. Keith, eds. *New Iberian World: A Documentary History of the Discovery and Settlement of Latin America to the Early 17th Century*. New York: Times Books, 1984.

Pascual Lizarraga, Jakue. *Movimiento de resistencia: Años 80 en Euskal Herria; Radios libres, fanzines y okupaciones*. Tafalla, Navarra: Txalaparta, 2019.

Picarol, Salvador. "Correspondencia—Demasié—Anuncios." *Ajoblanco* 4 (1975): 34.

Picarol, Salvador. "Correspondencia—Demasié—Anuncios." *Ajoblanco* 6 (1975): 37.

Picarol, Salvador. "Watergate en el Ayuntamiento?" *Ajoblanco* 7 (1975): 21.

Pinchevski, Amit. *Echo*. Cambridge, MA: MIT Press, 2022.

Piqueras, José Antonio. *Negreros: Españoles en el tráfico y en los capitales esclavistas*. Madrid: Catarata, 2021.

Poch Olivé, Dolors, ed. *El español en contacto con las otras lenguas peninsulares*. Madrid: Iberoamericana Editorial Vervuert, 2016.

Pons, Ventura, dir. *Ocaña, retrat intermitent*. Producciones Zeta, 1978.

Prado, Emili. *Las radios libres: Teoría y práctica de un movimiento alternativo*. Barcelona: Editorial Mitre, 1983.

Pratt, Mary Louise. *Imperial Eyes: Travel Writing and Transculturation*. New York: Routledge, 2008.

Pratt, Mary Louise. *Planetary Longings*. Durham, NC: Duke University Press, 2022.

Público. "El saludo fascista y el 'cara al sol' cierran la concentración en Madrid contra el referéndum catalán." September 30, 2017. https://www.publico.es /politica/saludo-fascista-y-cara-al.html.

Puig, Toni. "Yo también soy travesti." *Ajoblanco* 19 (1977): 15.

Pujol, Anton. "The Cinema of Ventura Pons: Theatricality as a Minoritarian Device." *Journal of Adaptation in Film and Performance* 3, no. 2 (2010): 171–84.

Quera, Jordi. *Anem d'excursió per Catalunya: 30 excursions per a nois i noies; Les excursions de Cavall Fort*. Barcelona: Cim Edicions, 2014.

Quijano, Aníbal. "Coloniality and Modernity/Rationality." *Cultural Studies* 21, no. 2–3 (2007): 168–78. https://doi.org/09502380601164353.

Rabasa, José. "Thinking Europe in Indian Categories, or, 'Tell Me the Story of How I Conquered You.'" In *Coloniality at Large: Latin America and the Postcolonial Debate*, edited by Mabel Moraña, Enrique Dussel, and Carlos A. Jáuregui, 43–76. Durham, NC: Duke University Press, 2008. https://doi.org/10.1515 /9780822388883-020.

Radano, Ronaldo, and Tejumola Olaniyan, eds. *Audible Empire: Music, Global, Politics, Critique*. Durham, NC: Duke University Press, 2016.

Radiomai. "Manifiesto de Villaverde." 1983. Accessed February 6, 2025. https:// radiomai.com/manifiesto-de-villaverde/.

Radiotelevisió Espanyola (RTVE). *Les aventures d'en Massagran*. 1983.

Rama, Ángel. *La ciudad letrada*. Hanover, NH: Ediciones del Norte, 1984.

Rancière, Jacques. *The Politics of Aesthetics: The Distribution of the Sensible*. Translated by Gabriel Rockhill. London: Continuum International, 2011.

Rappaport, Joanne, and Tom Cummins. *Beyond the Lettered City: Indigenous Literacies in the Andes*. Durham, NC: Duke University Press, 2012.

Reguero Jiménez, Núria, and Eloi Camps Durban. "Medio siglo de radios libres en Cataluña: La sostenibilidad del sector hoy." *Quaderns del CAC* 50 (2024): 63–73.

Remedi, Gustavo. "The Production of Local Public Spheres: Community Radio Stations." In *The Latin American Cultural Studies Reader*, edited by Ana del Sarto, Abril Trigo, and Alicia Ríos, 513–34. Durham, NC: Duke University Press, 2004.

Renan, Ernest. "What Is a Nation?" In *Nation and Narration*, edited by Homi K. Bhabha, 8–22. New York: Routledge, 1990.

Resina, Joan Ramon. *Barcelona's Vocation of Modernity: Rise and Decline of an Urban Image*. Palo Alto, CA: Stanford University Press, 2008. https://doi.org/10.11126/stanford/9780804758321.001.0001.

Resina, Joan Ramon. "Double Coding of Desire: Language Conflict, Nation Building, and Identity Crashing in Juan Marsé's *El amante bilingüe*." *Modern Language Review* 96, no. 1 (2001): 92–102. https://doi.org/10.2307/3735718.

Resina, Joan Ramon. "It Wasn't This: Latency and the Epiphenomenon of the Transition." *The Ghost in the Constitution: Historical Memory and the Memory of Denial in Spanish Society*. Liverpool: Liverpool University Press, 2017.

Riba, Pau. *Los '70 a destajo: Ajoblanco y libertad*. Barcelona: Editorial Planeta, 2007.

Riles, Annelise. "The View from the International Plane: Perspective and Scale in the Architecture of Colonial International Law." In *The Legal Geographies Reader*, edited by Nicholas Blomley, David Delaney, and Richard T. Ford, 276–84. Malden, MA: Blackwell, 2001.

Rivera Cusicanqui, Silvia. "*Ch'ixinakax utxiwa*: A Reflection on the Practices and Discourses of Decolonization." *The South Atlantic Quarterly* 111, no. 1 (2012): 95–109. https://doi.org/10.1215/00382876-1472612.

Robbins, Dylon. *Audible Geographies in Latin America: Sounds of Race and Place*. London: Palgrave, 2019.

Robertson, Pamela. *Guilty Pleasures: Feminist Camp from Mae West to Madonna*. Durham, NC: Duke University Press, 1996.

Robinson, Dylan. *Hungry Listening: Resonant Theory for Indigenous Sound Studies*. Minneapolis: University of Minnesota Press, 2020.

Roca, Josep. *Això es Ràdio PICA*. Barcelona: Laertes, 1997.

Rodrigo y Alharilla, Martín. "Cataluña y el colonialismo español (1868–1899)." In *Estado y periferias en la España del siglo XIX: Nuevos Enfoques*, edited by Salvador Catalayud, Jesús Millán, and María Cruz Romeo, 315–56. Valencia: Universitat de València Servicio de Publicaciones, 2009.

Rodrigo y Alharilla, Martín, and Lizbeth Chaviano Pérez, eds. *Negreros y esclavos: Barcelona y la esclavitud atlántica (siglos XVI–XIX)*. Barcelona: Icaria, 2017. https://doi.org/10.17502/mrcs.v9i1.453.

Rodríguez, Alexis. "Jóvenes latinos y geografías nocturnas." In *Jóvenes latinos en Barcelona: Espacio público y cultura urbana*, edited by Carolina Recio, Carles Feixa, and Laura Porzio, 205–14. Barcelona: Anthropos, 2006.

Rodríguez, Anto. *¡Eres tan travesti! Breve historia del transformismo en España*. Barcelona: Egales, 2024.

Rodríguez, Clemencia. *Fissures in the Mediascape: An International Study of Citizens' Media*. New York: Hampton Press, 2001.

Rodríguez, Richard. *Hunger of Memory: The Education of Richard Rodríguez: An Autobiography*. Boston: D. R. Godine, 1981.

Roig, Montserrat. *L'òpera quotidiana*. Barcelona: Edicions 62, 2022. Kindle.

Roure, Sebastià. *Quadre damunt quadre*. Barcelona: Roure Edicions, 2024.

Rovisco, Maria. "A New 'Europe from Below'? Cosmopolitan Citizenship, Digital Media and the *Indignados* Social Movement." *Comparative European Politics* 14, no. 4 (2016): 435–57. https://doi.org/10.1057/cep.2015.30.

Ruiz, Rafico. *Slow Disturbance: Infrastructural Mediation on the Settler Colonial Resource Frontier*. Durham, NC: Duke University Press, 2021.

Rull, Xavier. "El model de llengua del pop-rock en català durant la seva eclosió (1989–1993)." *Catalan Review* 30 (2016): 257–74. https://doi.org/10.3828/CATR.30.13.

Sagarra, Josep Maria de. *Vida privada*. Barcelona: Proa, 2007. Originally published 1932.

Said, Edward. *Orientalism*. London: Routledge, 1978.

Salvans, Aleix. *Gestió del caos: Escenes de la contracultura a Catalunya 1973–1992*. Barcelona: Angle Editorial, 2018.

Salvattore—Ràdio PICA. *Història-anàlisi de les ràdios lliures a Catalunya i el seu futur*. Roure Edicions, 2016. www.lahaine.org/mm_ss_est_esp.php/historia -analisi-de-les-radios-lliures-a-2.

Sánchez Fuarros, Íñigo. "Music and Migration in Multicultural Spain." In *Made in Spain: Studies in Popular Music*, edited by Sílvia Martínez and Héctor Fouce, 144–53. New York: Routledge, 2013.

Saunders, Robert A. "Media and Terrorism." In *The Routledge History of Terrorism*, edited by Randall D. Law, 428–41. New York: Routledge, 2015.

Schafer, R. Murray. "Radical Radio." In *Radiotext(e)*, edited by Neill Strauss and Dave Mandl, 291–98. Los Angeles: Semiotext(e), 1993.

Schmurz, Timothy. "58 de julio: Ascensión de la Santa Schmurz." *Ajoblanco* (1978): 41.

Schwartz, Hillel. "Whistling for the Hell of It." In *The Routledge Companion to Sounding Art*, edited by Marcel Cobussen, Vincent Meelberg, and Barry Truax, 341–50. New York: Routledge, 2017.

Seed, Patricia. *Ceremonies of Possession in Europe's Conquest of the New World, 1492–1640*. Cambridge: Cambridge University Press, 1995.

Segato, Rita. *La nación y sus otros: Raza, etnicidad y diversidad religiosa en tiempos de políticas de identidad*. Buenos Aires: Prometeo Libros, 2007.

Snyder, Bob. *Music and Memory: An Introduction*. Cambridge, MA: MIT Press, 2001.

Soley, Lawrence. *Free Radio: Electronic Civil Disobedience*. Boulder, CO: Westview, 1999.

Spivak, Gayatri Chakravorty. "Can the Subaltern Speak?" In *Marxism and the Interpretation of Culture*, edited by Cary Nelson and Lawrence Grossberg, 271–316. Urbana: University of Illinois Press, 1988.

Steingo, Gavin, and Jim Sykes, editors. *Remapping Sound Studies*. Durham, NC: Duke University Press, 2019.

Sterne, Jonathan. *The Audible Past: Cultural Origins of Sound Reproduction*. Durham, NC: Duke University Press, 2003.

Sterne, Jonathan. "Hearing." In *Keywords in Sound*, edited by David Novak and Matt Sakakeeny, 65–77. Durham, NC: Duke University Press, 2015.

Sterne, Jonathan. "Sonic Imaginations." In *The Sound Studies Reader*, edited by Jonathan Sterne, 1–17. London and New York: Routledge, 2012.

Stoever, Jennifer. "Splicing the Sonic Color Line: Tony Schwartz Remixes *Nueva York*." *Social Text* 28, no. 1 (102) (2010): 59–85. https://doi.org/10.1215/01642472-2009-060.

Strange, Hannah. "Barcelona Attack Suspect Younes Abouyaaqoub Shot Dead Wearing Suicide Belt by Spanish Police." *The Telegraph*, August 21, 2017.

Surribas Balduque, Mariona. "La ficción como amparo legal del arte: Titiriteros, raperos y libertad de expresión en la España pos 15-M." *Journal of Spanish Cultural Studies* 21, no. 3 (2020): 411–28. https://doi.org/10.1080/14636204.2020.1801300.

Taylor, Adam. "Anti-Muslim Views Rise Across Europe." *Washington Post*, July 11, 2016. https://www.washingtonpost.com/news/worldviews/wp/2016/07/11/anti-muslim-views-rise-across-europe/.

Taylor, Diana. *The Archive and the Repertoire: Performing Cultural Memory in the Americas*. Durham, NC: Duke University Press, 2003.

Thacker, Eugene. "Biomedia." In *Critical Terms for Media Studies*, edited by W. J. T. Mitchell and Mark B. N. Hanssen, 117–30. Chicago: University of Chicago Press, 2010.

Thompson, Emily. *The Soundscape of Modernity: Architectural Acoustics and the Culture of Listening in America*. Cambridge, MA: MIT Press, 2004.

Tofiño, Iñaki. *Guinea, el delirio colonial de España*. Barcelona: Edicions Bellaterra, 2023.

Tonkiss, Fran. "Aural Postcards: Sound, Memory and the City." In *The Auditory Culture Reader*, edited by Michael Bull and Les Back, 303–9. Oxford: Berg, 2003.

Torrent, Joan. "La campana de Gràcia." *Destino* 1729 (1970): 29–31.

Truax, Barry. "Acoustic Space, Community, and Virtual Soundscapes." In *The Routledge Companion to Sounding Art*, edited by Marcel Cobussen, Vincent Meelberg, and Barry Truax, 253–64. New York: Routledge, 2017.

Trueba, David. "Algarabía." *El País*, September 12, 2012.

Tsuchiya, Akiko. "Monuments and Public Memory: Antonio López y López, Slavery, and the Cuban-Catalan Connection," in "Unstable Foundations: Reconsidering Nineteenth-Century Monuments and Memories." Special issue, *Nineteenth-Century Contexts* 41, no. 5 (2019): 479–500. https://doi.org/10.1080/08905495.2019.1657735.

Tufekci, Zeynep. *Twitter and Tear Gas: The Power and Fragility of Networked Protest*. New Haven, CT: Yale University Press, 2017.

TV3. "Wert el colonitzador." *Polònia*, May 12, 2012. www.ccma.cat/tv3/alacarta/programa/wert-el-colonitzador/video/4371030/.

Ugarte Ballester, Xus. "El *Polònia* de TV3: Tastets paremiològics amb versions en francès i castellà." *Catalonia* 8 (2011): 19–25. https://doi.org/10.4000/11sr6.

Vasallo, Brigitte. *PornoBurka: Desventuras del Raval y otras f(r)icciones contemporáneas*. Cardedeu: Ediciones Cautivas, 2013.

Vázquez Montalbán, Manuel. *Barcelonas*. Translated by Andy Robinson. New York: Verso, 1992.

Vázquez Montalbán, Manuel. *Crónica sentimental de España*. Barcelona: Editorial Lumen, 1971.

Vázquez Montalbán, Manuel. *Crónica sentimental de la Transición*. Barcelona: Editorial Planeta, 1985.

Venegas, José Luis. *Sublime South: Andalusia, Orientalism, and the Making of Modern Spain*. Evanston, IL: Northwestern University Press, 2018.

Venegas, José Luis. "Uneven Souths: The Mediterranean Dimension of Spain's Southern Regionalism." *Comparative Literature Studies* 58, no. 3 (2021): 532–56. https://doi.org/10.5325/complitstudies.58.3.0532.

Vialette, Aurélie. "An On-Screen Trial: Resistance to Corruption or Audience Manipulation in *Ciutat Morta*?" *Letras Femeninas* 42, no. 1 (2016): 101–14. https://doi.org/10.14321/letrfeme.42.1.0101.

Vila, F. Xavier. "¿Quién habla hoy en día el castellano en Cataluña? Una aproximación demolingüística." In *El español en contacto con las otras lenguas peninsulares*, edited by Dolors Poch Olivé, 135–58. Madrid: Iberoamericana Editorial Vervuert, 2016.

Vila-Matas, Enrique. *Desde la ciudad nerviosa*. Madrid: Alfaguara, 2004.

Vila-San-Juan, Morrosko, dir. *Barcelona era una fiesta (Underground 1970–1983)*. Generalitat de Catalunya, Institut Català de les Indústries Culturals (ICIC), Monseiur Alain, and Séptimo Elemento, 2010.

Vilardell, Laura. *Books Against Tyranny: Catalan Publishers Under Franco*. Nashville, TN: Vanderbilt University Press, 2022.

Vilarós, Teresa. *El mono del desencanto: Una crítica cultural de la Transición española (1973–1993)*. Madrid: Siglo XXI de España, 1998.

Vilaseca, Stephen Luis. *Barcelonan Okupas: Squatter Power!* Madison, NJ: Fairleigh Dickinson University Press, 2013.

Villagrasa, Fèlix. *Una història de Ràdio PICA: 25 anys a contrapèl*. Barcelona: Mar Blava, 2006.

Viñuela, Eduardo. "Popular Music in *Televisión Española*: Cultural Practices, Consumption, and Spanish Identity." In *Made in Spain: Studies in Popular Music*, edited by Sílvia Martínez and Héctor Fouce, 178–86. New York: Routledge, 2013.

Walden, Joshua S. *Sounding Authentic: The Rural Miniature and Musical Modernism*. New York: Oxford University Press, 2014.

Warner, Michael. "Publics and Counterpublics." *Public Culture* 14, no. 1 (2002): 40–90.

Westergaard, Marit. "Microvariation in Multilingual Situations: The Importance of Property-by-Property Acquisition." *Second Language Research* 37, no. 3 (2021): 379–407.

Westerman, Frank. "The Man Stuffed and Displayed Like a Wild Animal." *BBC News*, September 16, 2016. https://www.bbc.com/news/magazine-37344210.

Whittaker, Tom. *The Spanish Quinqui Film: Delinquency, Sound, Sensation*. Manchester: Manchester University Press, 2020.

Wolfe, Patrick. "Settler Colonialism and the Elimination of the Native." *Journal of Genocide Research* 8, no. 4 (2006): 387–409. https://doi.org/10.1080 /14623520601056240.

Wolfson, Todd. *Digital Rebellion: The Birth of the Cyber Left*. Chicago: University of Illinois Press, 2014.

Woods-Peiró, Eva. *White Gypsies: Race and Stardom in Spanish Musicals*. Minneapolis: University of Minnesota Press, 2012.

Woolard, Kathryn A. *Double Talk: Bilingualism and the Politics of Ethnicity in Catalonia*. Palo Alto, CA: Stanford University Press, 1989.

Yarza, Alejandro. *The Making and Unmaking of Francoist Kitsch Cinema: From* Raza *to* Pan's Labyrinth. Edinburgh: Edinburgh University Press, 2018.

Yúdice, George. *The Expediency of Culture: Uses of Culture in the Global Era*. Durham, NC: Duke University Press, 2003.

Zanón, Carlos. "Jordi Cuixart: Parlar fins a l'afonia." *El País*, March 10, 2018.

Zeuske, Michael. "Capitanes y comerciantes catalanes de esclavos." In *Negreros y esclavos: Barcelona y la esclavitud atlántica (siglos XVI–XIX)*, edited by Martín Rodrigo y Alharilla and Lizbeth J. Chaviano Pérez, 63–100. Barcelona: Icaria Editorial, 2017.

Index

Page numbers followed by f refer to figures.

possession, 130, 168–69, 178, 203, 213; aurality of, 178; capitalist, 211; global economic, 204; narrative, 208. *See also* dispossession

Pratt, Mary Louise, xi–xii, 5, 30–31, 242n3

precariousness, 3, 52, 158, 163; of dispossession, 196; of life, 199, 203; of voice, 223

presence, 15, 111, 207, 223; accent and, 101, 104; of acoustic body, 123; Arabic, 172; of the body, 85, 87, 90, 92, 95, 97–98, 103–4, 122, 179, 182, 194; counterculture and, 126, 131; hauntological, 166; immigrants' embodied, 75; metaphysics of, 82, 201; Muslim, 171; Ocaña's, 78, 95, 97–100; online, 137–38; radio, 134; sonic, 202, 221; sound as, 190; temporal, 73; visual, 112; vocal, 85; of the voice, 90, 92, 109, 192

progressive rock, 124, 151

pronunciation, 35, 50–52, 72, 86, 246n59; accent and, 33, 78, 253n74; Andalusian, 85, 250n11; Catalan, 34, 48, 101–2, 108, 117, 147, 150, 253n74; of English, 227–28; errors, 101; of Spanish, 116, 228, 251n33. *See also* accent; dialect; lisp; mispronunciation; *seseo*; *yeísmo*

protest, 4, 27, 122, 157–58, 172–74, 181, 183, 193–99, 206, 210–14, 223; antigentrification, xii; aural imaginaries of, 184; Bio-lentos and, 210–11; La Campana de Gràcia and, 167; counterculture and, 144; COVID-19 and, 265n80; Diada, 170, 187; of military conscription, 143; *nova cançó* movement as, 133; *okupa*, xi, 168, 197; September 11, 104; sonic mattering maps and, 201; sound of, 28, 177–79, 185, 202–3, 232; terrorism and, 219. See also *cacerolazo*

publics, 35, 158, 181, 196, 221; globalized, 31; North American, 61; online, 192; young, 41. *See also* counterpublics

public sphere, 24, 198, 232; Catalan, 34, 63, 132; in Catalonia, 196; Francoist, 247n85; globalized, 14; masculine, 24; mediatic, 184, 190; mediatized, 75; networked, 198; representation in, 181; sonic agency and, 3

Puig, Toni, 76, 97

punk, 27–28, 130, 135–38, 143–45, 149–51, 154, 163–64, 168, 173, 177; bands, 28, 143, 145, 149, 155, 157, 259n83; culture, 16, 126, 131, 145; ethos, 143, 197; *meneo barcelonés* and, 166; Movida Madrileña and, 254n3; movement, 148, 154, 197; music, 16, 123–24, 159, 258n77; Ràdio PICA and, xi, 16, 142–43, 147–48, 153–54, 161, 198; sound, 159; squats, 232; tradition, 209; women and, 155–56. *See also* fanzines; hardcore

punks, 135, 137, 143, 148; Barcelona, 149, 200, 205; working-class, 2

puntos de escucha, 86, 100

queerness, 9, 25, 28, 77. *See also* gender; Ocaña, José Pérez; trans*; Vasallo, Brigitte

Quijano, Aníbal, 126, 239n19

race, 6, 35–36, 66, 185, 232; accent and, 63, 75; animality and, 177; binary construct of, 116; colonial imaginaries of, 27; echoic memories of, 37, 59–60, 65, 219; ideologies of, 74; linguistic soundings of, 4; Massagran series and, 42, 67; sonic production of, 231; Spain and, 79, 89, 171; stereotypes about, 8; voice and, 57, 79

racism, 36, 58, 67, 89, 116, 200, 218, 247n88; class, 35; state, 185

radio, 28, 122, 126–27, 129–30, 141–43, 152–53, 158–59, 213, 215, 229–30, 260n97; airwaves, 168; broadcasts, 247n85; Catalan, 9, 34, 39, 64, 133–34; clandestine, 169, 257n68; commercial, 140–41; community, 225; frequencies, 254n3; illegal, 123; licenses, xi, 130, 255n26; mobile, 162; pirate, 139–40, 257n65, 257n67; programs, 127, 129, 255n18; punk, 154; sound and, 31; Spanish, 31, 155; stations, 2, 64, 136–37, 169; transmissions, 11, 130; voice and, 33; waves, 189. *See also* Catalunya Ràdio; free radio; La Campana de Gràcia

sound art, 3, 28, 132, 134, 160; experimental, 124, 126, 128, 136, 166

sound artists, 27, 89, 126, 152

sound culture: experimental, 123; studies, 233

soundscape, 16, 99, 149, 171, 237n1 (intro.), 240n41; of *algarabía*, 62; of Barcelona, xii, 9, 27, 63, 75, 111, 115, 121, 125–27, 177, 223, 239n28; Castilian, 100, 132; Catalan, 3, 34, 64–66, 74, 83, 85, 91, 100, 109–10, 132, 147, 243n14; cosmopolitan, x–xi; democratic, 28; everyday, 102; linguistic, x, 80, 147, 239n28; musical, xi; normative, 168

sound studies, 9, 224, 226

sound theory, 87, 125

Spain, 2–3, 28, 36–37, 60, 66–67, 72, 75, 118, 130–32, 149, 174, 194, 196, 227; Africa and, 1, 26, 33, 42–43, 45, 57, 59, 61, 89, 243n22 (*see also* Equatorial Guinea; Morocco); antiausterity protests in, 198; *apertura* and, 213; Arab, 94; Catalonia and, 5–7, 10, 13, 38, 42, 83–85, 114–15, 173, 178, 185, 190–92, 194, 202–3, 246n59, 249n10, 264n50; coloniality and, 10; colonies of, 1, 40; comics in, 258n74; Constitution of, 186; Constitutional Court, 183, 185, 263n38; democracy and, 24, 123, 134, 169; Europe and, 2, 6, 33, 71, 106, 169; experimental film in, 89–90; expulsion of Muslims from, 171–72; foreign investment in, 207; gender and, 252n33; hackers in, 191; imaginaries of, 100; immigrants in, 119, 265n64; immigration and, 78, 219–20; Islamic practices of submission in, 241n68; language and, 14, 41, 106, 188; leftists in, 163; Ley de Memoria Democrática, 179, 259n91; media distribution in, xi; music and, 124, 126–27, 135 (see also *nova cançó*); NATO and, 154; normativity and, 77; obligatory military service in, 165; plurilingual culture of, 238n5; polyglot space of, 27; post-Franco, 151; protopunk in, 256n49; public space in, 199; race and, 89; radio in, 130, 136–40, 155, 169, 255n26 (*see also* free radio; Radio Gladys Palmera;

Ràdio PICA; radio: pirate); Reconquista, 171–72, 218; Republicanism in, 209, 211; resistance against, 16; revolt against, 128; Second Republic, 34, 79, 88; *senyera* and, 262n2; slave trade and, 248n97; sonic color line in, 218–19; southern, 76, 78, 84, 93, 147; stage traditions in, 88; V de Vivienda movement in, 197; voice and, 86. *See also* Andalusia; *crónicas*; *El País*; Franco, Francisco; Madrid; Rajoy, Mariano; *Requerimiento*; *seseo*; *yeísmo*

Spanish Civil War, 76, 83–84, 110, 151, 168, 219, 238n4; Durruti and, 145; Transition and, 132

Spanish language, ix–xi, 8, 10, 12, 14, 18–19, 26, 29, 136, 190, 227–29, 246n59, 267n108; *el andaluz* (*see* aural imaginaries), 78, 80, 85, 87, 90, 108, 110, 250n11; Arabic and, 115; barbarisms, 2; in Barcelona, 240n37; cultural production in, 225; decolonizing, 226; dominance of, 34; films in, 77; immigrants and, 116, 118–19; lyrics in, 134, 147, 256n40; newspapers, 215; as official language, 238n5; public sphere and, 63; puns in, 245n59; sound of, 21–22, 211; southern-accented, 84; *xarnego*-accented, 107. See also *seseo*; *yeísmo*

Spanish monarchy, 18, 42, 148

Spanish music industry, 3, 16, 129–30, 136

Spanishness, 78, 80, 87–89, 115, 130, 172–73

Spanish state, 42, 114, 130, 192, 264n46; Ajuntament and, 132; Barcelona terror attack and, 266n86; Castilian Spanish and, 188, 263n42; Catalonia and, x, 190; criminality and, 264n47; undemocratic posture of, 194; voting mechanisms of, 191

speech, 34, 47, 53–54, 83, 88, 109–10, 141, 222; accent and, 51–52, 87, 228; of *altres catalans*, 253n74; Andalusian, 108; Barthes's city and, 122; Catalan, 2–3, 51, 64, 243n14, 246n59; conquistador's, 23; everyday, 136, 114, 154; free, 169; freedom of, 223; human, 217; local, 44; market concepts and, 168; nation and, speech (continued)

35; nonvocalized vocalizations and, 50; Ocaña's, 85; performative, 19; performativity of, 103; politically valued, 24; the radiophonic and, 128; voice and, 81, 87; writing and, 82. *See also* accent; bilingualism; polyglot sound; pronunciation

Spivak, Gayatri, 21, 226

squatters, 3, 9, 136–37, 144, 154, 197, 199, 202. See also *okupa*; *okupes*; Vilaseca, Stephen

Star (magazine), 124, 129, 143, 258n76

Sterne, Jonathan, xiii, 54, 202, 223, 226, 231

subaltern, 20–21

subjectivity, 4; body and, 92; Catalan, 102; geographical, 8; Indigenous, 17, 47; national, 17; postcolonial understanding of, 86; racialized, 55; shared, 189; sound and, 83; vococentric, 80, 85, 92; voice and, 55, 81–82, 86–87, 121; Western, 55, 92

Subterranean Kids, 135, 143, 149

Subtítulo, 173–74

television, 31, 133–34, 141–42, 154–55, 212, 215; cartoons, 28; Catalan culture and, 38; Catalan language and, 2, 34, 39, 64, 77, 108; Massagran and, 36–37, 39, 59, 63, 64–68; programming, 34, 64; Spanish, 152; stations, 13, 130, 265n65

terrorism, 112, 119, 174, 180, 185, 214–15, 219–21; acoustic, 125; anarchist, 238n4; biomediated hearing of, 223; global, 214, 220; musical, 90; poetic, 173

textuality, 20, 190. *See also* entextualization

Thacker, Eugene, 184

Till, Emmett, 217

timbre, 23, 221; of Black voice, 79; of voice, 20, 22, 36, 58, 61, 97, 191

Tofiño, Iñaki, 36, 243n22

tone, 22–23, 60–661; ironic, 150; language and, 8, 18; musical, 16, 152; of voice, 17, 20, 22, 36, 61, 183, 192. *See also* sleight of ear

Tonkiss, Fran, 35–36, 121–22

tourism, 6, 36, 45, 57, 65, 112, 114–15, 131–32, 196, 214; Blackness and, 37; direct action against, xi; English and, 118; investment in, 9; local sentiment around, 205, 212; migration and, 213; resistance to, 265n80; terrorism and, 119

trans*, 9, 28, 80, 132

transculturation, xii, 30

Transition, 35, 38, 57, 83, 132, 173, 178, 259n85; accent and, 101; capitalist ownership and, 138; Catalan identity and, 27; cinema of, 77; early days of, 28; gender and, 101; historical memory and, 95; immigration and, 219; Ràdio 4 and, 258n71; Ràdio PICA and, 154; sexuality during, 95; Spanish Civil War and, 132; tourism and, 9; violence and, 103; wake of, 178

travesti, 76, 97–98, 249n5. *See also* drag; gender; Ocaña, José Pérez

travestí, 76, 97, 249n5

TV3, 13, 34, 116, 264n50

Último Resorte, 135, 149, 254n5

unintelligibility, 174, 191

United States, xii, 17, 29, 87, 227–28, 230–31; Black voice and, 58; bombing of Belgrade, 229; comics and writers in, 134; counterculture, 143; Dakota Access Pipeline protests, 202; Más Voces and, 137; New Journalism, 237n4; Nicaragua and, 205; Occupy Wall Street, 198; Platja de Pals and, 140; punk in, 130; punk and experimental musicians and, 2–3; racial identity and, 79

Valencia, 27, 47, 51, 83, 239n28; Radio Klara, 256n45

Valencian language, 27, 246n59

Valtònyc, 173, 262n8

Vasallo, Brigitte, 83, 111, 221; *PornoBurka*, 80, 111–21, 233

Vázquez Montalbán, Manuel, 88, 112, 154, 244n38

V de Vivienda, 197

Venegas, José Luis, 7, 171

Verdaguer, Jacint, 41, 44, 133
Via Catalana, 188–90, 193, 198
Vibraciones (magazine), 124, 260n94
victimhood, 180
Vila, Xavier, 63, 240n37
Vilarós, Teresa, 250n20, 252n63
Vilaseca, Stephen, 202, 204
vocal chords, 28, 75, 80, 91, 94, 101, 123
vococentrism, 28, 80–81, 177, 264n50
voice, 5, 12, 49, 54, 61, 64, 67, 82–83, 92–96, 110, 134, 150, 177–79, 181–84, 188–95, 198, 213, 217–18, 223, 242n76; accent and, 4, 15, 51, 74, 79, 96–98, 121, 219; African, 48; Andalusian, 88, 91; aural imaginaries of, 75, 196; aurality of, 22–23, 25, 116, 221; bare life and, 182; Black, 58–59; body and, 81, 90, 233; Catalan, 34, 36, 38, 64, 72, 107; collective, 181, 186, 189, 195; constructs of, 19, 26, 57, 78, 103, 196, 223, 232; COVID-19 and, 265n80; deboned, 55, 92, 99–100, 109; delivery of, 246n67; democracy and, 181; of democracy, 169; democratic, 174, 179, 184, 186, 195, 201–2, 221; dispossession and, 11; echoic memory of, 221; embodied, 205; emplaced hearing of, 106; emplacement of, 85, 87; English, 119; geographical, 77; grain of the, 8, 22, 99, 181–82; identitarian reflection of, 259n85; immigrant, 85, 104, 108, 114, 122; language and, 18, 20; language as, 25–26; of the law, 20–21, 207, 212; limits of, 159; Master's Voice, 182, 193–95; narrative, 101; non-Catalan, 36, 72; nonvoice, 179, 221; Ocaña and, 79, 81–82, 90–91, 97, 100; one's own, 37; of others, 32; pitch of, 20, 60–61; poetic, 151; political representation and, 221; politics of, 180; public, 210, 229; queering of, 78; radio, 155, 169; shadow, 166; silenced, 168, 184; silencing of, 3, 53, 183; silent, 203; as sound, 18–19, 22–23; sound and, 11–12, 17; sound of, 4, 7–9, 20, 22, 35, 55, 78–80, 86–87, 92–93, 100, 114, 185–86, 194, 208, 219, 253n76, 264n50; state and, 263n42; subaltern, 20–21; subjectivity and, 4, 55, 81–82, 86, 121; toothless, 28, 52, 55–58, 60, 67, 81, 103; volume of, 20, 191; Western notion of, 54, 121. *See also* democratic voice; emplaced voice; timbre; tone
voicing, 9, 49, 87, 90, 265n80

Warner, Michael, 181–82, 221
Washington, DC, xii, 229
West, the, 18, 82; aurality and, 214; colonial aurality of, 23; identity in, 182; immigrant voices and, 27; national identity and, x; Orientalism and, xiii; voice and, 221
whistling, 216–17
whiteness, 68–69, 79, 89
workers, 128, 132, 154, 249n10; associations, 128; Catalan, 44; movements, 104; textile, 72; unions, 84. *See also* anarchism
world-system, 6–7
writing, 20, 54; accent and, 88; of Catalan nation, 36; ethnographic, 31; speech and, 122; travel, xi, 30; voice and, 59, 82, 182

xarnego, 65, 84–85, 105, 107, 113, 117, 253n76; accent, 108, 116; culture, 250n29

yeísmo, 84, 251n33

Zeleste club, 135, 147, 149, 151, 256n48, 259n92, 260n94
zorongo, 93–94, 252n57

www.ingramcontent.com/pod-product-compliance
Lightning Source LLC
Chambersburg PA
CBHW020458270326
41926CB00008B/660